AF324102

Advances in Geophysical Methods Applied to Forensic Investigations

Giovanni Leucci

Advances in Geophysical Methods Applied to Forensic Investigations

New Developments in Acquisition and Data Analysis Methodologies

 Springer

Giovanni Leucci
Institute of Cultural Heritage Sciences
National Research Council
Lecce, Italy

ISBN 978-3-030-46241-3   ISBN 978-3-030-46242-0   (eBook)
https://doi.org/10.1007/978-3-030-46242-0

This Springer imprint is published by the registered company Springer Nature Switzerland AG
The registered company address is: Gewerbestrasse 11, 6330 Cham, Switzerland

Great spirits have always encountered violent opposition from mediocre minds

Albert Einstein

Preface

With the term "Forensic Geophysics" means *"the application of non—invasive geophysical methods to study, locating and mapping invisible objects that often are buried and hidden both in the subsoil, under the water, in the walls, etc."* (Larson et al. 2011; Dick et al. 2015; Pringle et al. 2016; Barone 2017; Khaldaoui et al. 2017).

The potentials of geophysical methods are manifold; for example, they aid the location and recovery of human remains and clandestine burial, as they can, non-invasively, very rapidly survey extensive areas.

The law agencies around the world have, in the last ten years, successfully used geophysical methods to help the forensics searches in the civil and criminal investigation.

In the criminal investigation, the search for hidden weapons, explosives, drugs, human remains can be made using several geophysical methods, as also several geophysical methods can be used for the search of water leaks (cause of damage to buildings and roads), the analysis of the conservation state of the large infrastructures (bridges, viaducts, highways, tunnels) which can cause major damage to men such as the Genoa bridge. However, at present, geophysics is underutilised, and common techniques may not be optimal for some specific targets or sites. Therefore, more controlled research is needed to understand the applicability of these methodologies better when searching for forensic applications in various environments, materials, and soil types.

There is a significant gap in knowledge of the best methods for collection, analysis, and interpretation of geophysical data for forensics applications. It should be noted that as long as there is a physical contrast between a target and the background materials, the items that should be solved are related to the question: which geophysical method helps to highlight the searched target better?

This book has the ambition to provide a general introduction to the more used geophysical methods in forensics investigation. Physical and mathematical principles are easily explained. The field acquisition procedures, data processing and interpretation, are illustrated to guarantee the best understanding of the results by a non-expert. Also, the fields of application of the various methods are well explained. It also introduces new suggestions for better data acquisition, processing and interpretation. It represents a useful guide for all those who approach these methods for the first time since the text widely demonstrates that there is no need for extensive math skills for a general understanding of forensics geophysics methods.

Although it must be emphasised that to understand the more advanced data processing and interpretation methodologies in-depth, reasonable mathematic and physic knowledge are required; the approach used in this book employs physics and mathematics as simple as possible and reduces the physical and mathematical analysis to transparent cases.

However, the user employing that approach to forensic geophysics methods should know the more advanced techniques of analysing and interpreting geophysical data because they can greatly increase the amount of useful information obtained from the data.

Therefore, the approach used in the book will enable the reader to assess the scope and importance of the advanced techniques of analysis without entering into the details of their implementation.

More examples are taken from cases history in which geophysical methods are applied to help the investigators to understand the causes of an accident better or to highlight the illegal aspect.

Through the use of these examples, the reader can understand how to design and perform a geophysical survey for forensic purposes. The reader is guided to the choice of the most suitable method related to a particular type of problem which arises and is guided to the type of data acquisition and processing that best enables the best possible result.

The most innovative data acquisition and processing systems are described that enable rapid reconnaissance of the surface sub-levels of the earth, even over areas of considerable extension, yielding highly detailed evidence, even in very challenging cases.

It is hoped that the book will serve as a text for students in geophysics, architecture, engineering, and legal disciplines and can also be a useful guide for specialists who want to increase their knowledge related to this newborn discipline.

Lecce, Italy   Giovanni Leucci

References

Barone, P. M. (2017). Forensic geophysics. In: R. M. Di Maggio & P. M. Barone (Eds.), *Geoscientists at crime scenes: A companion to forensic geoscience* (pp. 175–190). Basel, Switzerland: Springer International Publishing.

Dick, H. C., Pringle, J. K., Sloane, B., Carver, J., Wisneiwski, K. D., Haffenden, A., et al. (2015). Detection and characterisation of Black Death burials by multi-proxy geophysical methods. *Journal of Archaeological Science, 59*, 132–141. https://doi.org/https://doi.org/10.1016/j.jas.2015.04.010.

Khaldaoui, F., Djeddi, M., Zagh, A., & Naa, A. (2017). Use of near-surface geophysical methods for forensic investigations. In *Proceedings of International Conference on Engineering Geophysics*, Al Ain, United Arab Emirates, Oct 9–12, 2017. https://doi.org/10.1190/iceg2017-037.

Larson, D. O., Vass, A. A., & Wise, M. (2011). Advanced scientific methods and procedures in the forensic investigation of clandestine graves. *Journal of Contemporary Criminal Justice, 27*(2), 149–182. https://doi.org/10.1177/1043986211405885.

Pringle, J. K., Jervis, J. R., Roberts, D., Dick, H. C., Wisniewski, K. D., Cassidy, N. J., & Cassella, J. P. (2016). Long-term geophysical monitoring of simulated clandestine graves using electrical and ground penetrating radar methods: 4–6 years after burial. *Journal of Forensic Sciences, 61*(2), 309–321. https://doi.org/10.1111/1556-4029.13009. ISSN 1556-4029. PMID 27404604.

Introduction

The application of geophysical science in the forensic scope is relatively recent. Geophysical methods have been successfully utilised to assist Police and Law Enforcement Investigation Teams in Forensic searches and criminal investigations around the world.

Large areas of ground can be searched quickly and non-destructively, minimising the impact on the environment. This technology can greatly improve the efficiency of a search operation by reducing the amount of unnecessary excavation, and consequence repair, hence reducing costs and time on site.

Geophysical methods are related to the measurements of several physical parameters. Every geophysical method measures a particular physical parameter. The estimated of a particular physical parameter is successively used to investigate the subsoil (to obtain information about buried features or differences in characteristics related to the hidden objects) or to evaluate materials (evidence of discontinuities) without destroying the serviceability of the surveyed part.

Geophysical methods provide knowledge of unforeseen highly variable subsurface ground and investigated materials conditions. They are assisting in investigation phases to resolve many many mysteries related to all that is hidden.

In the case of forensic, the application of geophysical methods provides the possibility of obtaining indirect information on the presence of hidden objects and contributes to resolving more question related to civil cases to help the judge's decision.

Geophysics provides, in fact, an efficient and cost-effective means of collecting information about hidden objects. Various geophysical techniques are used to determine for example buried bodies, arms deposit, the extent of groundwater contaminant plumes, location of voids related to clandestine excavations, and the presence of buried material, such as steel drums, or illegal dumps. The common sense in investigations suggests the use of multiple methods to refine to provide useful data results. In the selecting the more useful geophysical method, one should be considered: to have clear and define what are the objective of the investigation; to have apriori information about the site geology; make a preliminary site inspection, if possible, to determine both the site access and the presence of features that may interfere with the instrument and the survey procedure; and determine the survey costs.

The variation of the measured physical parameters is related to the investigated materials or subsoil: a perfectly homogeneous medium done the same value of the measured physical parameter; if in a certain position of the subsoil, there is a body with different physical properties compared to the surrounding material, the instrument records a value **anomalous** (i.e. a variation with respect to the reference value relative to a homogeneous situation).

The methods discussed in this book include gravimetry, time-domain reflectometry, magnetometry, ground-penetrating radar, resistivity, spontaneous surface potential, seismic. These methods and their applications for forensic sciences are summarised in Table 1.

Gravimetry method measurements take into consideration the measurement of the acceleration of gravity. The gravity field allows successively to obtain the position of the bulk density anomaly that, as will see in the following chapters, produces an anomaly zone. The amplitude of this anomaly help in understanding the subsoil characteristics (i.e. presence of voids, etc.). Also, gravity methodology applies to any problem involving mass variations.

Magnetic is a method that allows surveying large areas in a short time. It is used successfully for underground iron and steel objects (i.e. tanks, illegal dumps, and hidden metal arms). Nowadays it is also used to locate buried human bodies by mean a gradiometry configuration. Proton precession magnetometers are used for both the Earth's total magnetic field and local magnetic gradients.

Ground-penetratingRadar (GPR) is used to locate buried objects. It can locate every shallow buried object (voids, metal and non-metal, human rests, etc.). GPR data can be displayed in realtime with a computer display unit and therefore leads to easy and fast interpretation in the field. For this reason, this method is the most used in forensics.

Table 1 Main field of application of geophysical methods in order of importance for forensic purpose

Field of application	More appropriate geophysical methods
Urban disaster causes, illegal dumps	(1) ground-penetrating radar (2) electrical resistivity tomography (3) self-potential (4) magnetic (5) microgravimetry
Hidden bodies, arms, clandestine graves	(1) ground-penetrating radar (2) electrical resistivity tomography (3) magnetic
Food sophistication	(1) time-domain reflectometry
Analysis of the quality of construction materials	(1) ground-penetrating radar (2) electrical resistivity tomography (3) self-potential (4) seismic tomography

Seismic techniques provide detailed information about the geomechanical properties of the investigated materials. The method allows the study of the reflection, refraction, and direct seismic waves that determine the seismic wave velocities distribution by either direct (in tomography), refracting (in refraction survey) or reflecting (in reflection survey) waves of material units with different seismic velocities or impedance.

Resistivity technique provides detailed information about resistivity contrast occurs in the subsurface between, for example, buried human bodies and the surrounding environment. Electrical apparatus provides two current electrodes (that inject current in the material) and two potential electrodes (that measure the potential at the material surface). The electrical current will crowd into the more conductive layers and will rarefy in the more resistive layers. The measured potential at the surface will reflect these path differences and will provide a distribution of the electrical resistivity in the investigated materials.

Spontaneous Potential is a passive electrical method. It measures natural voltage and can be used as a surface technique. In surface, techniques are important to use a non-polarising probe. The resolution of the method is a function of the spacing of the probes. Surface SP can be used to investigate chemical variations associated with the corrosion of metal bars in the concrete structures, subsurface water movement and landslides, shafts, and tunnels and sinkholes. It is also possible that SP can be used to map contaminant plumes.

Time-domain reflectometry (TDR). In this technique, the reflections that result from a signal travelling through a transmission environment is measured. In this case, electromagnetic pulse travel through the medium and the measure is a comparison of the reflections from the "unknown" transmission environment to those produced by a standard impedance. It can be used in several fields.

Generally, it is used to determine the characteristics of electrical lines, to characterise and locate faults in metallic cables, but it can also be used to locate leaks in pipes, to individuate the corrosion in metal bars, and to individuate the food sophistication.

Geophysical methods are often used in combination. Thus, for example, the search for arms deposits takes place at an early stage with the use of GPR and magnetic methods. The ambiguities resulting from the results of a single method can be removed by considering the results obtained by using a second method. For example, the reflections in a GPR survey due to the presence of a wall or a buried human rests could be similar (a hyperbole shaped reflection). By integrating the GPR survey with an electrical survey, this ambiguity can be solved considering that relatively high resistivity values could be associated with the wall, while relatively low resistivity value could be associated to the earth-filled human buried.

It is important to stress that, although an interpretation of the results of the, here described, geophysical methods requires relatively advanced mathematical treatments, initial information, as will be shown in the book, can be obtained from the simple observation of the acquired data.

Despite this, geophysical surveying, as will be shown in this text, is a valuable tool for the investigation. However, be careful not to create "false myths" and to the belief that geophysics is the only way to conclude an investigation, or that they are the only ones able to give answers. This is not the case, or rather it should not be; analyses, tools and the contribution of knowledge scientific should help the professionals working in the field to facilitate them in conducting investigations, but they are not and should not be perceived as the solution to the question.

Contents

Chapter 1
Short Note About Geophysical Data Analysis

Abstract The processing of geophysical data generally involves improvement of the signal, by filtering out induced noise, and removal of artefacts created during data collection, which could otherwise be interpreted as a hidden structure of forensic interest. As will underline in the book—for a broad understanding—the analysis of geophysical data and its interpretation require a good level of knowledge of basic mathematics. For this reason, this chapter will provide some basic concepts that could help in geophysical data analysis. The here used approach is an attempt to enables readers to approach to the data analysis without going into the intricacies of mathematics. However, it is important to underline that the success of data processing largely depends on the quality of the acquired data. If the quality of acquired data is very low, processing might not be able to improve it sufficiently for the interpretation and their use to help the investigators. Unsatisfactory quality of raw data implies to consider the repetition of the survey.

Keywords Basic mathematics · Data digitalisation · Fourier transforms · Spectral analysis

1.1 The Continuous Signal Definition

Is important to underline that the physical parameters that can be measured in the geophysical surveys take values that vary according to the spatial position and/or time. The variation of the physical parameter is presented as continuous signal time or spatial dependent. The continuous-time or spatial dependent signal can be defined as a complex function f(t) or f(s) of a real variable f: R → C, where R is the domain of the real numbers and C is the complex numbers codomain.

The variables t and s are independent and typically interpreted as time or space (Titchmars 1937; Cooper and McGillem 1974; Bracewell 1986).

The subclass of periodic signals is characterized by the relationship f(t + T) = f(t), where T > 0 is the period. The aperiodic signals did not satisfy the periodicity condition.

© Springer Nature Switzerland AG 2020

G. Leucci, *Advances in Geophysical Methods Applied to Forensic Investigations*,

https://doi.org/10.1007/978-3-030-46242-0_1

1.2 The Fourier Transform

The Fourier Transform (FT) is the most extensively used transform and has an important part in the theory of many branches of science. In particular, in its discrete form, the FT plays a key role in signal processing and digital linear signal filtering.

Given a generic signal $s(t)$ that can represent a continuous function of a real variable t, its Fourier transform is defined by the integral (Bracewell 1986):

$$S(f) = \int_{-\infty}^{+\infty} s(t)e^{-i2\pi ft}\,dt$$

where i is the imaginary unit, while the inverse Fourier transform (IFT) can be written as:

$$s(t) = \int_{-\infty}^{+\infty} S(f)e^{i2\pi ft}\,df$$

When t is a time variable, f represents the temporal frequency. It is customary to substitute the angular frequency $\omega = 2\pi f$ in the above relations obtaining:

$$S(\omega) = \int_{-\infty}^{+\infty} s(t)e^{-i\omega t}\,dt$$

and

$$s(t) = \frac{1}{2\pi} \int_{-\infty}^{+\infty} S(\omega)e^{i\omega t}\,d\omega$$

In the case of a spatial variable, for example, x, the spatial frequency, fx, takes the place of f and the (angular) wavenumber, kx, that of the angular frequency ω. A bit of confusion is sometimes generated by the use of 'k' for spatial frequency (the counterpart of the temporal frequency f) instead of angular wavenumber. Moreover in electrical engineering and seismic processing (Robinson et al. 1986; Yilmaz 1987) an opposite sign convention is generally used for time and spatial variables: in the forward transform the argument in the exponent is negative if the variable is time and positive if the variable is space; of course, the converse is true for the inverse transform. The function $S(\omega)$ can be expressed in its complex form as:

$$S(\omega) = S_r(\omega) + iS_i(\omega) \ [\text{real part} + \text{imaginary part}]$$

or, in the exponential form, as:

$$S(\omega) = Ae^{i\varphi(\omega)}$$

where

$$A = \sqrt{S_r^2(\omega) + S_i^2(\omega)} \text{ and } \varphi(\omega) = arctang\frac{S_i(\omega)}{S_r(\omega)}$$

A are the *amplitude* (or *magnitude*) and *phase spectra*, respectively.

A well-known property is that the function bandwidth-duration product cannot be less than a certain minimum value. The uncertainty relation mathematically expresses this:

$$\Delta t \Delta f \geq \frac{1}{4\pi}$$

or equivalently:

$$\Delta t \Delta \omega \geq \frac{1}{2}$$

where $(\Delta t)^2$ is the variance of $s(t)$ and $(\Delta f)^2$, $(\Delta \omega)^2$ the variances of the corresponding power spectra.

Some of the main properties of the FT are summarised in Table 1.1.

Instead of continuous functions, often we must deal with digital signals such as discrete-time functions or *time series*. In general a digital signal is an ordered *sequence* of numbers; in some cases it comes from a uniform sampling operation that extracts from a continuous function $x(t)$ the values corresponding to integer multiples of a time interval Dt (sampling interval): xn = x(n) = x(n t) = x(n Δ t). If the sequence has non-null values only within a range of N samples, its *Discrete Fourier Transform* (DFT) is:

Table 1.1 Fourier transform theorems (Bracewell 1986)

Theorem	Time domain	Frequency domain
Similarity	$s(at)$	$\frac{1}{|a|}S\left(\frac{\omega}{a}\right)$
Shift	$s(t-a)$	$e^{-ia}\ S(\omega)$
Addition	$s(t) + g(t)$	$S(\omega) + G(\omega)$
Multiplication	$s(t)g(t)$	$S(\omega) * G(\omega)$
Convolution	$s(t) * g(t)$	$S(\omega)G(\omega)$
Autocorrelation	$s(t) * s^*(-t)$	$|S(\omega)|^2$
Derivative	$ds(t)/dt$	$i\omega S(\omega)$
Parseval's Theorem	$\int_{-\infty}^{+\infty} |s(t)|^2 dt = \int_{-\infty}^{+\infty} |s(\omega)|^2 d\omega$	

$$X(v) = \frac{1}{N} \sum_{n=0}^{N-1} x(n) e^{-i2\pi\left(\frac{v}{N}\right)n}$$

and the *Inverse Discrete Fourier Transform (IDFT)* is:

$$x(n) = \sum_{v=0}^{N-1} X(v) e^{i2\pi\left(\frac{v}{N}\right)n}$$

It is worth noting that both sequences are periodic of period N. Thus, as for $x(n)$, only N values are taken for the DFT, with $n = 0, 1, \ldots, N - 1$. It has to be remembered, however, their cyclic dependence on n and n. (For computer implementation, for example, to avoid starting by 0, both integers are increased by one unity). Taking into account their cyclic nature, another way of representing the discrete Fourier pair is:

$$X(v) = \frac{1}{N} \sum_{n=\frac{1}{2}N}^{\frac{1}{2}N-1} x(n) e^{-i2\pi\left(\frac{v}{N}\right)n}$$

$$x(n) = \sum_{v=\frac{1}{2}N}^{\frac{1}{2}N-1} X(v) e^{i2\pi\left(\frac{v}{N}\right)n}$$

and v/N may be identified with frequency measured in cycles per sampling interval over the range—$(1/2)N \le n \le (1/2)N$; thus the frequency measured in Hz is

$$f_v = \frac{v}{N\Delta t}$$

Moreover, considering that, apart for $X(0)$, a half number of samples are the complex conjugate of the other half, only $N/2 + 1$ samples are needed.

The frequency $f_{Ny} = 1/2\Delta t$ is the '*Nyquist frequency*'. It is the maximum frequency contained in the digital signal corresponding to a period of only two samples.

The sampling theorem by Shannon states that a (band-limited) function must be sampled at least twice per period of its highest frequency to recover the continuous signal from the sampled version. If this criterion is not met, the signal is undersampled, and its frequency spectrum is said to be *aliased*. In this case the single periods of the periodic spectrum overlap. If we consider as the DFT of the digital signal only one period of this periodic sequence, it comes out that high-frequencies beyond the Nyquist frequency (also said the '*alias*' or '*folding*' frequency) are folded back inside the band limits and become disguised (aliased) as lower frequencies (Fig. 1.1). The discrete transform became popular after the introduction in 1965 by Cooley and Tukey of an algorithm, known as *Fast Fourier Transform* or more concisely *FFT*,

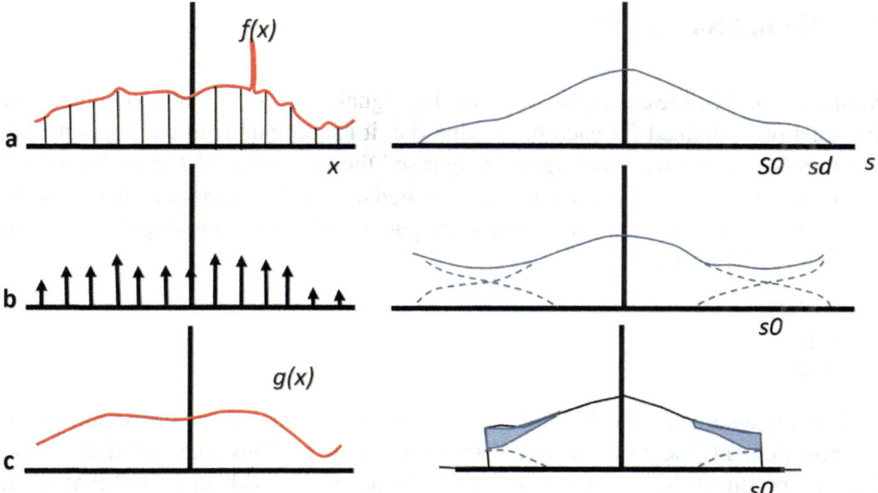

Fig. 1.1 Aliasing due to undersampling of a continuous function $f(x)$ (**a**) as overlapping of the periodized spectrum near the Nyquist frequency s_0 (**b**), or as wrap-round in the single-period DFT (**c**) (After Bracewell 1986)

that reduces the computational cost from N^2 to $N\log_2 N$ total number of operations, and is particularly efficient for large N.

Summarise spectral analysis is one technique used for characterising and analysing sequenced data. Sequenced data are observations that have been taken in one, two, or three-dimensional space, and/or time. In spectral analysis is important the decomposition of a sequence into oscillations of different lengths or scales. Therefore to understand the relationship between a continuous-time signal and the sampled version of that signal, the discrete Fourier transform represents one of the best methods that can be used. The FT allows constructing the so-called power spectrum of a signal. This important parameter describes the dependence of the signal's power with the frequency. When the signal phase characteristics have little meaning, the power spectrum of the signal can be done useful information. By dividing the signal into several possibly overlapping segments and averaging the spectrum obtained from each segment, a smoothed power spectrum can be obtained. The result is a frequency curve. In this case, the result emphasises the broadband or general characteristics of a signal's spectrum but loses some of the fine detail. The FT can be extended easily to two or three-dimension. Very important is the 2D Fourier Transform.

1.3 Signal Sampling

A function of the time can represent an analog signal it enjoys the following features: the function is defined for each time value (i.e. it is continuous in the domain); the function is continuous. Wanting to "vulgarise" the properties of the analog signal consequent to the two characteristics above-cited it could be stated that "it is possible to draw the temporal course of a signal analogue without ever removing the pen from the sheet …" (Fig. 1.2).

$$a = f(t)$$
$$t \in R$$
$$a \in R$$

That the analogue signals, in particular, can be considered to correspond to the two requirements mentioned above derive from some obvious considerations. First, they are produced by generators operating in the real world, in a field that is, in which time appears to be "dense". In the real world, time is great it continues, and it is always possible to imagine that, between two instants, however, close to one another find an intermediate instant in which the generator in question is producing a value signal finished. Secondly, it must be considered that there are no generators in the real world infinite power. This consideration leads to the conclusion that it is not possible to generate a signal that presents a finite variation in null time: even the fastest real transient presents a continuous signal evolution.

Unlike the analogue signal, the digital one consists of a "discrete-time" function e "Quantized". This function is, therefore: defined only in a countable set of "equispaced" instants; equipped with a codomain consisting of a discrete set of values (Fig. 1.3).

$$d = f(nT_c)$$
$$n \in Z \text{ (rational numbers)}$$

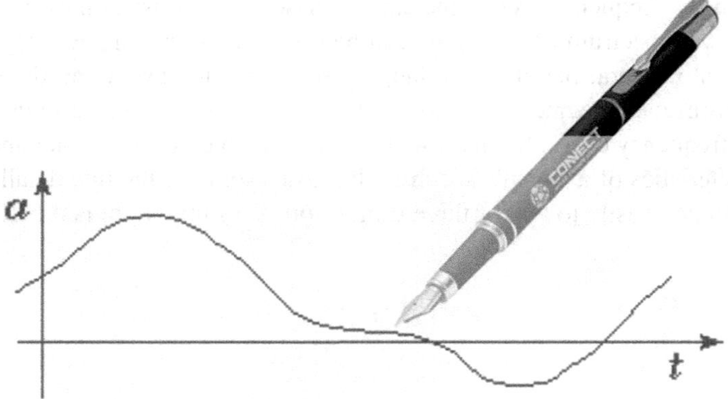

Fig. 1.2 The analog signal thinking as write without ever removing the pen from the sheet

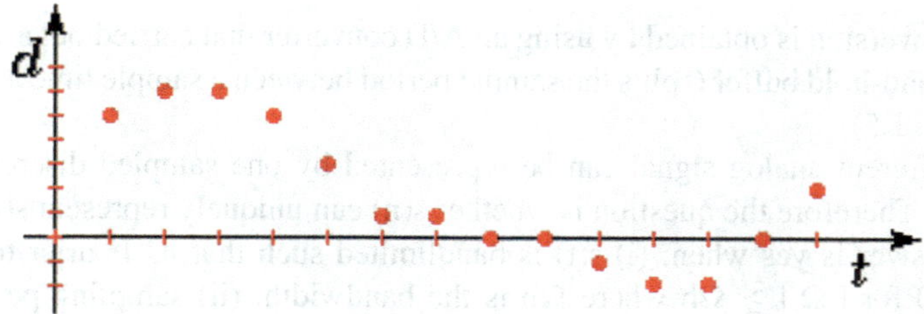

Fig. 1.3 The digital signal

$d \in Z$

The numerical signals, therefore, present two fundamental differences when compared with those analogs: are defined only at predetermined instants and between an instant and the next, they are "undefined", can assume only a predetermined number of values and pass, without continuity, from a value to the other "skipping" the possible values that might be found included among these.

Sampling an analog signal means taking a temporal sequence of values from this constituted by the succession of instantaneous values assumed by the signal in correspondence with details instants, called "sampling instants". The interval that separates two successive moments of sampling is called "sampling period" Tc and its reciprocal, indicated as fc, is called "sampling rate" (Fig. 1.5).

One might be led to think that sampling causes a reduction in content information of the analogue signal as the information on the value assumed by the signal is lost in all instances other than the sampling ones. Important: The sampling theorem ci says instead that, in ideal conditions, the execution of the sampling does not cause loss of information.

The sampling of the analog signal to a digital signal is regulated by the sample rate defined as the number of samples taken during one second. If s(t) is the expression of the analog signal and s(n) is the corresponding expression of the digital signal the conversion process can be written as

$$s(n) = s(t)|_{t=nT} = s(nT) \; n = \ldots - 1, 0, 1, 2, 3, \ldots$$

i.e. s(n) is obtained by extracting s(t) every T where T is known as the sampling period or interval (Fig. 1.4).

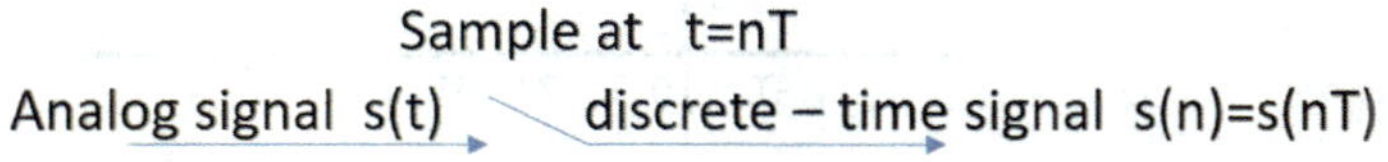

Fig. 1.4 Schematic representation of the conversion of the analogical signal to discrete-time sequence signal

The conversion is obtained by using an A/D converter that carried out a sampling a sample-and-hold buffer (splits the sample period between a sample time and a hold time) (Fig. 1.5).

The different analog signal can be represented by one sampled discrete signal (Fig. 1.6). Therefore the question is whether s(n) can uniquely represent s(t).

The answer is yes when: (i) s(t) is bandlimited such that its Fourier transform $S(j\Omega) = 0$ for $|\Omega| \geq \Omega b$ where Ωb is the bandwidth, (ii) sampling period T is sufficiently small.

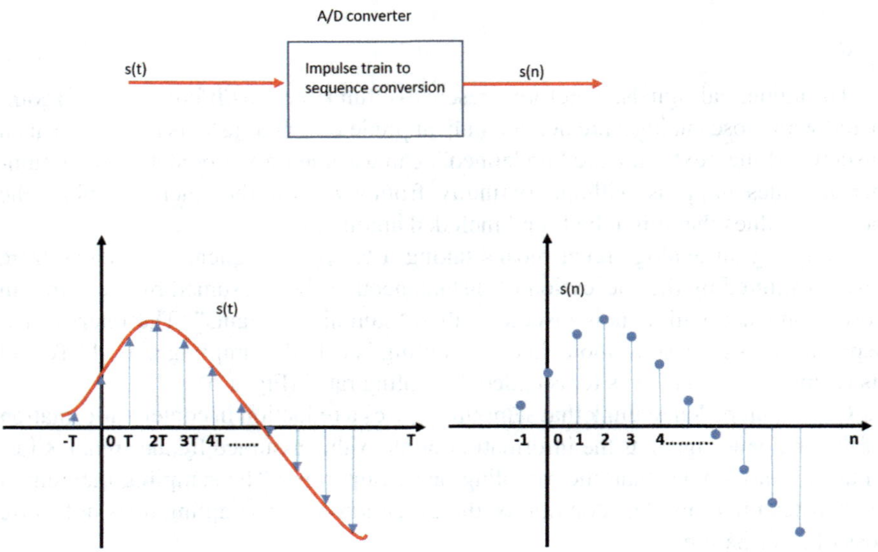

Fig. 1.5 Schematic diagram of A/D converter

Fig. 1.6 Different analog signals map to the same sequence

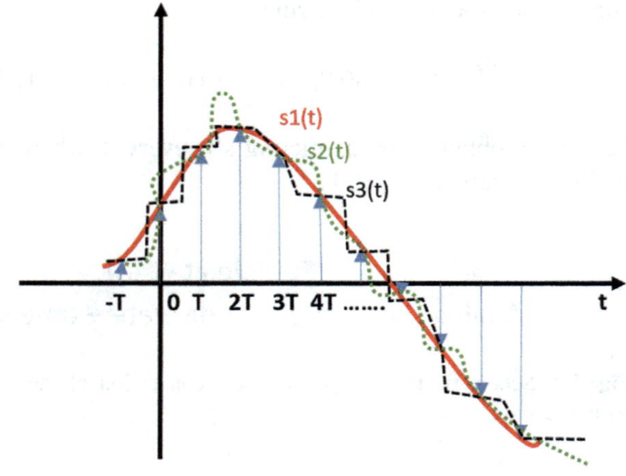

1.4   Important to Remember to Read This Book

Desired signal: is a signal in which the ratio signal/noise is infinite (i.e. it is an ideal signal that therefore is not affected by a component of noise).

Signal sampling: the process of obtaining a sequence of instantaneous values of a particular signal characteristic, usually at regular time intervals.

Sampling frequency: is the number of samples per second and the unit of the measure id the Hertz (Hz). Practically it is the frequency at which the analogic signal is converted in the digital signal.

Sampling period: is the inverse of the above-defined sampling frequency.

Nyquist interval: is the maximum time interval between equally-spaced samples of a signal that will enable the signal waveform to be completely determined.

Nyquist Sampling rate: the sampling frequency value. It is the twice of the maximal frequency of the acquired signal.

Aliasing: is an effect of violating the Nyquist-Shannon sampling theory. Aliasing can be prevented by using a steep-sloped low-pass filter with half the sampling frequency before the conversion (for more see Leucci, 2019).

References

Bracewell, R. N. (1986). *The fourier transform and its applications* (2nd ed.). New York: McGraw-Hill.

Cooper, G. R., & McGillem, C. D. (1974). *Continuous and discrete signal and system analysis*. New York: Holt, Rinehart and Winston.

Leucci, G. (2019). *Nondestructive testing for archaeology and cultural heritage a practical guide and new perspectives*. Berlin: Springer International Publishing.

Robinson, E. A., Durrani, T. S., & Peardon, L. G. (1986). *Geophysical signal processing*. Prentice-Hall International.

Titchmars, E. C. (1937). *Introduction to the theory of fourier integrals*. New York: Oxford University Press.

Yilmaz, O. (1987). *Seismic data processing*. SEG.

Chapter 2
Forensic Geosciences and Geophysics: Overview

Abstract In this chapter, the theoretical foundation of the geophysical methods applied in the field of Forensic Geosciences will be treated. After an introduction about state of the art, the Time Domain Reflectometry (TDR), Gravimetry, Magnetometry, Ground-Penetrating Radar (GPR), Electrical Resistivity Tomography (ERT), Self-Potential (SP), and seismic sonic an ultrasonic methods will be considered. Using practical examples, some important theoretical aspect of these methods will be explained as simply as possible.

Keywords Forensic geophysics state of the art · TDR · Gravimetry · Magnetometry · GPR · ERT · SP · Seismic sonic · Seismic ultrasonic · Background theory

2.1 Geophysical Methods in Forensic Geosciences: State of the Art

The forensic geoscience is defined as the set of the geo-discipline related to the Earth Science (e.g. geology, geophysics, geochemistry, etc.). As shown in Fig. 2.1 the forensic disciplines consider the application of wider environmental science. All these disciplines (geology, geophysics, environment, remote sensing, archaeology, engineering, botany, biology) are part of the so-called *"forensic geoscientific methods (FGM)"*. They are related to all aspects of earth materials, including rocks, sediments, soil, air and water, and with a wide range of natural phenomena and processes (Pye and Croft 2004). Since modern sediments and soil also often contain objects and particles of human origin, human-made materials such as brick, concrete, ceramics, glass and various other industrial products and raw materials are also sometimes of interest (Pye and Croft 2004). The numerous books and case study published in the last decade (Missiaen et al. 2010; Novo et al. 2011; Killam 2004; Pye and Croft 2004; Pye 2007; Chainey and Ratcliffe 2008; Jervis et al. 2009; Arosio 2010; Dionne et al. 2011; Schultz and Martin 2011) attest that the geoscience are increasingly used as a search tool by investigators. The FGM is currently considered an emerging discipline that provides important contributions in several fields of investigations

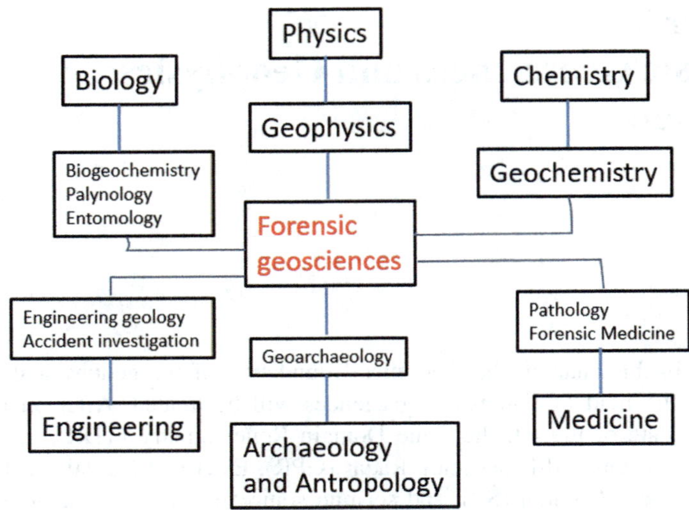

Fig. 2.1 Schematic representation of the relationship between scientific disciplines and subdisciplines with the forensic geoscience

such as environmental crimes (i.e. discovery of illegal dumps), humanitarian crimes (i.e. murders, concealment of corpses, counterfeit food, etc.) etc.

The first case of the application of the FGM is attributable to the English writer Arthur Canon Doyle (1859–1930), graduated in natural science from Cambridge University, the inventor of the great detective Sherlock Holmes. In his stories the investigator Holmes, by simple observation of shoes and clothes mud, can distinguish different soil type and understand the places where people have been (Murray and Tedrow 1975, 1992). More scientists developed individual theories and method using the FGM from the late 1800s to middle 1900s. The first school of Forensic Science and Criminology was realised in 1909 by Rodolphe Reiss (Lusan University). In the same period a private Locard Pollice Laboratory begun working in FGM applications.

The Aldermaston in the UK and FBI in the USA were the first government agencies that, between the year's thirties and seventies, considered the FGM (soil sciences, botanology, etc.) (Ruffell and McKinley 2008). Muarry and Tedrow (1975) write the first textbook about forensic geology. They in 1992 re-write a textbook considering a wider range of geoscience applications and development.

Successively Ruffell and McKinely (2005, 2008) and Pye (2007) done a comprehensive review of the geoforensic activities and a complete list of famous forensic geologists. The FGM includes more sub-disciplines related to the fact that exist a wide range of information and relations to other basic sciences (Fig. 2.1). An example is related to the forensic geology divided into large scales, that includes disciplines as geophysics, geomorphology and remote sensing, and small scales that include mineralogy and geochemistry (Ruffell and McKinley 2005). But it must be said that in the FGM research different methods are contemporaneously used eliminating thus the boundary between large and small scales.

The application of geophysical methods as support to the law to discover the buried truths define the forensic geophysics (FG). The FG is not formally defined or listed in the encyclopedic dictionary of exploration geophysics (Sherriff 1994). These methods could equally apply to both criminal and civil investigations. A little curiosity reported in Fennning and Donnelly (2004), affirms that the seismic method was used in the First World War to the location of enemy artillery guns.

In the last 30 years high-resolution surveys of the shallow subsurface using geophysical methods has used in the forensic field (Ruffell and McKinley 2008; Watters and Hunter 2004; Leucci et al. 2016). The applications of some of these methods and advances in geophysical instruments and computing technology are well described in Leucci (2019).

Geophysical methods can be divided into active methods and passive methods. In the active methods, an artificial source of particular physical parameters (i.e. electromagnetic waves, seismic waves, electrical current, etc.) is used, and their distribution will be measured. In passive methods, special instruments measure the natural distribution of some physical parameters (i.e. self-potential, gravity acceleration, magnetic properties, etc.) into the ground (Ruffell and McKinley 2008; Leucci 2019). In forensic studies, both active and passive methods are used. The search of hidden objects such as weapons or smuggled goods, illegal drums, human bodies, etc. is one of the most forensic applications of geophysical methods (Watters and Hunter 2004; Ruffell and McKinley 2008). Although for different purposes, the principles which apply in FG are similar to other geophysical researches (i.e. archaeo-geophysics, engineering geophysics, environmental geophysics, etc.).

The applicability of geophysical methods in forensic purpose can be summarised as seismic methods, magnetic methods, resistivity method, gravity method, self-potential method, the electromagnetic method that includes time-domain reflectometry, (Fennning and Donnelly 2004; Ruffell and McKinley 2008). It is clear that these methods are suitable for different case studies.

These methods have individual characteristics which are suitable for different case studies as their sensibility and reliability of resulted data differ in various situation and assorted materials. Each technique allows the built map of the subsurface that can vary in the way to visualise it and content (intended as a physical parameter).

The best methods for acquisition, analysis and interpretation of geophysical data for forensic applications depending on the type of target. Also, geophysicist experience may affect the interpretation of these data. Is well noted that if there is a strong geophysical contrast between a target and the background materials, the target can be identified (Leucci 2019).

Although any change in physical properties of the ground caused by a hidden object can be due to different causes, and it is important to be aware of these and their potential magnitude to assess the applicability of the technique to a particular site.

Ultimately, forensic geophysics, if well used, could have great potential to assist the police service in both the rapid detection and characterisation of every type of buried objects (such as will see in the next chapters of this book).

The application of geophysical methods in forensic investigations depending on (i) the target size; (ii) burial depth below ground level, (iii) background soil type, (iv) depositional environment (e.g. urban, rural, woodland, moorland), (v) vegetation type (e.g. grasses, woods, urban gardens), (vi) local climate (e.g. arid to humid), (vii) water table depth below ground level, and if the searcher is addressed to buried bodies, (viii) style of burial (e.g. for human bodies they are commonly wrapped, clothed or naked, or inserted into a box), (ix) time since burial, (the time of burial change the physical and chemical characteristics of the bodies) and many other specific variables. Each of these variables has a very important effect on the choice of the optimum technique and therefore on their subsequent success or failure in the results. For these reasons, it is extremely difficult to produce specific guidelines that can help in all forensic searches. Recent advances in the field equipment and data processing software has meant that some geophysical techniques are now relatively quick to collect and quantitatively analyse to pinpoint suspected locations for further investigation. In this book the following updated review of geophysical methods applied to forensic investigations and related active cases. It includes the following techniques: i) time-domain reflectometry (TDR), (ii) gravity, (iii) Magnetometry, (iv) Electrical methods (Resistivity), (v) Electrical passive methods (self-potential), (vi) Seismic tomography, (vii) Ground penetrating radar (GPR).

2.2 Forensic Geophysical Methods

This part of the book presents the background theory of the most commonly applied geophysical methods: time domain reflectometry (TDR), gravimetry, magnetometry, ground-penetrating radar (GPR), active and passive electrical (electrical-resistivity tomography and self-potential—SP), and seismic sonic and ultrasonic tomography.

Like all these methods, those described are nondestructive and useful for forensic applications. They allow a description of the subsurface and a discovery of hidden objects with saving time and money.

2.2.1 Time Domain Reflectometry Method

Time Domain Reflectometry (TDR) is based on the emission of electromagnetic (EM) waves and then measures the amplitudes of the reflections of the waves together with the time intervals between the waves are launched, and the reflections are detected. By measuring (i) the time interval between launching the electromagnetic waves and detection of the reflections, (ii) the amplitude of the reflections and (iii) the direction of the reflections and combining this with knowledge of the travelling EM waves velocity it is possible to determine position and size of the subject that created the reflections. The TDR methodology, thanks to its high versatility, is very often used for diagnostic and monitoring purposes in civil forensic. It is due to several reasons, first one the low costs and the possibility of carrying out real-time measurements in continuous mode. Is important to underline that TDR measurement can be controlled in remote and through the use of a simple series of multiplexers is possible to use simultaneously several probes with a single reflectometer. The reflectometer HL1500 can, for example, control, trough multiplexer systems, up to 512 probes.

Most applications of TDR were performed to the industrial field for health monitoring of electric wires in buildings, aircraft, and transportation systems (Cataldo et al. 2016). The TDR is also used for dielectric characterisation (i.e. dielectric properties of solid and liquid foods) of materials and moisture content measurements (Heimovaara et al. 2004). Other important applications of TDR in the forensic field are the monitoring liquefaction of soils, for the detection of organic pollutants in sandy soils, etc., investigation of contaminated land, and in particular of leachate contaminated soil (Cataldo and De Benedetto 2011).

In landslide monitoring, a coaxial cable grouted in the soil is used. It watching for reflections due to cable deformity induced by the ground deformation. The same configuration can be used in the slope underground movements with the scope to have geotechnical monitoring of road embankment in the landslide area (Cataldo and De Benedetto 2011). The basic theory of TDR is related to the microwave reflectometry study. In this case, a low-power electromagnetic signal is propagated into the system under test. With a specific data analysis of the reflected signal allows having information on the system under test.

Two main elements are involved in microwave reflectometry measurements: i) the instrument for generating/receiving the electromagnetic (EM) signal, and ii) the measurement cell, which includes the sensing element (or probe) and the system under test.

In TDR measurements, the EM signal (usually a step-like signal) propagates along the probe and successively through the system under test. For this purpose, a step-pulse generator (a sampling scope) and a signal analyser are used.

The step pulse generator produces a positive incident wave. The step pulse travels in the transmission line at a given velocity of propagation that is typical of the transmission line.

An impedance variation can be observed as a partial reflection of the propagating signal. If the load impedance is equal to the characteristic impedance of the line, no wave reflections will be observed, and the oscilloscope will display only the incident voltage step waveform. A variation of the load impedance causes a reflection event and the reflection coefficient in the time domain, ρ (t) can be obtained from the ratio between the amplitude of the reflected signal, v_{refl}(t), and the amplitude of the generated signal, v_{inc}(t):

$$\rho(t) = \frac{v_{refl}(t)}{v_{inc}(t)}$$

where $-1 \leq \rho$ (t) $\leq +1$.

The behaviour of $\rho(t)$ is strictly associated with the impedance variations along the electrical path travelled by the electromagnetic (EM) signal.

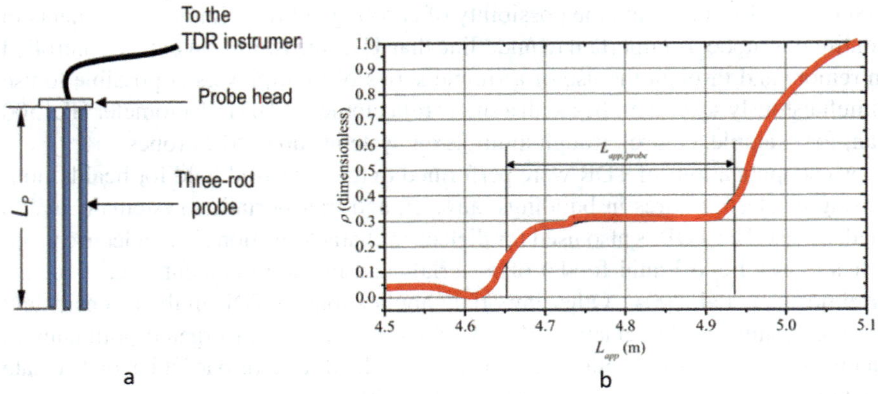

Fig. 2.2 **a** Schematic representation of TDR measurement probe; **b** example of TDR acquired waveform

Assuming the medium with a lossless conductivity, the TDR signal's propagation velocity, v, is related to the relative dielectric permittivity (ε_r) of the medium and to the relative magnetic permeability (μ_r), which is equal to 1 for most materials:

$$v = \frac{c}{\sqrt{\varepsilon_r \mu_r}}$$

where c is the velocity propagation of light in free space (c $\sim = 3 \times 10^8$ m/s).

Considering a typical three-rod probe, whose electrodes have length L a typical TDR waveform for such a probe, immersed in a generic homogeneous dielectric, is represented in Fig. 2.2.

The analysis of TDR waveform allows the evaluation of L_{app} (also called electric distance). L_{app} is the distance that the EM signal travel at velocity c, in the same interval of time. L_{app} can be directly associated with the dielectric characteristic of the medium. Typically the apparent distance is evaluated through the individuation of the beginning and end of the probe distance; thus, it can be associated with the physical length, L, through the following equation:

$$L_{app} = \sqrt{\varepsilon_{app}}L = \frac{c \times t_t}{2}$$

where t_t is the travel time (round-trip time taken by the signal to travel between the beginning and endpoints), and ε_{app} is referred to the apparent dielectric permittivity of the material in which the probe is inserted. The following equation gives a more rigorous expression for ε_{app}:

$$\varepsilon_{app}(f) = \frac{\varepsilon_r'(f)}{2}\left(1 + \sqrt{1 + \left\{\frac{\left[\varepsilon_r''(f) + \frac{\sigma_0}{2\pi f \varepsilon_0}\right]}{\varepsilon_r'(f)}\right\}^2}\,\right)$$

where $\varepsilon_0 \sim = 8.854 \times 10 - 12 Fm - 1$ is the dielectric permittivity of free space, f is the frequency, $\varepsilon_r'(f)$ describes energy storage, $\varepsilon_r''(f)$ accounts for the dielectric losses and σ_0 is the static electrical conductivity.

In a low-loss material (lossless conductivity), the imaginary part of the complex permittivity can be neglected. Additionally, considering low dispersive materials, the dependence of $\varepsilon_r'(f)$ on frequency can be considered negligible. Under these conditions, $\varepsilon_{app}(f)$ can be considered approximately constant:

$$\varepsilon_{app}(f) \cong \varepsilon_r'(f) \cong const$$

Based on the above considerations, the TDR dielectric measurements based on the observations of the points corresponding to the beginning and the end of the probe. To individuate these two points in the TDR waveform, different approaches based on the so-called tangent method can be used. Another method to esteem L_{app} is the derivative of the TDR waveform. The derivative of TDR waveform in fact exhibits prominent peaks in correspondence of the probe-beginning and probe-end sections.

It is also important to employ a probe with a well-known impedance profile, in this way, it is easier to discriminate and interpret the impedance variations due to the system under test.

For example, the coaxial probes are the most simple to design. They are widely used for monitoring and diagnostics on liquids.

A coaxial probe is composed of an outer cylinder (acting as the outer conductor) and a rod along the centerline of the cylinder (acting as a central conductor). The impedance profile can be determined from the transmission line theory:

$$Z(f) = \frac{60}{\sqrt{\varepsilon_r}}\ln\left(\frac{b}{a}\right)$$

where Z(f) is the frequency-dependent impedance of the probe filled with the considered material; $\varepsilon_r(f)$ is the relative dielectric permittivity of the material filling the probe (which is assumed lossless); b is the inner diameter of the outer conductor, and a is the outer diameter of the inner conductor.

Conversely for soil monitoring and granular materials monitoring the multi rod probe is a good solution. In this case for a two wire transmission line, the characteristic impedance can be derived from the previous equation, and it may be written as (O'Connor and Dowding 1999)

$$Z(f) = \frac{120}{\sqrt{\varepsilon_r}}\ln\left(\frac{2D}{d}\right)$$

where D is the distance between the centre of the conductors, and d is the diameter of each conductor. Unfortunately, for configurations different from a coaxial line, models are not always available (Cataldo et al. 2009), and the evaluation of the probe impedance is not straightforward, especially for unbalanced probes. The impedance (Zp) of the multi rod probes in the air also related the geometric characteristics can be:

$$Z_p = \frac{1}{2\pi(n-1)}\sqrt{\frac{\mu_0}{\varepsilon_0}}\,ln\left[H + \sqrt{H^2 - 1}\right]$$

where H is given by

$$H = \frac{\left(s^2 - a_0^2\right)^{n-1} - a_i^{2(n-1)}}{a_i^{n-1}\left\{(s + a_0)^{n-1} - (s - a_0)^{n-1}\right\}}$$

In the previous equations, n is the number of rods of the probe, a_i is the radius of the probe centre conductor, a_0 is the radius of the outer conductors, s is the distance between the middle of the centre conductor and the middle of the outer conductor, $\varepsilon_0 = 8.854 \times 10^{-12}$ F/m is the dielectric permittivity of free space, and $\mu_0 = 4\pi \times 10^{-7}$ H/m is the magnetic permeability of free space.

For three-rod probes, the above equation can be written as follows:

$$Z_{p,three-rod} = \frac{1}{4\pi}\sqrt{\frac{\mu_0}{\varepsilon_0}}\,ln\left(\frac{1 - \left(\frac{g}{s}\right)^4}{2\left(\frac{g}{s}\right)^3}\right)$$

where s is the centre to centre rod spacing, and g is the rod radius.

Figure 2.3 shows the schematisation of a reflectogram relative to a two bar probe which is partly in the air (medium with relative dielectric constant equal to $\varepsilon_r = 1$) and partly inserted in a different material ($\varepsilon_r > 1$).

From Fig. 2.3 is possible to note that the value of ρ (nominally between -1 and $+1$) varies with the dielectric characteristics of the medium in which the EM signal is propagated (vice versa, it remains constant in the sections in which there are no variations in the characteristics dielectric). The reflection coefficient, ρ, measured through the TDR technique, is directly linked to the electrical impedance (Z) of the transmission line along which the EM signal is propagating, according to the following relation:

$$\rho \cong \frac{Z - Z_0}{Z + Z_0}$$

where Z_0 is the reference impedance (typically 50 Ω), by inverting the above equation, it is, therefore, possible to obtain the impedance profile of the probe inserted in the material under examination, and to trace the dielectric (dielectric permittivity) and electrical (resistivity, conductivity) characteristics of the material under examination.

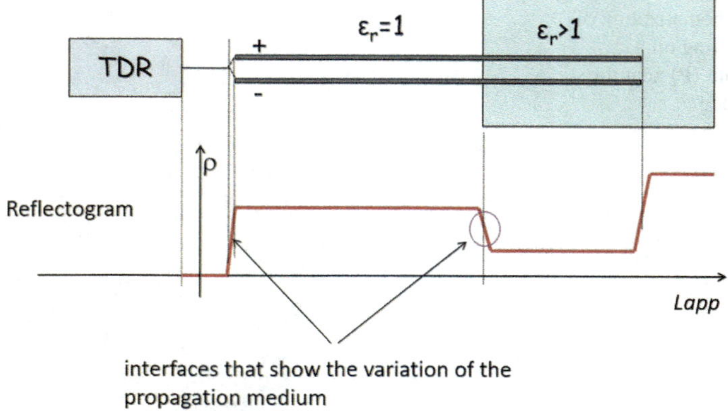

Fig. 2.3 Schematic of a sensitive two-wire element inserted in materials with different relative dielectric permittivity (ε_r). As the EM signal generated by the TDR travels through the probe, the corresponding reflectogram highlights the variations of ρ in correspondence with the different media

2.2.2 Gravimetry Method

The gravimetry method studies the mass distribution within the earth. The basic physical parameter that can be esteemed by gravimetry measurements is the bulk density (weight per unit mass). The basic theory of this method is related to the law of Newton. The law is defining the force of attraction between two masses within a certain distance. In this way, it is possible to define the acceleration that is directly proportional to the masses itself and therefore to the bulk densities of the two bodies. The method allows the measurements of the acceleration between a sample mass (contained in a measurements instrument) and masses in the subsurface. Starting from these measurements, the bulk density distribution in the subsoil can be derived. In the gravimetry measurements, there is some ambiguity because the observation in the field is affected not only by one mass but the sum of all masses. These ambiguities can be reduced by taking into consideration some known constraints.

The force of gravity is defined as the resultant of the forces that are independent of time acting on a mass body m placed near the Earth. These forces are:

(1) the Newtonian attraction force F_N due to the complex of the masses (solid, liquid and gaseous) that constitute the Earth whose intensity is expressed by:

$$F_N = \Gamma \cdot \frac{M \cdot m}{R^2}$$

where M is the mass of the Earth (5.98×10^{27} g); R the radius of the Earth (6378 km); Γ is the universal gravitational constant which in the cgs system is 6.66×10^{-8} dyna cm^2 g^{-2}.

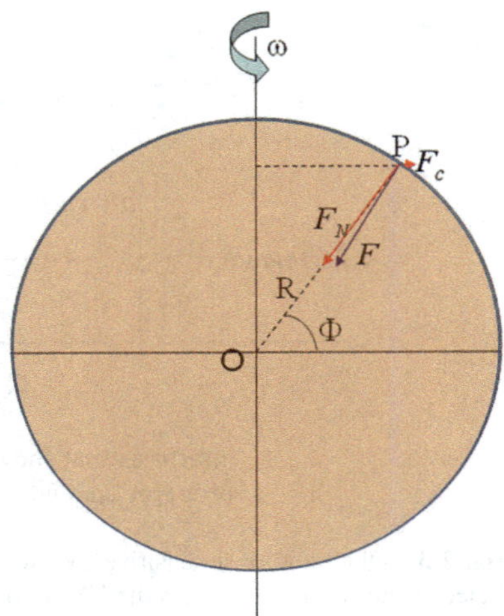

Fig. 2.4 Schematization of the forces acting on a material point (P) near the earth's surface

The Newtonian attraction force is directed towards the centre of the Earth according to the line joining the point P, considered of unitary mass, placed near the earth's surface and the centre O of the Earth (Fig. 2.4);

(2) the centrifugal force Fc, of intensity:

$$F_c = m\omega^2 \cdot d = m\omega^2 \cdot R \cdot \cos\Phi$$

where ω (7.29212 × 10⁻⁵ rad/s) is the angular velocity of the Earth; d the distance of P from the rotational axis of the earth; Φ the latitude. The centrifugal force is directed towards the outside normally to the rotation axis (Fig. 2.4). The angular velocity ω is expressed by:

$$\omega = \frac{2\pi}{T}$$

where T is the time taken by the Earth to make a complete rotation around its axis.

On the mass unit placed in the point P, other forces act due to the attraction of the other planets and in particular of the Sun and the Moon. These forces, which vary over time, give rise to tides.

The resultant of the two forces F_N and Fc is the force of gravity F; its direction coincides with that of the plumb line.

$$F = \Gamma \cdot \frac{M \cdot m}{R^2} - m\omega^2 \cdot R \cdot \cos^2\Phi$$

in terms of gravity acceleration, the above equation is written as follows:

$$g = \Gamma \cdot \frac{M}{R^2} - \omega^2 \cdot R \cdot \cos^2 \Phi$$

As is clear from the above equation, the acceleration of gravity increases as the move from the equator to the poles. The preponderant contribution for g is given by the Newtonian part g = 980 gal.

The "gal" is the unit of measurement of the acceleration of gravity introduced in honour of Galileo Galilei and, in the cgs system, 1 gal = 1 cm/s^2.

The force of gravity is conservative and as such, derives from a potential, known as the earth gravitational potential. The place of the points of equal potential value is a closed surface, called equipotential surface. The surface of the sea (purified from the tides, currents, waves, etc.) and thought prolonged below the continents, to form a closed surface, is an equipotential surface of the force of gravity (not taking into account the Moon and the Sun). This particular surface is called the geoid.

The geoid is assimilated to a rotation ellipsoid or, more correctly to a spheroid whose equation can be reduced to the simple form:

$$\zeta = a \cdot (1 - \alpha \cdot \sin^2 \varphi)$$

where ζ is the geocentric distance, φ the geographical latitude, a is the terrestrial equatorial radius and α the terrestrial crushing which results to be equal to:

$$\alpha = \frac{3\sigma}{2a^2} + \frac{\gamma}{2}$$

This crushing, initially derived from geodetic measurements, was redetermined by observations from satellites leading to the result $\alpha = 1/298$.

It should be pointed out that the term terrestrial spheroid means the revolution ellipsoid adopted to approximate the shape of the Earth; so we use to speak of a reference ellipsoid. The parameters that characterise the reference ellipsoid are the terrestrial equatorial radius a and the polar crushing α. Several ellipsoids are proposed to approximate the shape of the Earth, but what is normally used in geodetic and geophysical applications, is the international ellipsoid of reference taken by convention in 1924 by the II General Assembly of the Geodetic Union and International Geophysics (UGGI) with a = 6,378,388 m and $\alpha = 1/297$ (Fig. 2.5).

The gravity anomaly is defined, in the measurement point P, as the difference between the measured gravity in P and the normal or theoretical gravity (calculated on a model of Earth) in P. In formulas:

$$\Delta g(P) = g^k(P) - g_T(P)$$

where $g^k(P)$ indicates the gravity value observed at a point P and reduced for the tide and drift, $g^T(P)$ indicates the value of the gravity calculated, in the same point, based on the assumed Earth model.

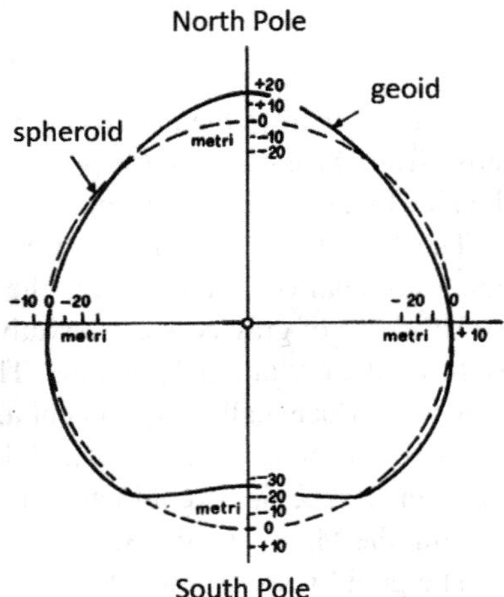

Fig. 2.5 Comparison between the surfaces of the geoid and the spheroid

Gravimetric measurements are generally performed on the topographic surface and in any case on surfaces that do not coincide with the surface of the ellipsoid on which the gravimetric effect of the Earth model is normally calculated. For this reason, it is necessary to make some reductions to bring the measured gravity values $g^k(P)$ and calculated $g^T(P)$ on the same surface to make them comparable. The use of corrections is still normal, and almost all gravimetric maps published are related to correct data. Undoubtedly the most interesting aspect of the use of correct data consists of their validity to illustrate more or less complex geological situations that with the only measured data, could not be highlighted. The classical formulation of the Bouguer anomaly includes a model, called "Bouguer model", in which only the masses inside the geoid are taken into consideration, on whose surface the observed values of gravity are reduced: the effect of the masses external is calculated and subtracted from the observed values. The biggest difficulty of this procedure lies in the estimation of the density of the masses exceeding the geoid; it is clear that, especially in the case of measurements carried out in mountainous areas or with uneven topography, the estimate of the average density of the masses between the topographical surface and the sea level may be affected by coarse errors.

The value of the gravity anomaly obtained from the reduction of the theoretical gravity value g^T, calculated on the ellipsoid, at the measurement point and compare this value with the observed one is numerically equal to the value of the **Bouguer anomaly**.

Instead, the classical Bouguer anomalies, by their very definition, do not allow for highlighting inhomogeneities in the density distribution located between the topographic surface and the sea level, an area of great interest, mainly but not only, for forensic, environmental, archaeological and engineering purposes. In the following, the anomalies, even if indicated as Bouguer's anomaly, will be referred to the physical surface of the Earth. Some corrections are made to reduce the theoretical gravity value to the measurement point.

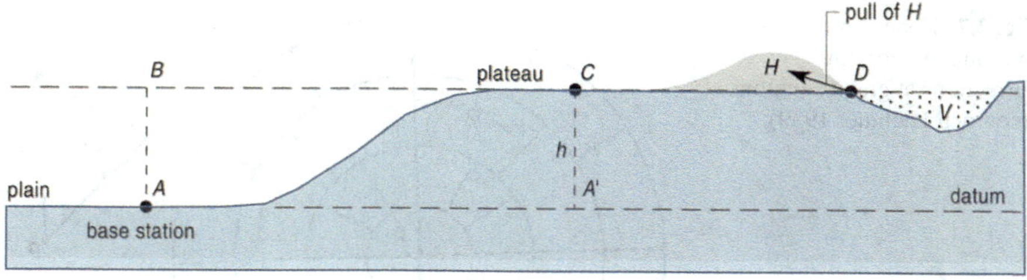

Fig. 2.6 Topographic corrections (modified Musset and Khan 2000)

Latitude correction. Latitude correction is related to: (i) the Earth's centrifugal force. This force must be added (vector sum) to the gravitational force (Newton's law). The correction involves a variation of the gravity force as a function of the radius of rotation. Hence the smallest gravitational force is on the equator (maximal centrifugal force), and the largest is on the pole; (ii) the gravitational force is also affected by the ellipsoidal shape of the Earth and this implies a decrease of the gravitational force on the equator.

The International Gravity Formula allows to remove both of these effects:

$$g_\lambda = 978,031 : 8(1 + 0 : 0053024 \sin^2\lambda - 0 : 0000059 \sin^2 2\lambda) \text{ mGal}$$

Free-air correction. It is a correction related to the topographic effect related therefore to the changes in the elevation considering only air between the instrument (gravity meter) and selected datum (leftmost part of Fig. 2.6).

To get the change in gravity acceleration with height, the free air correction will be:

$$F.A.C. = -\frac{2g}{R} = 0.3086 \text{ mGal/m}$$

An accuracy of 0.01 mGal needs a measure the elevation of the gravity meter with an accuracy of 3 cm.

Bouguer correction. The effect of rocks laying between the measured point and reference datum (Fig. 2.6 in the centre), ignored during the free-air corrections, is considered in the Bouguer correction. This correction removes this effect. Hence a slab with an average density of surrounding rocks added and the Bouger correction can be written as:

$$B.C. = 0.04192\rho \text{ mGal/m}$$

where ρ is the density of the Bouguer slab.

The free-air and Bouguer correction is often combined into an elevation correction

$$F.A.C - B.C = (0.3086 - 0.04192\rho) \text{ mGal/m}$$

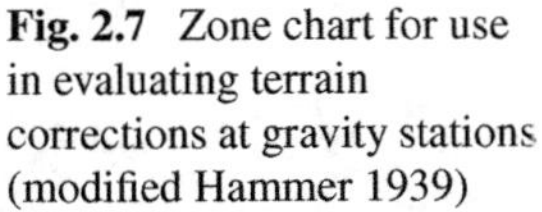

Fig. 2.7 Zone chart for use in evaluating terrain corrections at gravity stations (modified Hammer 1939)

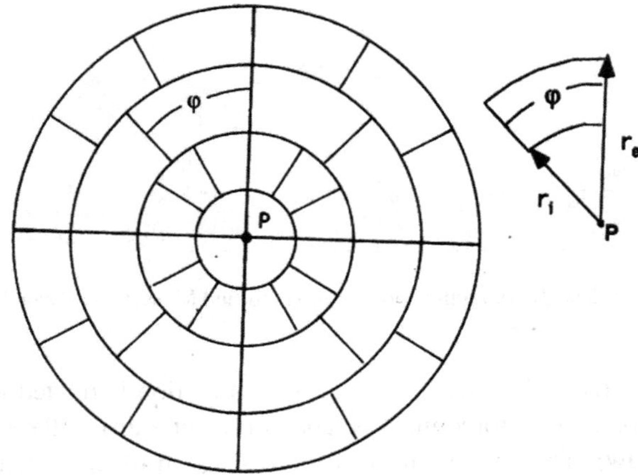

Terrain correction. Regarding the Fig. 2.6, right, is importantly considered the topographical correction. The hill above the Bouguer slab with its gravity force contributes to a decrease of the measured acceleration. Similarly, the valley also decreases the value, because in the computing of the Bouguer correction, the Bouguer slab (with a flat surface) has already subtracted and did not account for the missing masses of the valley. Hence the terrain correction is always added. This correction can be made dividend the surroundings areas of the gravity stations into zones, with average altitude in every zone and compute the gravity effect of the zones (e.g. Hammer 1939). The easiest way is to print the zoning chart (Fig. 2.7) into the transparent sheet and overlay it over the topographic map. Based on the DEM (digital elevation model of the terrain) for every sector the terrain correction is:

$$\mathrm{T.\,C.} = \Gamma \rho \phi [(r_e - r_i) + (r_i^2 + \Delta z^2)^{1/2} - (r_e^2 + \Delta z^2)^{1/2}]$$

where Γ is the Newton's constant and ρ is the density, Δz is equal to the difference between the station quota and the average share of the sector.

Tidal correction. The tidal correction is considered in order to keep into account the gravity is time-independent. Since the gravity effect of the Sun, Moon and large planets must be removed. This correction must be computed (e.g.) according to the Longman (1959).

Drift correction. Also, this correction must be considered to keep into account the gravity is time-independent. The correction removes the changes caused by the instrument itself.

In fact, if the gravimeter would be at one place and take periodical readings, the readings would not be the same. The drift is usually estimated from repeated readings on the base station. The measured data are then interpolated, e.g. by a third-order polynomial, and corrections for profile readings are found.

The final expression for the Bouguer anomaly becomes:

$$(\Delta g)(P) = g_H^k - (g_T - 0.3086H + 0.04192\rho H - T.C.)$$

where P is on the topographic surface or in any case on the surface defined by the measurements, the anomalies thus defined are referred to the physical surface of the Earth. It should also be noted that for the reduction of the plate and the topographic correction it is assumed that the masses comprised between the topographic surface and the geoid have constant density; any non-homogeneity or any eventual deviations between the density of the model and the real density are the sources of the anomalies. Inhomogeneity in the density distribution can occur at any depth; they will always affect the measurements taken on the whole Earth. The deeper the perturbing mass is, the more extensive is the surface effect. In general, a mass of the order of a few meters will cause extended anomalies some tens of meters. If the bodies have kilometres dimensions, the anomalies generated will be in the order of kilometres.

2.2.3 Magnetometry Method

The purpose of the magnetics methods is to study the subsoil from the point of view of its magnetic properties. The method as a lot common with gravimetry but it must also be said that there are also significant differences.

First one is that there are no magnetic monopoles, but dipoles (and higher orders— quadrupoles and more) are the principal units. Unlike the gravimetric field, the magnetic field of the Earth is less stable and this means that the magnetic field could change quickly. Another substantial difference is related to the fact that the magnetic results, displayed in the form of maps, are dominated mainly by local anomalies, and different rock types present, in many cases, a magnetisation quite large than in the case of densities. The magnetic methods are parts of the "family" of the potential field methods and therefore suffers from the non-uniqueness.

As a consequence of Ampere's law, the magnetic field is due to a flow of electrically charged particles (electrical current). A current $\mathbf{I}$ in a conductor of length Δl creates at a point P a magnetising field $\Delta \mathbf{H}$

$$\Delta \mathbf{H} = \mathbf{I}\,\Delta l \times \frac{r1}{4\pi r^2}$$

where the $\mathbf{H}$ is the magnetising field in amperes per meter, r and l are in metres, $\mathbf{I}$ is in amperes, and the directions are as shown in Fig. 2.8. Regarding the Fig. 2.8 the current flowing in a circular loop behaves like a magnetic dipole located in the centre of the loop. In the same way on a microscopic level, the electrons that orbit around an atomic nucleus are electrical charges that move circularly and cause therefore atoms to have magnetic moments. Also, the molecules have spin, which gives them magnetic moments; so a magnetisable body placed into the magnetic field undergoes

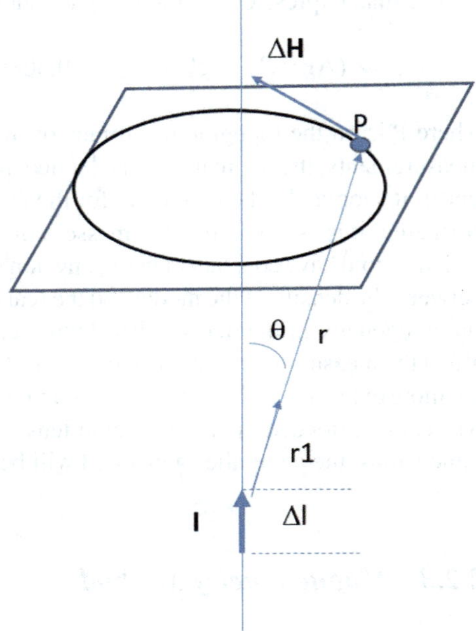

Fig. 2.8 Ampère's law. A current I through a length of conductor Δl creates a magnetising field ΔH at a point P (Telford et al. 1990)

a magnetisation by induction. The magnetisation is caused by reorientation of atoms and molecules so that their spins line up. The magnetisation is measured by the magnetic polarisation M (also called magnetisation intensity or dipole moment per unit volume). The lineup of internal dipoles produces a field M, which is added to the magnetisation field **H**. The SI unit for magnetisation is ampere per meter (A/m). For low magnetic fields, **M** is proportional to **H** and is in the direction of **H**. The degree to which a body is magnetised is determined by its magnetic susceptibility μ, which is defined by

$$\mathbf{M} = \mu\mathbf{H}$$

The magnetic susceptibility is the basic rock physical parameter determining the applicability of a magnetic survey. Amounts and susceptibilities of magnetic minerals determine the overall magnetic response of rocks in them.

The magnetic induction **B** is the total field, including the effect of magnetisation:

$$\mathbf{B} = \mu_0(\mathbf{H} + \mathbf{M})$$

where μ_0 is the permeability of free space which has the value of $4\pi10^{-7}$ Wb/Am (Weber per ampere meter). SI unit for B is the tesla (1 T = 1 N/ampere-meter = 1 Wb/m^2). Although the interest is for the Earth's field **H,** in the magnetic field prospection, the quantity measured is **B,** but usually, $\mu = 1$ and the maps of **B** can be treated as maps of **H**.

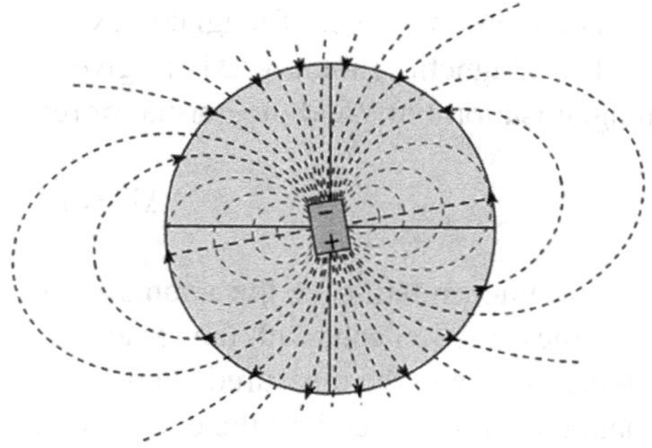

Fig. 2.9 Magnetic field of the Earth

The Earth's magnetic field has three components:

1. The main field is originating within the Earth's interior and changing relatively slowly (Fig. 2.9).
2. A small field (compared to the main field) which varies relatively rapidly and originates outside of the Earth:

The changes in the geomagnetic field caused by external sources have lower amplitude than changes stemming from the internal changes; however, their period is much shorter and thus could seriously affect the magnetic survey. The periodical variations are the Solar diurnal variations with a period of 24 h and amplitude of several tens of nT and Lunar variations with a period of 25 h and amplitude of 2 nT. Next, there are emphshort period variations with periods of a tenth of seconds up to tens of minutes with amplitudes from a tenth of nT up to tens of nT. These could be periodical or random and are mostly effects of the Solar activity. The most important are magnetic storms. They are the effect of increased Solar activity, could appear several times per month and last even for several days. The amplitude of the storms could be several thousands of nT and have a random fluctuation. Effects of these variations could be easily removed from the measured magnetic data in a similar way as in the gravity prospection—using a base station and subtracting the base-station data from the measured ones. However, the magnetic storm has such a high amplitude and random course that it is best to avoid measurements during the storm.

3. Spatial variations of the main field, caused by the inhomogeneities of the Earth's crust. They are the local variations of the main field originate from the magnetic minerals (that vary in concentration) in the near-surface rocks.

The anomalies could have very different amplitudes; exceptionally, they could even double the Earth's main field. They are usually localised, and hence the magnetic maps are often hard to read, compare with the Bouguer anomaly map. The sources of magnetic anomalies could not be very deep since temperatures below about 40 km should be above the Curie point, the temperature at which rocks lost their magnetic properties (about 550 °C). Thus local anomalies must be associated with features in the upper crust.

These are the target of magnetic exploration.

The magnetic anomaly ΔF is given by the difference between the measured magnetisation Fmis and the normal or reference F_N:

$$\Delta F = \text{Fmis} - F_N$$

The interpretation of the anomalies is complex for two reasons: the field has variable direction and, with the same source, the anomaly changes shape depending on the place and the measured component. In the following considerations, only the magnetisation induced by the current earthly magnetic field is considered, and it is assumed that the magnetic axis of the source is always parallel to the direction of the external field.

The magnetic interpretation of real data, where the assumption of a purely induced magnetism is not always true, is very complicated.

Before talking about the interpretation of magnetic anomaly is important to introduce a speech about the magnetic properties of the earth materials.

The magnetisation of the rocks can be considered as a vectorial sum of two vectors. The magnetisation presents only if an external field is applied that ceases when the external field is removed (induced magnetisation) and magnetisation present even without the external magnetic field (the so-called natural remanent magnetisation). For example, effusive rocks have had the remanent magnetisation often much stronger than the induced one.

According to their behaviour, when placed into an external magnetic field, the materials could be divided into two main groups—diamagnetic and paramagnetic (Fig. 2.10). Atoms dominate diamagnetic material with orbital electrons oriented to oppose the external field—the susceptibility is negative (see Table 2.1). Diamagnetic materials are graphite, quartz, feldspar, marble, salt, etc. Atoms of paramagnetic materials have non-zero moments without the presence of an external field, and magnetic susceptibility of such materials is positive. The direction of the magnetisation of individual atoms is randomly oriented, and their vector sum is non-zero but weak. In the presence of the external field, the magnetic atom slightly aligns, forming a weak

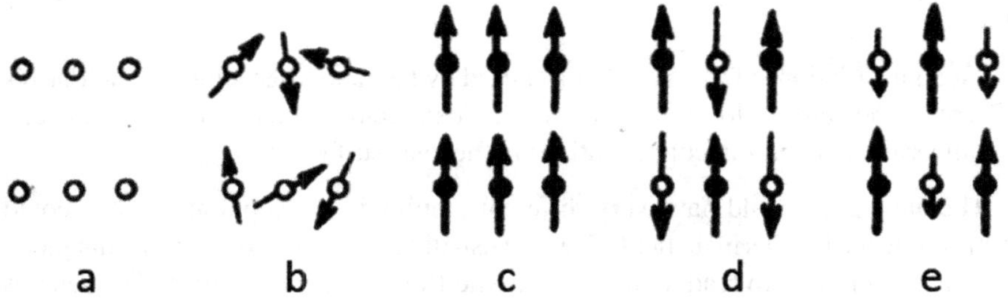

Fig. 2.10 Magnetic Schematic diagram showing the orientation of magnetic moments in the crystal lattice of different materials: **a** diamagnetic, **b** paramagnetic, **c** ferromagnetics, **d** antiferromagnetic and **e** ferromagnetic (Mareš and Tvrdý 1984)

Material	Susceptibility (SI $\times 10^6$)
Air	0
Ice	−9
Freshwater	0
Seawater	0
Topsoil	0.1–10
Coal	0–1000
Dry sand	30–1000
Wet sand	30–1000
Gravel	20–5000
Clay	10–500
Weathered	10–10,000
Salt	−10
Shale	0–500
Siltstone	10–1000
Sandstone	20–3000
Chalk	0–1000
Limestone	10–1000
Slate	0–2000
Graphitic	10–1000

magnetisation—an induced magnetisation. When the external field is removed, the magnetisation ceases. The magnetic effect of diamagnetic and most paramagnetic substances is weak. Certain paramagnetic materials (iron, nickel, cobalt) could have such strong magnetic interactions that the magnetic moments in large regions—domains—align. This effect is called ferromagnetism and is about 106 times the effect of diamagnetism and ferromagnetism. The ferromagnetism decreases with increasing temperature and ceases when the temperature exceeds the Curie point. Some materials have domains further divided into subdomains with opposite orientation and the overall magnetic moment nearly cancels. These materials are called antiferromagnetic, and their susceptibility is low. The common example is hematite.

The last group have subdomains also aligned in oppositions; however, their net magnetic moment is non-zero. This could be either because one orientation of subdomains has a weaker moment or that there is less domain with one of the orientations. Such substances are called ferrimagnetic. Magnetite, titanomagnetite, oxides of iron and iron and titanium are examples of the first type of minerals. The second group is represented by pyrrhotite. There is a direct proportionality between the induced magnetisation and the susceptibility and concentration of magnetic minerals present in the rocks. The dipoles orientation is the same as the geomagnetic field (in general the external field). However, the measured magnetisation is not always of this direction. Responsible for this phenomena is the remanent magnetisation. The remanent magnetisation is present even if we remove the external magnetic field. The most common types of remanent magnetisation are described below.

When a magnetic material is cooled below the Curie temperature in the presence of the external magnetic field, it is the so-called thermoremanent magnetisation and the direction of the dipole depends on the direction of the external field at the time and place where the rock cooled. Detrital magnetisation has fine-grained sediments. When magnetic particles slowly settle, they are oriented into the direction of an external field. Various clays exhibit this type of remanence.

Chemical remanent magnetisation is created during a grown of crystals or during alteration of existing minerals. The temperature must be low (below the Curie point). This type might be significant in sedimentary or metamorphic rocks.

Isothermal remanent magnetisation is the residual left following the removal of an external field. Its amplitude is low unless it was created within a very large magnetic field like during the lightning strike. Viscous remanent magnetisation is produced by long exposure to an external field. It grows with a logarithm of time. It is common for all rock types, the direction is usually close to the direction of the present magnetic field, is quite stable and amplitude could be up to 80% of the induced magnetisation. Dynamic remanent magnetisation is created when a rock is exposed to variate pressures within a magnetic field. The pressures could be of various types ranging from tectonic or seismic pressures up to hammer strikes.

2.2.4 The Electrical-Resistivity Active Method

The electrical resistivity method studies the electrical properties of the subsoil or, in general, the ground materials. It is used to underline the horizontal and vertical discontinuities in terms of some physical, electrical parameters (electrical resistivity and conductivity).

The electrical resistivity method employs an artificial source (continuous or very low-frequency electrical current) which is introduced into the ground or the material through a pair of electrodes (so-called current electrodes). Another pair of electrodes (so-called potential electrodes) measure the electrical potential. Therefore a current flow through a medium consisting of materials with different individual resistivities where the resistivity ρ of a material is a measure of how the material allows or not allow the flow of electrical current. Resistivities values of the materials vary in a range illustrated in Table 2.2.

As shown in Table 2.2 the great variation in the resistivity values of different materials makes the resistivity measurements of unknown material is very useful in identifying the materials.

To have better results in the field measurements is important to combine the resistivity of a material with the so-called apriori information about the site. The information about the investigated sites (geology, hydrogeology, etc.) allows identifying well the materials that constitute the various underground layers.

Table 2.2 Resistivity and conductivity values for some types of rocks (modified Loke 2001)

Material	Resistivity (ohm, m)	Conductivity (mS/m)
Igneous and metamorphic rocks		
Granite	$5 \times 10^3 - 10^6$	0.001–0.2
Basalt	$10^3 - 10^6$	0.001–1
Slate	$6 \times 10^2 - 4 \times 10^7$	$2.5 \times 10^{-5} - 1.7$
Marble	$10^2 - 2.5 \times 10^8$	$4 \times 10^{-6} - 10$
Quartzite	$10^2 - 2 \times 10^8$	$5 \times 10^{-6} - 10$
Sedimentary rocks		
Sandstone	$8 - 4 \times 10^3$	0.25–125
Shale	$20 - 2 \times 10^3$	0.5–50
Limestone	$50 - 4 \times 10^2$	2.5–20
Soils and water		
Clay	1–1000	1–1000
Alluvium	10–800	1.25–100
Groundwater (fresh)	10–100	10–100
Sea water	0.2	5000

The basic theory of the method is related to Ohm's law. The resistivity is related to the resistance of an ideal cylinder of length L and cross-sectional area S of uniform composition by:

$$R = \rho(L/S)$$

Ohm's law affirms that $R = V/I$, where V is the potential difference between the ends of the cylinder and I is the total current flowing through the cylinder. Therefore the resistivity of the material can be obtained by:

$$\rho = (S/L)\,(V/I) = R_{app}K$$

The ratio S/L is the geometric factor (related to the geometry of the cylinder), and R_{app} is the apparent resistance. It is important to underline that in nature, the geometry is not as simple as the above considerate uniform cylinder. To resolve this, a good starting point is shown in Fig. 2.11a that consider a current flowing radially away from a single electrode at positive potential located on the surface of the ground.

As shown in Fig. 2.11a the lines of current and the equipotential surfaces are perpendiculars. This can explain considering a "voltage drops" (potential local gradients) that drive the current according to the simple scalar form of Ohm's law given by $I = V/R$. In this case, the resistance of the medium above the ground (in this case, air) is "infinite" so that the ground forms a Dirichlet-type boundary (Jackson 1975).

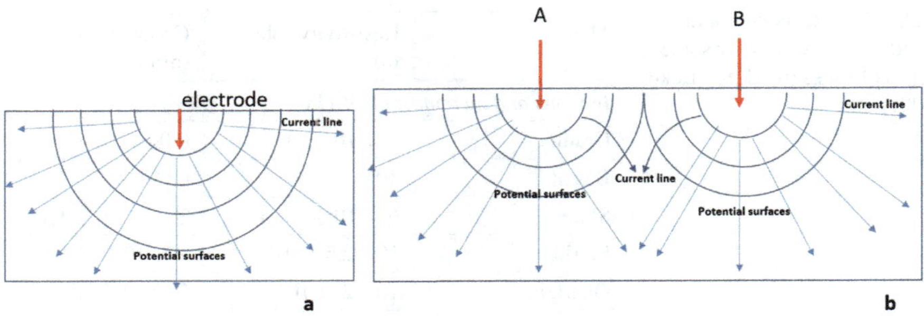

Fig. 2.11 Schematization of current flow and equipotential surfaces in material: **a** in a level field with homogeneous subsurface structure, **b** between the two current electrodes A and B in a level field with homogeneous subsurface structure

But in a real case, the current electrodes, A and B, are two (Fig. 2.11b). The "effective depth" of penetration is related to the spacing between the two current electrodes. The spacing of the order of a meter implies that a total current will flow no more than a few meters from the surface. The spacing of the order of a kilometre most of the current will penetrate very deeply into the underlying material (Leucci 2019). Is important to underline here that the total resistivity measured at the ground surface in field studies of multilayer systems is a weighted average of the resistivities of the various materials that the current encounters and therefore it is not the true resistivity of the underlying material (apparent resistivity).

The effective depth of the boundary between the two layers can be esteem using the "cumulative resistivity" method. This method employs a plot of the sum of the apparent resistivities, $\Sigma \rho_{app}$, versus the effective depth (Robinson and Coruh 1988).

The electrodes array and spacing is, therefore related to the depth of investigation (Roy and Apparao 1971). Using Table 2.3 (Loke 1999) is possible to esteem the average depth of investigation. It is an average depth because of the values shown in the Table 2.3 have been calculated for a homogeneous medium that is an ideal case. In real cases, the noise related to the particular condition of the investigated material can affect an increase or decrease of the observed penetration depth (for more see Leucci 2019).

The most commonly used conventional arrays in forensic applications include Wenner (alpha), Schlumberger, and dipole-dipole. These arrays with their corresponding geometric factor are illustrated in Fig. 2.12.

The choice of a particular array that can be used in field procedures depends on several factors, such as the geological structures to be delineated, objects to be evidenced, heterogeneities of the subsurface, sensitivity of the resistivity meter, the background noise level and electromagnetic coupling. Other important factors are related to the subsurface variations of the resistivity: some arrays are sensitive to vertical variations and some arrays to lateral variations in the resistivity. As seen above the depth of investigation, and the horizontal data coverage and signal strength of the array are the other two important factors (Leucci 2019).

Table 2.3 Average depth of investigation of the various electrode arrays (Loke 1999)

Electrodes array	Zmean/a	Zmean/L
Wenner alpha	0.519	0.173
Wenner beta	0.416	0.139
Wenner gamma	0.594	0.198
Dipole-dipole		
n = 1	0.416	0.139
n = 2	0.697	0.174
n = 3	0.962	0.192
n = 4	1.22	0.203
n = 5	1.476	0.211
n = 6	1.73	0.216
Wenner-Schlumberger		
n = 1	0.52	0.173
n = 2	0.96	0.186
n = 3	1.32	0.189
n = 4	1.71	0.19
n = 5	2.09	0.19
n = 6	2.48	0.19

Fig. 2.12 Electrode arrays with their corresponding geometric factors

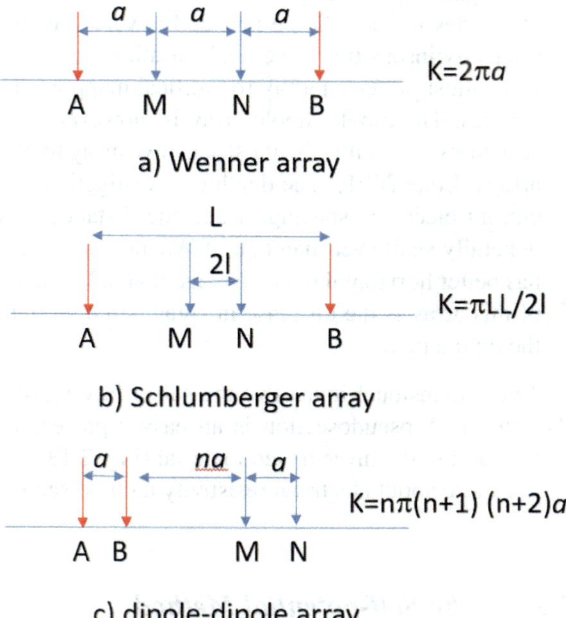

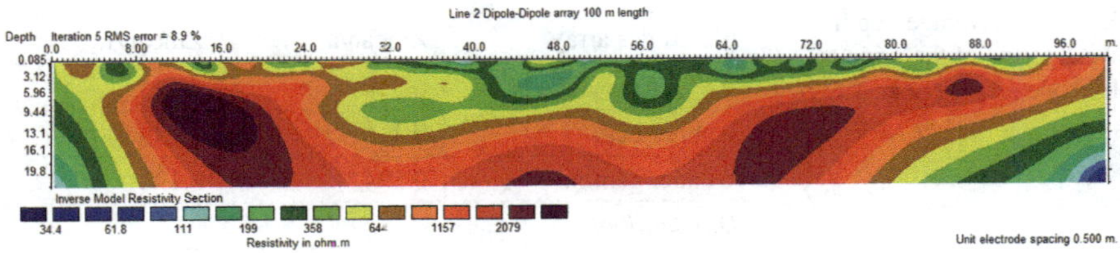

Fig. 2.13 Pseudosection visualization on a computer during ERT data acquisition

About these last factors that influenced the choice of the best array is important to underline that:

the conventional Wenner and Schlumberger arrays are relatively sensitive to vertical variations in the subsurface resistivity below the centre of the array but less sensitive to horizontal variations in the subsurface. The arrays have moderate depths of investigation and generally strong signal strength, which is inversely proportional to the geometric factor used in calculating the apparent resistivity values. The major limitation of these arrays is the relatively poor horizontal coverage with increased electrode spacing. Wenner array is preferred for surveys in a noisy site because of its high signal strength; however, the array is less sensitive to 3D structures (Loke 2001);

the dipole-dipole array is the most sensitive to resistivity variations below the electrodes in each dipole pair and is very sensitive to horizontal variations but relatively insensitive to vertical variations in the subsurface resistivities. Thus, it is the most preferred array for vertical mapping structures like tombs, walls, and cavities. The dipole-dipole array is, however, very poor in mapping horizontal structures. Also, it is the most sensitive array to 3D structure among the common arrays (Loke 2001). The depth of investigation of the array depends on both the current electrode spacing, *a* and the distance between the two dipoles; and is generally shallower than that of Wenner array. However, the dipole-dipole array has better horizontal data coverage than a Wenner array. The major disadvantage of this array is the decrease in signal strength with increasing distance between the dipole pair.

Two-dimensional images of the subsurface resistivity variation are called *pseudosections*. A pseudosection is an easy representation of an electrical-resistivity distribution in the investigated material (Fig. 2.13).

For more about electrical resistivity method see Leucci (2019).

2.2.5 *The Self-potential Method*

Like gravity and magnetic methods, the self-potential (SP) method is part of the family of potential field methods. It is a passive electrical method because it involves the measurement of the electric potential at a set of measurement points called self-potential stations. In the last years, the SP method has had an increasing number of applications. They vary from mapping fluid flow in the subsurface (to detecting preferential flow paths in earth dams and embankments) to detecting metal bar corrosion (helping in understand the conservation state). Three main mechanisms lead the SP:

the electrokinetic coupling (streaming potential) (Reynolds 2011);
chemical potentials gradients of ionic species (membrane or diffusion potentials), the potential redox gradients (electro-redox) (Corry 1985; Naudet et al. 2004; Naudet and Revil 2005).

All these mechanisms are related to potential chemical gradients of charges carriers creating polarisation in the porous media. Therefore The SP anomalies are associated with charge polarisation mechanisms occurring at depth.

Particularly the electrokinetic effect is related to the electric field associated with the flow of electrical charges (i.e. the groundwater flow). The parameters associated with this effect are the electrokinetic coupling "C" associated with the potential electrical field ($\nabla\varphi$) and hydraulic pressure difference (∇p).

In this case the electrical current density ($\mathbf{J}$) can be considered as vectorial sum of the driving current ($\mathbf{J}s$) and the conduction current ($\sigma\nabla\varphi$):

$$\mathbf{J} = -\sigma\nabla\varphi + \mathbf{J}s$$

where $\mathbf{J}s = C\ \nabla p$ (electrokinetic coupling = driver current density + pressure difference) and $C = -(L/\sigma)$.

φ is electrical potential (volt); p is the fluid pressure (Pascal); σ is the electrical conductivity (Siemens/meter); C is the electrokinetic coupling coefficient (volt/Pascal); L is an electrokinetic coupling term (Ampere/Pascal meter).

Using the above equations combined with the continuity equation (ex-prime the conservation of the charge), the electrical potential can be written as

$$\nabla^2\varphi = (1/\sigma)\nabla\mathbf{J}s - (\nabla\sigma/\sigma)\nabla\varphi$$

The first term is the primary source of polarisation (related to the current density) while the second term is the secondary source of polarisation (related to electrical resistivity heterogeneity).

Unlike the electrokinetic effect, electrochemical sources can be attributed to several phenomena. One of these is due to a concentration gradient between two regions (the diffusion of ions). This source of current is divergence-free because it is balanced by a current flowing throughout the Earth conductivity structure.

The redox processes, occurring in ore bodies and contaminant plumes, is another electrochemical source mechanism. Also, in this case, the total electric current density is divergence-free.

The redox potential can generate large SP anomalies in the surrounding conductive medium. Strong negative SP anomalies with magnitudes, usually reaching a few hundred millivolts have been reported over ore deposits for more than 50 years and explained thanks to a geobattery model (Bigalke and Grabner 1997).

2.2.6　Seismic Tomography Method

"Major" infrastructures (bridges, viaducts, tunnels, etc.) whether they are famous or so-called "minor" (places, houses, historical building, etc.), architecture represent an important topic of investigation in civil and criminal cases. To avoid disasters like that of the Morandi bridge in Genoa, these infrastructures must be monitored and preserved as much as possible. Seismic methods are useful for monitoring the physical-mechanical characteristic of the infrastructures.

The basic theory of seismic method involves the generation of a short pulse of seismic wave energy and recording the time of the arrival of the same seismic pulse at certain distance locations.

Source of seismic energy could be the impact of a mass which is detected by sensitive seismometers operating with electronic amplifiers and a suitable recorder. Using the measurements of the seismic wave travel time is possible to study the reflection, absorption, diffusion, and refraction wave that allow the seismic interpretation. Here quite different principles are involved than the methods above described that include gravity, magnetic, and electrical techniques. 2D and 3D seismic reflection, seismic refraction, seismic tomography, are used for infrastructure analysis. In this paragraph, some important considerations will be done about the seismic tomography method.

Generally, in seismic surveys, waves are emitted by a controlled source and propagate through the subsurface. Several waves will be backscattered at the surface due to refraction or reflection at boundaries within the surveyed materials. Receivers distributed along the surface detect the motion caused by these returning waves and measure the arrival times of the waves at various distances from the source. The travel times are then converted into velocities, and the spatial distribution of velocities can be systematically mapped.

The velocity value of the waves carries information on the type of sediment or rock or materials that they are crossing. This method is important not only for structural information but also for the physical characterisation of layers and thus is very useful in forensic investigations (Reynolds 2011). Seismic waves include body waves that travel three-dimensionally through solid earth volumes and surface waves that travel near the surface of the earth volume (Fig. 2.14). Surface waves are categorised further as Love waves and Rayleigh waves. For the seismic investigation of the ground, the use of body waves is standard. Here P-waves) and shear waves (S-waves); the difference is in the particle motion of the wave propagating through the underground material (Fig. 2.14). The velocities of these waves depend on the elasticity and density of the underground material and can be expressed by:

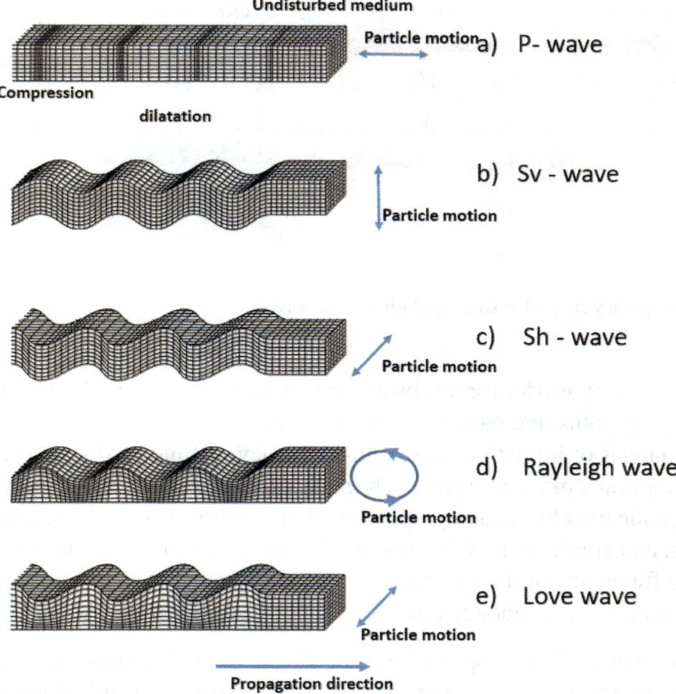

Fig. 2.14 Seismic waves and particle motion: **a** compressional P-wave travelling in a block of material, **b** vertical polarised and **c** horizontal polarised shear wave travelling in a block of material, **d** Rayleigh wave travelling in a section of the earth's surface, **e** Love wave travelling along a section of the earth's surface (modified Steeples 2005)

$$Vp = \sqrt{\frac{k + \frac{4}{3}\mu}{\rho}}$$

$$Vs = \sqrt{\frac{\mu}{\rho}}$$

with the elastic constants k (= bulk modulus) and μ (= shear modulus) and the mass density ρ of the material through which the wave is propagating (for more see Leucci 2019).

Tomography, which has been used in science and technology as the object's internal-structure imaging technique, can map the physical quantity distribution within the object non-destructively. It is an alternative way for quality inspection of many materials.

Seismic tomography is one of the main techniques to constrain the two and three-dimensional distribution of physical properties that affect seismic-wave propagation: elastic, anelastic, and anisotropic parameters, and density. The term tomography comes from the Greek tomos, which means 'slice'. The basic idea of tomography

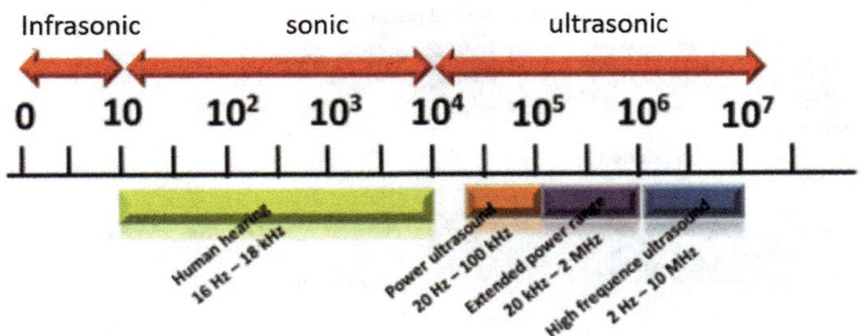

Fig. 2.15 Frequency range for sonic and ultrasonic waves

is to find a velocity model for a subsurface volume consistent with measured travel times along ray paths that pass through the volume.

In the analysis of infrastructures, high-frequency seismic waves are used. So one talk about sonic and ultrasonic pulse. The sonic traveltime tomography is very similar to the ultrasonic traveltime tomography for what concerns basic theories, the purpose of the tests, and application of the method. The difference between the two methods regards the frequency of the emitted wave (Fig. 2.15).

The seismic wave frequency must be selected considering:

(1) The thickness of the inspected object: higher frequency implies faster decays; this means that it becomes impossible to inspect structures with a large thickness.
(2) The assumed size of the defect: higher frequency, imply smaller wavelength (as in the equation below) and the smaller defect is possible to determine (Reynolds 2011):

$$\lambda = v/f$$

λ—wavelength; v—ultrasonic wave propagation velocity; f—ultrasonic wave frequency.

Incorrect assignment of the frequency of seismic wave leads to mistakes in the interpretation of the results of the inspection.

The seismic signal frequency is, therefore related to the concept of seismic resolution. Seismic resolution is a measure of how large an object needs to be seen by seismic methods.

The vertical resolution is derived from the wavelength of seismic source, and layers can be discerned when their thickness is below 1/4 of the wavelength.

The horizontal resolution is derived from the Fresnel zone, the part of a reflector covered by the seismic signal at a certain depth. On a buried horizon, all features with a lateral extent exceeding the Fresnel zone will be visible. The radius of this zone is often taken as the horizontal resolution for unmigrated seismic data. As with the wavelength, the Fresnel-zone size also increases rapidly with depth (Reynolds 2011).

Table 2.4 P- and S-wave velocities in various types of geological materials (Zanzi 2004)

Material	v_P (m/s)	v_S (m/s)
Air	330	–
Dry sands	400–1200	100–500
Saturated sands	1500–4000	400–1200
Clay	1100–2500	200–800
Marne	2000–3000	750–1500
Arenaria	3000–4500	1200–2800
Limestone	3500–6000	2000–3000
Gypsum	2300–2600	1100–1300
Shale	4500–5500	2000–3100
Dolomite	3500-6500	1900–3600
Water	1450	–
Granite	4500–6000	2500–3300
Basalt	5000–6000	2800–3400
Coal	2200–2700	1000–1400
Ice	3400–3800	1000–1900

$$\text{Fresnel zone} = v(d/f)^{0.5}$$

where f is the seismic frequency, v is the seismic velocity, and d is the depth in time.

Therefore resolution depend also on the seismic wave velocity in the investigated materials.

Table 2.4 shows some values of Vp and Vs for a wide variety of materials (Zanzi 2004).

Sonic (S) and ultrasonic (US) can be used direct tomography (DT), semi-direct tomography (ST), indirect tomography (IT) transmission method, or with tomographic (T) mode, depending on the objective of the investigation. Problems that can be analysed are listed in Table 2.5, with an indication of the most appropriate acquisition mode.

Table 2.5 Target of sonic and ultrasonic measurements

Target	Acquisition type
Detection of inhomogeneities (e.g., a variation of material texture, repair interventions, presence of different materials)	DT, T
Detection of multiple leaves and measurement of the thickness of each leaf	T
Detection of detached external leaves	DT, T
Detection of voids or chimney flues	DT, T
Evaluation of the effectiveness of repair interventions (e.g., grout injections, repointing, etc.)	DT, T, ST
Detection of damaged portions of the structure or crack patterns	DT, IT, T

2.2.7 Ground-Penetrating Radar: Fundamental Principles

Is well known that Ground-Penetrating Radar (GPR) method is one of the most widely used for the exploration of the shallow subsurface, especially for civil engineering, geological and environmental or archaeological applications. From a few years, it is also applied in criminal and civil forensic. Its ability to provide, easily and quickly, high-resolution (almost) continuous information on the uppermost few meters (or tens of meters) of the earth heavily contributed to the increasing popularity of this method and its expanding role among the shallow geophysical techniques in the last two decades. Nevertheless, the same reasons, easiness and speed, could make this method highly subjected to misuse. The knowledge of its power and limitations is related to the knowledge of its fundamental principles. This allows a successful application of this (as well as any other) geophysical technique. Since it is possible, concerning the nature of the practical problem to solve, to develop a suitable field and post-acquisition procedures for the specific problem at hand. Both theoretical bases and practical guidelines, as well as numerous case histories on GPR studies in various fields, can be found in the recent literature: books (Conyers and Goodman 1997; Leucci 2015, 2019), geophysical handbooks (Reynolds 2011), Proceedings and Special Issues of geophysical journals (as those devoted to the biennial International Conference on GPR held since 1986) and numerous research papers. Although in earlier times GPR data were generally used and interpreted as they were collected (raw data), they are now routinely subjected to digital data-processing, interpretation and display techniques aimed to further enhance the visibility of meaningful signals in the raw data and to help to understand their three-dimensional relationships. Due to the close kinematic similarity with seismic reflection methods, most of the processing and visualisation techniques currently available in GPR processing software are a direct adaptation of seismic ones. The physical bases and mathematical foundations underlying these techniques are therefore available from seismic literature (e.g. Yilmaz 1987). Nevertheless, although without presuming to furnish a deep examination and an exhaustive treatment of the theoretical and practical aspects of the GPR method, the main basic principles of GPR data, needed for the comprehension of the methodology are concisely exposed in the following pages.

The GPR technique is similar in principle to the seismic reflection technique, but instead of mechanical waves, it uses high frequency (10–2500 MHz) electromagnetic pulses to explore the underground. A radar wave, emitted by a transmitting antenna placed directly above the ground surface, propagates in the ground and is partially reflected by any change in the electrical properties of the subsoil. The reflected energy is then detected by the receiving antenna (Fig. 2.16).

Georadar antennas have a relatively large frequency band, whose width is approximately equal to the *centre-frequency*, that is the frequency around which most of the pulse energy is concentrated. Most GPR equipment uses dipole antennas (identified by their centre-frequency or by the *pulse width*, approximately corresponding to the reciprocal of the centre-frequency) arranged either in *monostatic* or in *bistatic* configurations. In the first case, the same antenna is used for transmission and receiver,

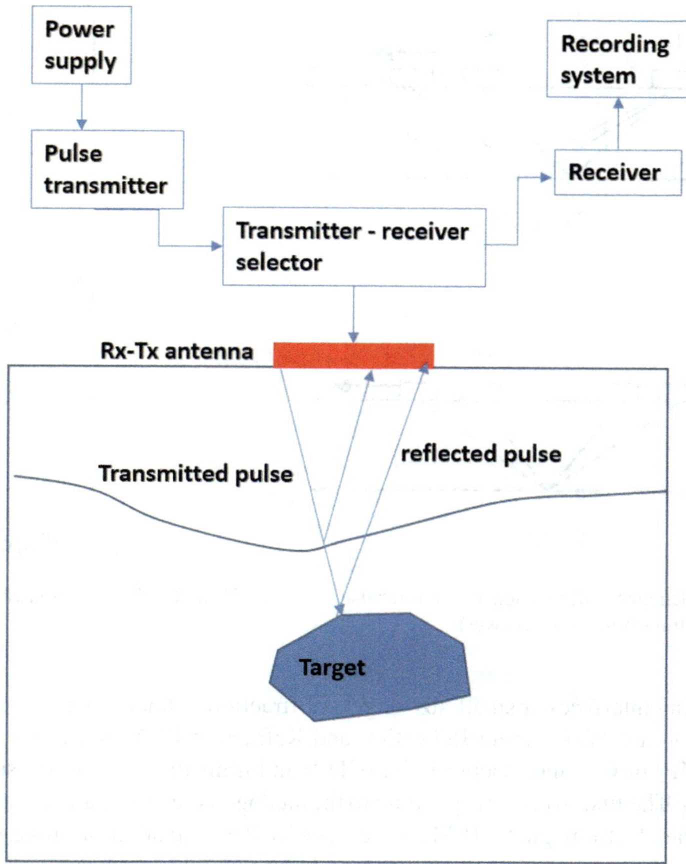

Fig. 2.16 Sketch of the basic components of a GPR system and principle of operation

while in the second situation there is a constant, small offset between the two antennas, that can be placed either in separated cases (as for the low-frequency antennas) or inside the same box (as for the higher-frequency ones). Generally, the offset is sufficiently small that it can be practically neglected and the last arrangement could be considered nearly monostatic.

For both arrangements, the usual data acquisition is the *reflection mode*, performed either as *continuous profiling* (moving the antennas along the profile at a slow, near Tx Rx constant towing speed) or as *stationary point collection* (shifting them stepwise). GPR data, properly amplified, are then recorded and displayed as a two-dimensional section with the antenna positions (or midpoint positions for bistatic systems) in the horizontal axis and the two-way travel time in the vertical axis. This section can be considered a normal-incidence time section (corresponding to the zero-offset section of the seismic reflection), where the two-way time is plotted vertically below the midpoint position even if the actual ray path is slanted, as for the reflection

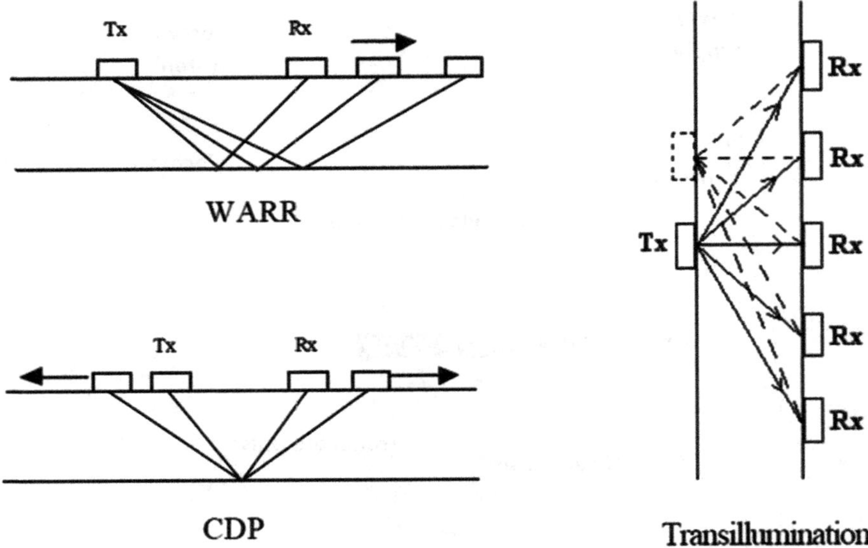

Fig. 2.17 Schematic illustration of data acquisition in the WARR, CDP and transillumination modes (Tx: transmitter, Rx: receiver)

from dipping interfaces or small-size targets (diffraction). Other less common modes of operation are Wide-Angle Reflection and Refraction (WARR), Common Mid-Point (CMP) or Common Depth Point (CDP) and transillumination measurements (Fig. 2.17). The first two, corresponding to the analogous seismic gathers, are mainly used for the electromagnetic (EM) wave velocity determination or, more rarely, in multi-folds surveys, whereas the last is used in tomographic studies.

The interpretation of GPR sections, with the aid of the EM wave velocity field (assumed or determined from in situ measurements), can provide high-resolution information on depth and geometry of reflectors and diffractors and on near-surface earth properties and structure, mainly about its dielectric characteristics. Ground-penetrating Radar is, indeed an EM method operating in a (high) frequency range, where displacement currents dominate over conduction currents, so that the main EM parameter influencing propagation, velocity and attenuation of GPR waves is the dielectric permittivity of the media in which they propagate, whereas the electrical conductivity mainly contributes to EM wave attenuation, and magnetic permeability has generally negligible influence for most common geologic materials. In the theory of GPR Maxwell's equations provide the starting point to understand how electromagnetic fields can be used in geophysical exploration to obtain information on the electric and magnetic properties of the earth, to which the EM field is related by means of empirical relationships known as constitutive equations (Keller 1987; Ward and Hohmann 1987). These relations allow the description of the behaviour of EM waves in a medium using three constitutive parameters, that in general are tensor quantities, but under the assumption of isotropy and homogeneity can be considered scalars: the *electric permittivity*, ε, the *electric conductivity*, σ, and the *magnetic permeability*, μ. A useful approximation, in the case of a homogeneous isotropic medium, is represented by the *damped plane-wave* solution of the scalar wave equation. In this case, each component of the electric (E) and magnetic (H) field at a distance z and time t is related to the corresponding fields at $z = 0$ and $t = 0$ ($E0$ and $H0$) by the expressions:

$$E = E_0 e^{-\alpha z} e^{i(\omega t - \beta z)}$$

$$H = H_0 e^{-\alpha z} e^{i(\omega t - \beta z)}$$

where

$$\alpha = \pm \omega \sqrt{\varepsilon \mu} \left\{ \frac{1}{2} \left[\sqrt{1 + \left(\frac{\sigma}{\omega \varepsilon}\right)^2} - 1 \right] \right\}^{1/2}$$

$$\beta = \pm \omega \sqrt{\varepsilon \mu} \left\{ \frac{1}{2} \left[\sqrt{1 + \left(\frac{\sigma}{\omega \varepsilon}\right)^2} + 1 \right] \right\}^{1/2}$$

α is called absorption constant and β is called the phase constant. While

$$\tan \delta = \sigma / \varepsilon \omega$$

is the *loss tangent* and

$$v = \omega / \beta$$

is the *propagation velocity* in the medium, and

$$\lambda = 2\pi / \beta = v/f$$

is the *wavelength* and is inversely proportional to the frequency f.

The constitutive parameters ε and σ are, in general, complex and have in-phase (d.c.) components, ε' and σ', and out-of-phase (high frequency) components, ε'' and σ'' (Turner and Siggins 1994):

$$\varepsilon = \varepsilon' - i\varepsilon''$$
$$\sigma = \sigma' - i\sigma''$$

At most radar frequencies the out-of phase component of the conductivity (σ'') is generally negligible, while the out-of phase component of the permittivity (ε'') is not. Moreover most geological materials, which are best suited for GPR investigations, are low-loss ($\tan\delta \ll 1$), non magnetic media ($\mu \cong \mu_0$). In this case approximated expressions for α and β can be write as:

$$\alpha = (Z_0/2)[s'/(k')^{1/2}]$$
$$\beta = (w/c)(k')^{1/2}$$

where ω is the radia frequency,
$\mu = \mu_0\mu_r = (4\pi)\ 10^{-7}$ H/m ($\mu_r = 1$)
$\varepsilon = \varepsilon_0\ \varepsilon_r = 8.85\ 10^{-2}\ \varepsilon_r = \varepsilon_r/(36\ \pi\ 10^9)$ F/m
$\omega = 2\ \pi\ f$
$c = 1/(\varepsilon_0\mu_0)^{1/2} = 3 \times 10^8$ m/s is the electromagnetic velocity in free space
$Z_0 = (\mu_0/\varepsilon_0)^{1/2} = 376.8$ O is the free space intrinsic impedance
$K' = \varepsilon'/\varepsilon_0$ is the real part of the relative permittivity (or dielectric constant) of the medium.

From the above equations results that for the materials with conductivities below 50 mS/m the electromagnetic wave velocity of propagation depends exclusively on the real part of the dielectric constant and is not frequency-dependent (Fig. 2.18):

$$v = \frac{c}{\sqrt{\varepsilon'_r}}$$

And the medium attenuation can be approximated by:

$$\alpha \cong \frac{1.69x\,10^3\sigma}{\varepsilon_r}$$

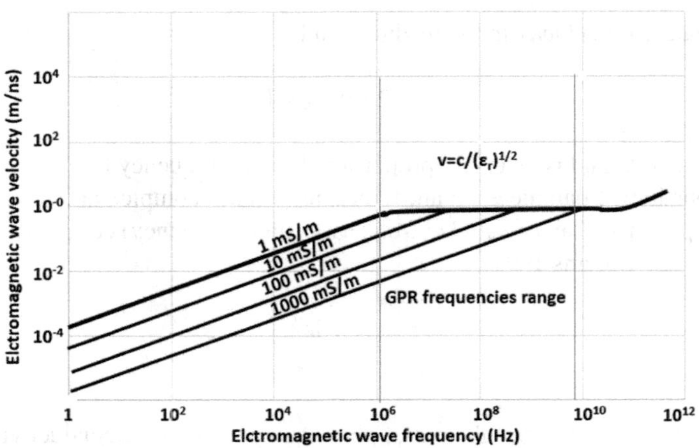

Fig. 2.18 Electromagnetic-wave velocity of propagation trend as a function of frequency [(Davis and Annan 1989) modified]

The dielectric constant varies from its space value of 1 to a maximum of 80 for water, whose presence, therefore, strongly influences the dielectric constant of rock- (or soil-) water mixtures. It is also clear that GPR is a method suited for sounding dielectric low-loss materials. Attenuation increases as the conductivity of the ground increases. Materials having high conductivities, as water-saturated clay or saltwater, rapidly dissipate the radar energy and restrict the investigation depths. The amplitude of radar waves is further reduced by spherical spreading losses, reflection and transmission at discontinuities as well as by small scale heterogeneity scattering which, in turn, increases with increasing frequencies. For these reasons, the penetration capability of GPR decreases as the centre frequency of the antenna increases. When a wave arrives at a boundary separating two media with different EM characteristics, energy is partially reflected and partially transmitted. For normal incidence and in the case of non-magnetic low-loss materials, the *amplitude reflection coefficient, R* can be expressed either in terms of the radar wave velocity in the two layers (v_1 and v_2):

$$R = (v2 - v1)/(v2 + v1) = \frac{\sqrt{\varepsilon_1} - \sqrt{\varepsilon_2}}{\sqrt{\varepsilon_1} + \sqrt{\varepsilon_2}}$$

In contrast to the penetration depth, which decreases as the frequency increases, radar resolution increases with higher frequencies. As in the seismic case, the resolution is a crucial point both in defining the acquisition geometry and interpreting georadar data. Resolution relates to how close two points can be, yet still, be distinguished.

Also for GPR is possible to consider:

1. *the vertical resolution*, which relates to the (minimum) depth separation between two boundaries to give separate reflection events;
2. *the horizontal resolution*, which refers to how close two reflecting points can be situated horizontally, yet be recognised as two separate points rather than one.

Resolving power with a radar system is determined by the bandwidth, which, in turn, for most GPR systems, is considered about equal to the centre (or dominant) frequency. Therefore resolution requirements become a major factor in selecting the operating frequency for the radar. Reflections from two boundaries, separated by a distance Dz, are separated for high centre frequency pulses and are merged together for low centre frequency pulses. The acceptable threshold for vertical resolution generally is a quarter of the dominant wavelength (Sheriff 1994), although this criterion is subjective and depends on the noise level in the data. The above criterion implies that the minimum depth separation (Δz) is:

$$\Delta z = \lambda/4$$

The Fresnel zone width is a measure of horizontal resolution (Leucci 2019). Two reflecting points separated by a distance less than the *first Fresnel zone radius (r)* are considered indistinguishable as observed from the earth's surface. The first Fresnel zone radius is:

$$r = [(\lambda d/2)^{1/2}]$$

and, in addition to velocity and frequency, is also depth-dependent. Since the Fresnel zone generally increases with depth, the spatial resolution also deteriorates with depth.

Among other constraints (Conyers and Goodman 1997), in a GPR survey, the central frequency of the antenna is chosen to obtain a viable compromise between the desired penetration depth and vertical resolution. Moreover, the lateral resolution is important in planning the acquisition geometry, and in particular the spatial sampling along the survey line (inline spacing) and the distance between consecutive lines (crossline spacing). The latter requirement is seldom fulfilled due to time and positioning problems. For more about GPR, see Leucci (2019).

References

Arosio, D. (2010). A microseismic approach to locate survivors trapped under rubble. *Near Surface Geophysics, 8,* 623–633.

Bigalke, J., & Grabner, E. W. (1997). The geobattery model: A contribution to large-scale electrochemistry. *Electrochimica Acta, 42*(1997), 3443–3452.

Cataldo, A., De Benedetto E., & Cannazza G. (2016). *Advances in reflectometry Sensig for industrial applications.* Morgan & Claypool Publisher. https://www.morganclaypool.com.

Cataldo, A., & De Benedetto, E. (2011). Broadband reflectometry for diagnostics and monitoring applications. *IEEE Sensors Journal, 11*(2), 451–459.

Cataldo, A., Monti, G., De Benedetto, E., Cannazza, G., Tarricone, L., & Catarinucci, L. (2009). Assessment of a TD-based method for characterization of antennas. *IEEE Transactions on Instrumentation and Measurement, 58*(5), 1412–1419.

Chainey, S., & Ratcliffe, J. (2008). *GIS and crime mapping* (p. 428). Chichester, UK: Wiley Ltd.

Conyers, L. B., & Goodman, D. (1997). *Ground penetrating radar: An introduction for archaeologists.* Walnut Creek: AltaMira Press.

Corry, E. (1985). Spontaneous polarization associated with porphyry sulphide mineralization. *Geophysics, 50,* 1020–1034.

Davis, J. L., & Annan, A. P. (1989). Ground-penetrating radar for high resolution map-ping of soil and rock stratigraphy. *Geophysical Prospecting, 37*(5), 531–551.

Dionne, C. A., Schultz, J. J., Murdock, R. A., II, & Smith, S. A. (2011). Detecting buried metallic weapons in a controlled setting using a conductivity meter. *Forensic Science International, 208,* 18–24.

Fenning, P. J., & Donnelly, L. J. (2004). Geophysical techniques for forensic investigation. In: K. Pye & D. J. Croft (Eds.), *Forensic geoscience: Principles, techniques and applications.* Special Publications of the Geological Society of London (vol. 232, pp. 11–20).

Hammer, S. (1939). Terrain corrections for gravimeter stations. *Geophysics, 4,* 184–194.

Heimovaara, T. J., Huisman, J. A., Vrugt, J. A., & Bouten, W. (2004). Obtaining the spatial distribution of water content along a TDR probe using the SCEM-UA bayesian inverse modeling scheme. *Vadose Zone Journal, 3,* 1128–1145.

Jackson, J. D. (1975). *Classical electrodynamics.* New York: Wiley.

Jervis, J. R., Pringle, J. K., & Tuckwell, G. W. (2009). Time-lapse resistivity surveys over simulated clandestine graves. *Forensic Science International, 192,* 7–13.

Keller, G. V. (1987). Rock and mineral properties. In M. N. Nabighian (Eds.), *Electromagnetic methods in applied geophysics* (vol. 1, chap. 2).

Killam, E. W. (2004). *The detection of human remains* (p. 268). Springfield, Illinois, USA: Charles C Thomas Publishers.

Leucci, G. (2015). *Geofisica Applicata all'Archeologia e ai Beni Monumentali* (p. 368). Dario Flaccovio Editore: Palermo.

Leucci, G. (2019). *Nondestructive testing for archaeology and cultural heritage a practical guide and new perspectives*. Berlin: Springer International Publishing.

Leucci, G., De Giorgi, L., Gizzi, F., & Persico, R. (2016). Integrated geo-scientific surveys in the historical centre of Mesagne (Brindisi, Southern Italy). *Natural Hazard, 3*, 1–21. https://doi.org/10.1007/s11069-016-2645-x.

Loke, M. H. (1999). Time–lapse resistivity imaging inversion. In *Proceedings of the 5th Meeting of the Environmental and Engineering Geophysical Society European Section, Em 1*.

Loke, M. H. (2001). Electrical imaging surveys for environmental and engineering studies. A practical guide to 2-D and 3-D surveys. RES2DINV Manual, IRIS Instruments. www.iris-instruments.com.

Longman, I. M. (1959). Forumlas for computing the tidal accelerations due to the moon and the sun. *Journal of Geophysical Research, 64*(12), 2351–2355.

Mareš, S., & Tvrdý, M. (1984). *Introduction to applied geophysics*. Springer.

Milsom, J., & Eriksen, A. (2011). *Field geophysics*. Chichester, UK: Wiley.

Missiaen, T., Söderström, M., Popescu, I., & Vanninen, P. (2010). Evaluation of a chemical munition dumpsite in the Baltic Sea based on geophysical and chemical investigations. *Science of the Total Environment, 408*, 3536–3553.

Murray, R. C., & Tedrow, J. C. F. (1975). *Forensic geology* (217 pp). New Brunswick, New Jersey: Rutgers University Press.

Murray, R. C., & Tedrow, J. C. F. (1992). *Forensic geology* (2nd ed., 203 pp). Englewood Cliffs, New Jersey: Prentice Hall Inc.

Musset, A. E., & Khan, M. A. (2000). *Looking into the earth: An introduction to geological geophysics* (pp.139–198). London: Cambridge University Press.

Naudet, V., & Revil, A. (2005). A sandbox experiment to investigate bacteria-mediated redox processes on self-potential signals. *Geophysical Research Letters, 32*(11), L11405. https://doi.org/10.1029/2005GL022735.

Naudet, V., Revil, A., Rizzo, E., Bottero, J.-Y., & Bégassat, P. (2004). Groundwater redox conditions and conductivity in a contaminant plume from geoelectrical investigations. *Hydrology and Earth System Sciences, 8*(1), 8–22.

Novo, A., Lorenzo, H., Ria, F., & Solla, M. (2011). 3D GPR in forensics: finding a clandestine grave in a mountainous environment. *Forensic Science International, 204*, 134–138.

O'Connor, K. M., & Dowding, C. H. (1999). *Geomeasurement by pulsing TDR cables and probes*. CRC Press.

Pye, K. (2007). *Geological and soil evidence* (p. 335). Boca Raton, USA: CRC Press.

Pye, K., & Croft, D. J. (2004). *Forensic geoscience: Principles, techniques and applications* (vol. 232, 318 pp). London: Geological Society of London Special Publication.

Reynolds, J. M. (2011). *An Introduction to applied and environmental geophysics*. Chichester: Wiley.

Robinson, E. S., & Coruh C. (1988). *Basic exploration geophysics*. New York: Wiley.

Roy, A., & Apparao, A. (1971). Depth of investigation in direct current methods. *Geophysics, 36*, 943–959.

Ruffell, A., & McKinely, J. (2005). Forensic geoscience: Applications of geology, geomorphology and geophysics to criminal investigations. *Earth-Science Review, 69*, 235–247.

Ruffell, A., & McKinely, J. (2008). *Geoforensics*. Wiley Ltd.: New York.

Schultz, J. J., & Martin, M. M. (2011). Controlled GPR grave research: Comparison of reflection profiles between 500 and 250 MHz antennae. *Forensic Science International, 209*, 64–69.

Sheriff, R. E. (1994). *Encyclopedic dictionary of exploration geophysics*. Tulsa: Society of Exploration Geophysics.

Steeples, D. W. (2005). Shallow seismic method. In Y. Rubin, S. Hubbard (Eds.), *Hydrogeophysics* (pp. 215–221). Berlin: Springer.

Telford, W. M., Geldart, L. P., Sheriff R. E., & Keys, D.A. (1990). *Applied geophysics* (2nd ed). Cambridge University Press.

Turner, G., & Siggins, A. F. (1994). Constant Q attenuation of subsurface radar pulses. *Geophysics, 59,* 1192–1200.

Ward, S. H., & Hohmann, G. W. (1987). Electromagnetic theory for geophysical exploration. In M. N. Nabighian (Ed.), *Electromagnetic methods in applied geophysics* (vol. 1, chap. 4).

Watters, M., & Hunter, J. R. (2004). Gephysics and burial: Field experience and software development. In K. Pye & D. J. Croft (Eds.), *Forensic geoscience: Principles, techniques and applications* (vol. 232, pp. 21–33). London: Special Publications of the Geological Society of London.

Ylmaz, O. (1987). *Seismic data processing, society of Exploration Geophysicists.*

Zanzi, L. (2004). Appunti di sismica di esplorazione e georadar, CUSL.

Chapter 3
Forensic Geophysics Instrumentation and Data Acquisition

Abstract Nowadays, in the forensic science geophysical methods should detect, characterise, and discriminate hidden structures to obtain useful information and provide accurate and efficient results for the investigative purpose. In this field in comparison with other direct procedures (such as direct excavations, drill cores, etc.), geophysical methods minimise time and cost factors and maximise the amount of data, information, and knowledge that can be obtained. This chapter introduces a new "not standard" procedures to acquire the geophysical data, new mode to process, and interpret them. A framework for identify the basic problem and successively chose the appropriate solution with data processing and the relationship between the data, the information, and the apriori knowledge is presented. In the book and particularly in this chapter attempts to respond to the question of how data, information, and knowledge can be enhanced for crime site characterisation. The latter item discussed in this chapter is related to the advanced visualisation techniques and how they can help and address classical crime and civil problems.

Keywords Geophysical data processing · Geophysical instrumentation

3.1 TDR Instrumentation Enhancement

Time-domain reflectometry is an electromagnetic investigation technique that, thanks to its high flexibility, finds application in the most varied sectors, for example in the field of hydrogeology, environmental monitoring and soil monitoring and consequently in the forensic science. TDR method allows:

- measurement of water content in the soil (O'Connor and Dowding 1999),
- measurement of electrical conductivity of the soil (Robinson et al. 2003; Huisman et al. 2008),
- detection of the presence of pollutants (Kim et al. 2010),
- liquefaction of the subsoil (Scheuermann et al. 2010),
- Characterisation of soil constituents (Friel and Or 1999).

For remote control of liquids, solids, granules or composites:

- detection of the number of substances (liquid, granular, etc.) (Nemarich 2001),

G. Leucci, *Advances in Geophysical Methods Applied to Forensic Investigations*,
https://doi.org/10.1007/978-3-030-46242-0_3

- detection of parameters associated with the quality of substances (Cataldo et al. 2010a, b),

In structural deformation monitoring applications:

- Localisation and characterisation of stresses and deformations present (Scheuermann and Huebner 2009), with the application also in rock strata, for example, to predict landslides (Kane et al. 2010).

In the field of telecommunications and electrical plant engineering:

- spatial localisation of cable faults (Smith et al. 2005)
- characterisation of electronic devices (Cataldo et al. 2009a, b, c).

In the field of leakage research in underground water and sewage pipelines (Cataldo et al. 2014).

As a common denominator, at the base of all TDR measurement systems related to the before-mentioned applications, there are electrical impedance measurements (Z).

As described in Chap. 2, the direct output of a TDR measurement is a parameter called reflection coefficient in the time domain (ρ) which is intrinsically linked to electrical impedance.

TDR-based measurement systems guarantee an optimal trade-off between measurement accuracy, implementation costs and ease of use. They are characterised by a modular structure that allows them to be adapted to specific application needs. Furthermore, the possibility of using specially designed and custom-made probes for the desired application allows the TDR systems to be adapted to the specific operating conditions (for example, probes can be made to resist in aggressive environments or probes for non-invasive monitoring, etc.).

Another very useful feature in TDR systems is the possibility of remote automation and control, via GSM (Global System for Mobile Communications)/GPRS (General Packet Radio Service) network (or satellite network) or the internet. Finally, through multiplexing systems, it is possible to simultaneously control/use up to 512 independent probes with a single measuring instrument, allowing a considerable reduction in the implementation costs of a TDR measurement system.

TDR is a survey and diagnostic technique that exploits the interaction between the electromagnetic field (EM) and matter. Typically, in TDR measurements, an EM signal is propagated through a probe (sensing element or sensing element) inserted or placed in proximity to the material/system to be analysed. The variations in electrical impedance (Z) encountered by the EM signal along its own "path" cause the reflection of part of the signal itself. This can be due, for example, to the variation of the dielectric characteristics of the system. The analysis of the reflected signal allows information about the electrical characteristics of the investigated materials (dielectric permittivity, electrical conductivity, etc.), as well as on other properties (not necessarily of electrical nature) of the investigated material/system. From the hardware point of view, the fundamental elements of a TDR measurement system are two: reflectometer (it is the measuring instrument) and probe.

To these elements, a computer must be added to manage and/or process measurement data. Furthermore, TDR measurement system can be equipped with "accessories" dedicated to specific tasks, such as multiplexers (to use multiple independent probes simultaneously with a single measuring instrument), modules for remote management (e.g. Wi-Fi connection modules), etc.

From the software point of view, TDR instruments are generally supplied by the manufacturer already equipped with software for data acquisition. Usually, the "direct" output of a TDR measurement is a reflectogram, which is a curve that shows the reflection coefficient (ρ) as a function of time or, similarly, of the apparent distance travelled by the EM test signal (d_{app}). Therefore, to trace the information of interest on the system in question, it is necessary to perform processing on the measurement data acquired by the TDR instrument. It is advisable to develop ad hoc algorithms (suited to the specific application requirement), which automatically process the reflectogram and return the value of the quantity of interest in real-time.

The reflectometry: The reflectometer generally contains both the electronics used to generate the EM test signal and the electronics responsible for acquiring the reflected EM signal. The test signal generated by the reflectometer is, typically, a step voltage signal (such as that shown in Fig. 3.1), characterised by a rise time (rise time, t_r) which is directly associated with the frequency content (BW) of the signal itself according to the following relation:

$$BW \approx \frac{0.35}{t_r} \tag{3.1}$$

In the TDR instruments, t_r is in the order of picoseconds or nanoseconds: at lower rise times, they correspond to better performance and higher costs. In the TDR measures, it is very important to take into account the relationship (3.1); through appropriate processing of the TDR data (based on the well-known Fourier transform,

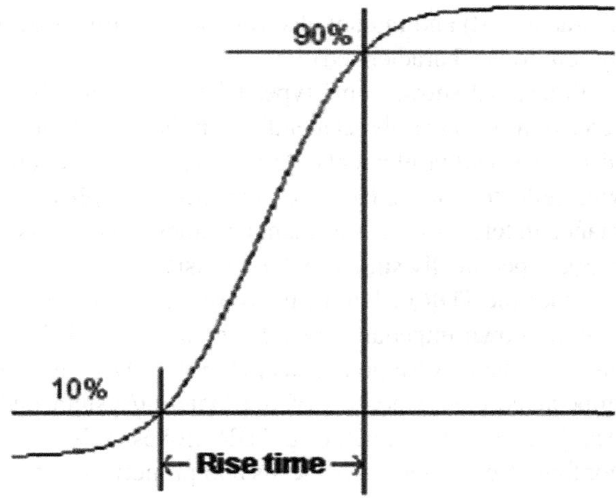

Fig. 3.1 Schematic of the EM test signal generated by a TDR instrument

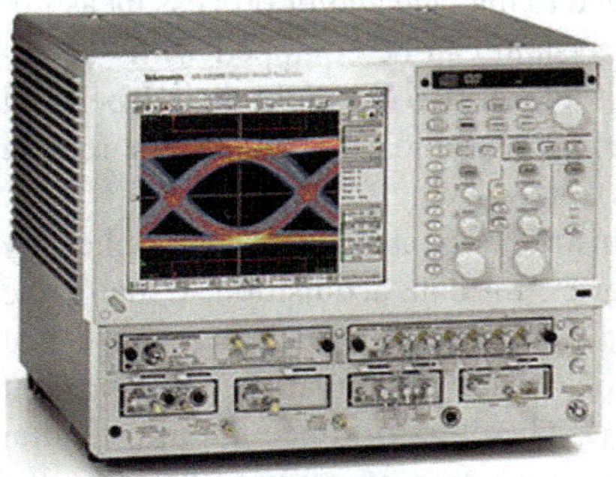

Fig. 3.2 Image of a bench reflectometer (model Tektronix DSA8200 with TDR modules) and a portable reflectometer (model HyperLabs HL8200)

see Chap. 1), it is possible to go back to information and quantities defined in the frequency domain. Therefore, knowing the BW of the reflectometer used allows defining the frequency band where it is possible to obtain significant results.

Different types of TDR instruments are available, and a general classification can be made by distinguishing between bench-top instruments (generally instruments with higher performances and dedicated to laboratory measurements) and portable instruments (generally suitable for being used for in situ measurements). Furthermore, ruggedised reflectometers are suitable for resisting in "hostile" environmental and/or operating conditions. The choice of which type of reflectometer to use depends on the specific application requirements. As an example, Fig. 3.2 shows the image of a bench reflectometer and a portable one.

The Probe: The probe (or sensing element) is the element responsible for the interaction between the EM test signal and the system to be analysed. A general classification of the probes can be made by distinguishing between guiding electromagnetic structures (for example, coaxial probes inserted in the system to be characterised) and radiant electromagnetic structures (i.e. antennas, placed near the system to be characterised).

Figure 3.3 shows some types of probes typically used for the dielectric characterisation of materials. Depending on the type of application, it is necessary to use different sensitive elements. Various types of probes are available (often sold together with reflectometers). However, for specific applications and to obtain better performance in terms of measurement accuracy, it is necessary to design and manufacture probes specifically suited to the forensic application.

Since the TDR technique measures variations in electrical impedance (Z), probes with a known impedance profile should be used. In this way, one can be sure that any impedance variations detected in the TDR measurements are attributable to the impedance characteristics of the material/system under examination. Even when designing and manufacturing TDR probes, it is essential to choose a geometrical configuration suited to the needs and properly size the probe.

Fig. 3.3 Examples of probes used in TDR measurements: **a** radiative probe used for non-invasive measurements; **b** coaxial probe used for measurements on liquid materials; **c** bar probes used for measurements on granular materials

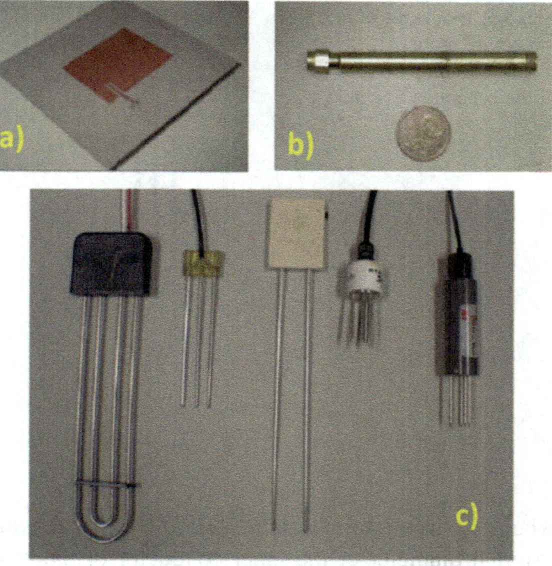

The configuration of a coaxial TDR probe is very simple to design and implement. It is a structure similar to that of a cylindrical condenser (Fig. 3.4): also, in this case, the material to be characterised is made to fill the space present between two concentric conducting cylinders (for this reason, this type of probe is mainly used for measurements on liquids).

The impedance profile of a coaxial probe in the air (i.e. not inserted in any material) can be easily obtained from the transmission line theory:

$$Z = 60 \cdot \ln \frac{b}{a} \tag{3.2}$$

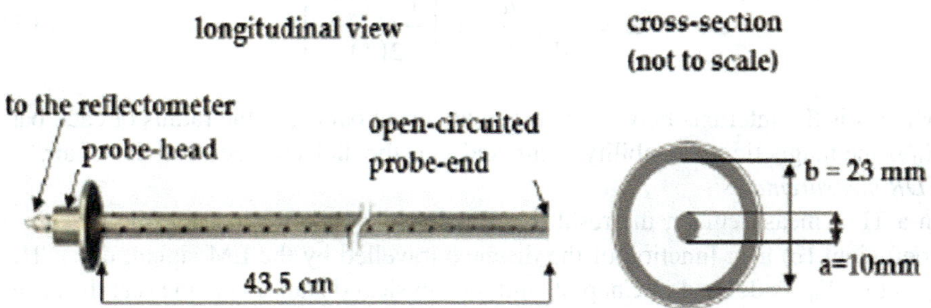

Fig. 3.4 Longitudinal and cross-section views of a coaxial probe, in which the diameters of the conductors have been chosen to guarantee an electrical impedance of the probe (in the air) equal to 50 Ω

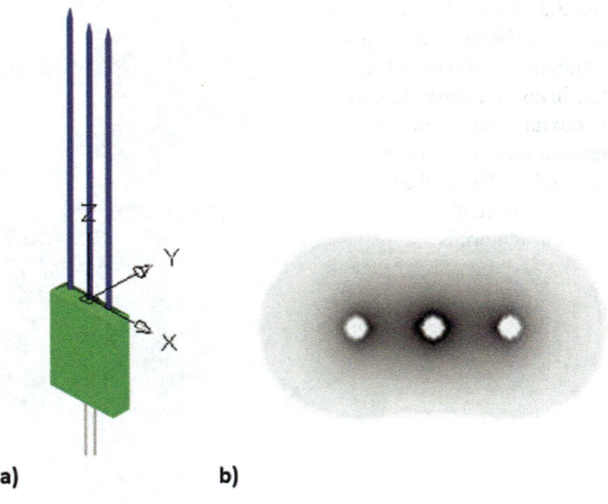

Fig. 3.5 **a** Scheme of a three-wire probe; **b** corresponding distribution of the generated EM field

where b is the internal diameter of the outer conductor of the probe, and a is the external diameter of the inner conductor of the probe (as shown in Fig. 3.4). For measurements on granular materials (such as soils), a multi-bar (or bar or multi-wire) configuration is usually preferred for the probe. This geometry makes it easier to insert the probe into the material being examined. In particular, the three-bar probe is widely used (each bar is an electrode), since this configuration has an electromagnetic behaviour due to that of a coaxial-type propagation structure (Zegelin et al. 1989). Figure 3.5 shows the schematisation of a typical three-bar probe and the corresponding distribution of the EM field. For probes with geometries different from the coaxial one, in the literature, exact models for the impedance profile are not available. However, some approximate formulas describe the impedance of multi-wire probes, allowing for proper design. In particular, for the three-bar probes, the following simplification applies (Cataldo et al. 2010a, b):

$$Z = \frac{1}{4\pi}\sqrt{\frac{\mu_0}{\varepsilon_0}} \cdot \ln\left[\frac{1 - \left(\frac{g}{s}\right)^4}{2\left(\frac{g}{s}\right)^3}\right] \tag{3.3}$$

where s is the interaxis between two neighbouring bars; g is the radius of each bar; μ_0 is the magnetic permeability of air, and ε_0 is the dielectric permittivity of air.
TDR measurements:
In a TDR measurement, the result is represented by a reflectogram (the reflection coefficient (ρ) as a function of the distance travelled by the EM signal, d_{app}). The quantity d_{app} is defined in Chap. 2, and the physical distance (i.e. the real distance) is related to d_{app} through the dielectric constant of the medium as:

$$d = \frac{d_{app}}{\sqrt{\varepsilon_{eff}}} \tag{3.4}$$

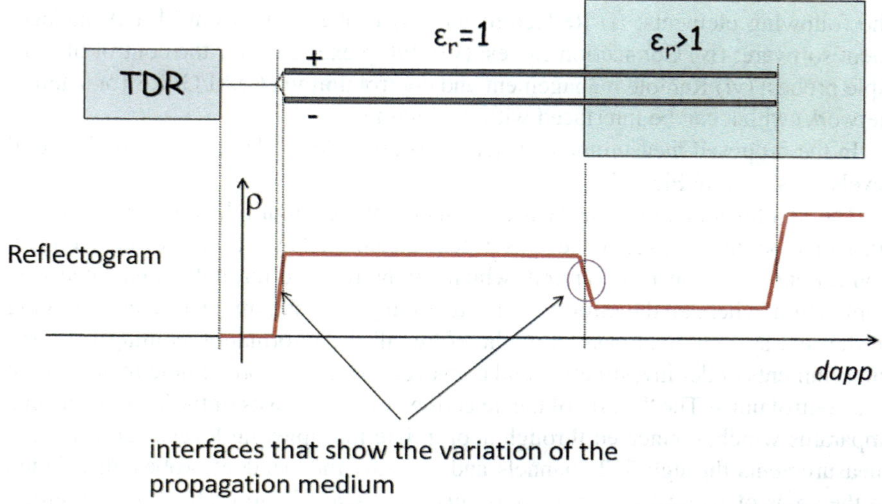

interfaces that show the variation of the
propagation medium

Fig. 3.6 Schematic of a sensitive two-wire element inserted in materials with different relative dielectric permittivity (ε_r). As the EM signal generated by the TDR travels through the probe, the corresponding reflectogram highlights the variations of ρ in correspondence with the different media

where ε_{eff} is the effective dielectric constant that can esteem from TDR measurements, it takes into account the dielectric characteristics of the medium in which the EM signal is propagated; and $d_{app} = (c \times tv)/2$ where $c \cong 3 \times 10^8$ m/s is the speed of light in a vacuum, t is the trave time in the medium and v is the EM wave velocity of propagation in the medium.

A schematisation of a reflectogram relative to a two bar probe which is partly in the air (medium with relative dielectric constant equal to $\varepsilon_r = 1$) and partly inserted in a different material ($\varepsilon_r > 1$) is shown in Fig. 3.6.

The electrical impedance (Z) of the transmission line along which the EM signal is propagating is related to the reflection coefficient, ρ, by the following relation:

$$\rho \cong \frac{Z - Z_0}{Z + Z_0} \tag{3.5}$$

where Z_0 is the reference impedance (typically 50 Ω). Using Eq. (3.5) is possible to obtain the impedance profile and the physical characteristic (dielectric permittivity, resistivity, and conductivity) of the probe inserted in the material under examination.

The TDR instrumentation enhancement:

In particular, the possibility of using different reflectometric instruments (with different performances) and different types of probes was evaluated, in order to identify the optimal configuration of the measuring apparatus, to guarantee a good trade-off between measurement accuracy and management/implementation costs, as well as the possibility of performing in situ measurements and controlling them remotely. The TDR instrumentation enhancement includes a modular system that includes the following elements: (i) Reflectometer; (ii) Probes; 9iii) Control and management software; (iv) Connection cables; (v) Multiplexer unit, for the control of multiple probes; (vi) Remote management and control unit via GSM/GPRS (or satellite network) which can be interfaced with the internet.

In the proposed measurement system, it is possible to identify two architectural levels, as shown in Fig. 3.7.

The first level includes the TDR instrument, probes, multiplexers and data transmission modules. The second level is instead constituted by the remote management, control and data transmission unit, which can be realised through a dedicated web application, where all the information was coming from the single devices belonging to the first system-level pass. It is, therefore, all the information coming from the environments under investigation and conveyed to one or more remote management and control units. The "heart" of the detection system consists of the TDR electronic apparatus which, connected through appropriate multiplexing levels, can carry out measurements through 512 channels and as many independent probes, distributed in the areas of interest covering areas very extensive. These probes can, therefore,

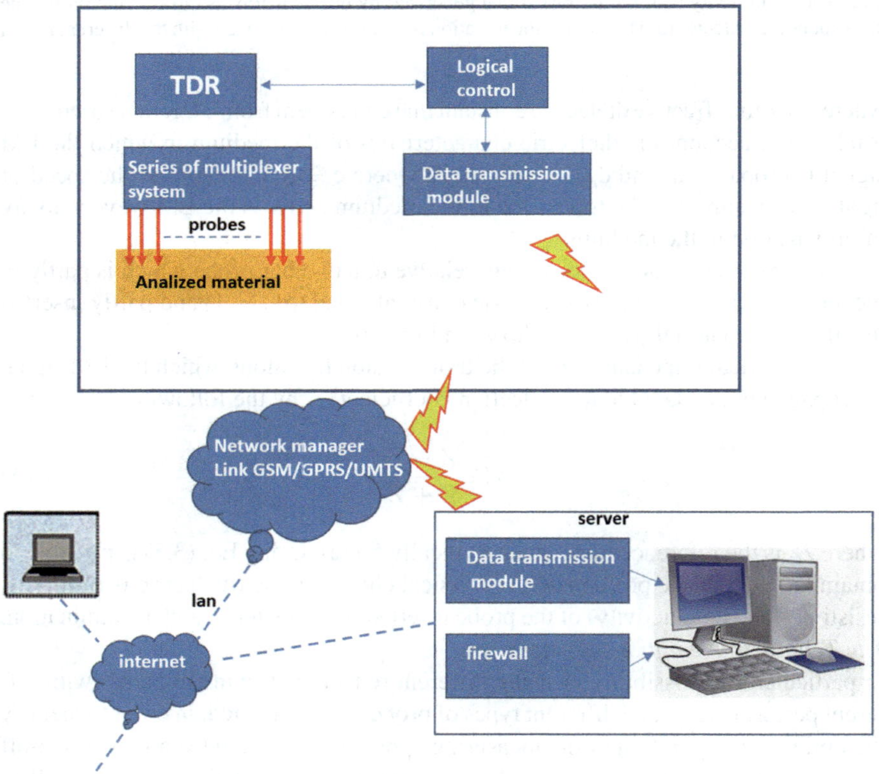

Fig. 3.7 General scheme of a TDR measurement system

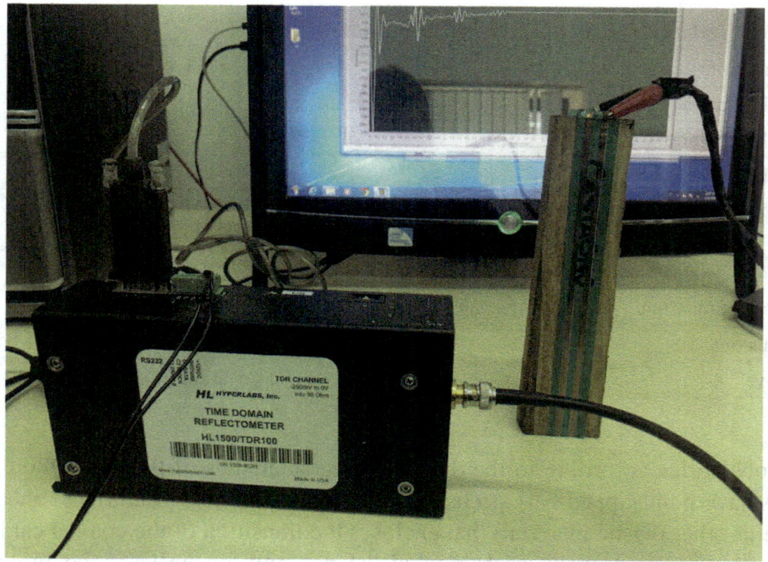

Fig. 3.8 Image of the TDR HL1500 instrument

be connected through one or more levels of multiplexing, a feature that allows a considerable reduction in instrumentation costs.

The choice of reflectometer: The identification of the optimal instrument was made taking into account the need to have a robust instrument (such as to withstand even in relatively hostile operating conditions, such as those of field measurements) and which would allow implementing the multiplexing mode of use. The instrument that is most suitable for the application considered is the HyperLabs HL1500 reflectometer (Fig. 3.8).

It is a compact instrument, which guarantees an optimal trade-off between cost, performance, ability to manage 256 channels simultaneously (and beyond). The HL1500 generates a step signal with a rise time of around 200 ps; therefore, its frequency content is approximately 1.75 GHz. The instrument works with a 12 V battery (however, the system can also be powered directly from the mains if it is available). Table 3.1 summarises the main technical specifications of the HL1500 reflectometer.

Another very useful feature of the identified tool is that its control system is "open", i.e. it is possible to develop and implement control software on the desired platforms (LabVIEW etc.). Furthermore, for this instrument, a series of bar probes suitable for measurements on soils with different characteristics are already commercially available. Finally, it is possible to equip this instrument with appropriate "accessories" that maximise the efficiency in the implementation of the measurement system.

Examples of probes: Typically for TDR measurements on soils, multi-bar probes are used: the probe acts as a waveguide, and the impedance measured (via TDR) along

Table 3.1 Technical specifications of the HL1500 instrument

The rise time of the incident signal	<200 ps
Signal duration	14 μs
Signal amplitude	250 mV
Resolution in the time base	12.2 ps
Maximum number of points that can be acquired for each waveform	2048
Output impedance	50 Ω ± 1%
Number of maximum automatic average	2048
Dimensions	235 mm × 110 mm × 55 mm
Supply	Batteria da 12 V

the probe varies according to the dielectric characteristics of the soil. Often these probes have point-tipped bars (to facilitate insertion into the ground) and ahead (in Teflon or other plastic material) that encloses the transition of the coaxial cable/bar electrical connection and keeps the mutual distance between the bars fixed. Figure 3.9 shows the schematisation of a three-bar probe.

For the HL1500 instrument, the Campbell Scientific company offers six types of three-bar probes, depending on the type of soil they are intended to inspect. For example, the CS605 probe is suitable for measurements on soils with low bulk electrical conductivity (≤ 1.4 dS/m). This probe has a coaxial cable RG58, and the bars have a diameter of 0.48 cm and a length of 30 cm; this probe must be connected to the reflectometer with a coaxial cable with a maximum length of 15 m. For longer cable lengths (up to 25 m), it is advisable to use the CS610 probe (also with 0.48 cm diameter bars and 30 cm lengths). The CS630 thread-pulling probe is suitable for measurements on soils with higher electrical conductivity (in any case less than 3.5 dS/m). This probe has bars with a length of 15 cm and a diameter of 0.318 cm; has

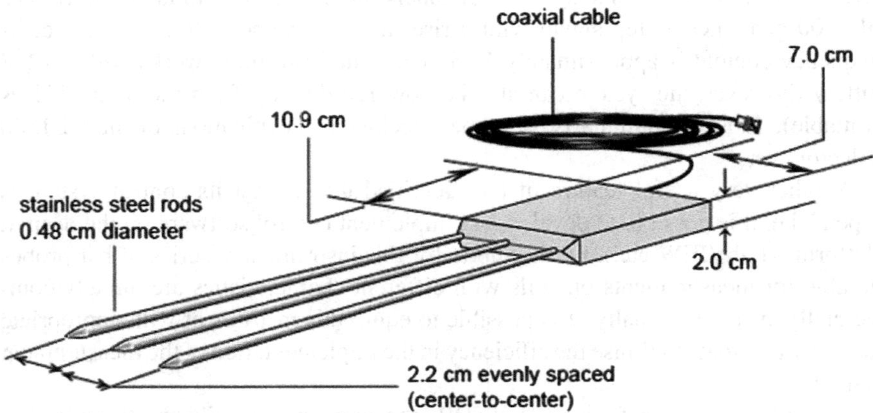

Fig. 3.9 Schematic of a trifle probe for TDR measurements in soils

an RG58 coaxial cable and must be used with cables up to 15 m long. For longer cable lengths (up to 25 m), the CS635 probe (which mounts an LMR200DB cable rather than RG58) is recommended. Finally, for soils with even higher conductivity (still less than five dS/m), the use of the CS640 probe is recommended (with bars of 7.5 cm length and 0.159 cm diameter, with RG58 cable and for use with long cables) up to 15 m). Alternatively, for cable lengths greater than 15 m (up to 25 m), the CS645 probe is recommended (which, unlike the CS640 probe, has an LMR200DB cable).

The probes described above generally have a length of less than 0.5 m; therefore, they provide a local measure of soil properties. If the area to be monitored is extensive and need to a "distributed" profile of the impedance of the soil, several probes must be used, be suitably located in the area of interest, as shown in Fig. 3.10.

As an alternative to this solution, to have a distributed measurement of the electrical impedance is possible to use the probe configuration as described in Cataldo et al. (2011, 2012) used to search for leaks in underground water pipes. The configuration described in Cataldo et al. (2011, 2012) foresees the use of a probe with two thread-like and parallel conductors (separated by a plastic sheath), to be placed in the ground, along the direction in which the electrical impedance profile is to be measured. Figure 3.11 shows a cross-section of the probe.

Figure 3.12 shows a diagram of the measurement system and the corresponding reflectogram in the presence of a variation in electrical impedance (due, for example, to a different composition of the ground in a certain area). The acquired reflectogram

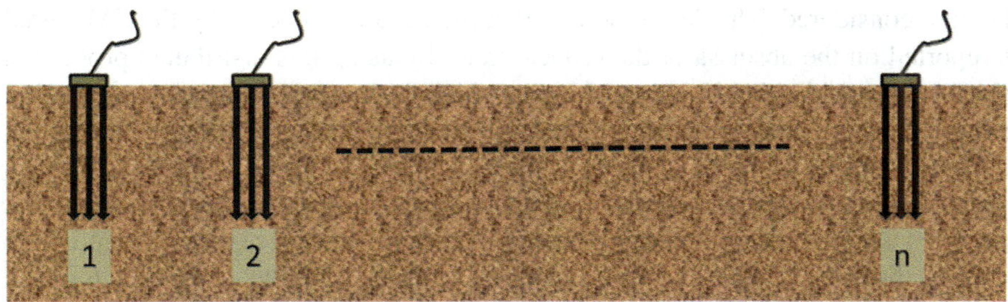

Fig. 3.10 Scheme of a hypothetical dislocation of a number n of three-bar probes for the reconstruction of the electrical impedance profile of an extended portion of soil (non-scaled scheme)

Fig. 3.11 Scheme of the cross-section of a distributed two-wire probe. The 4 mm size refers to a "typical" transverse dimension of this type of probe

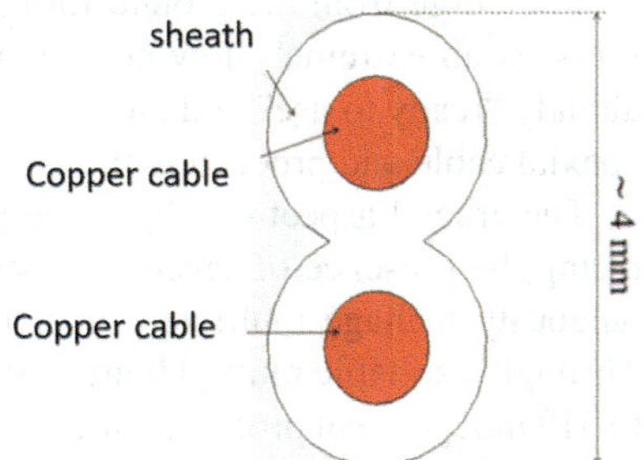

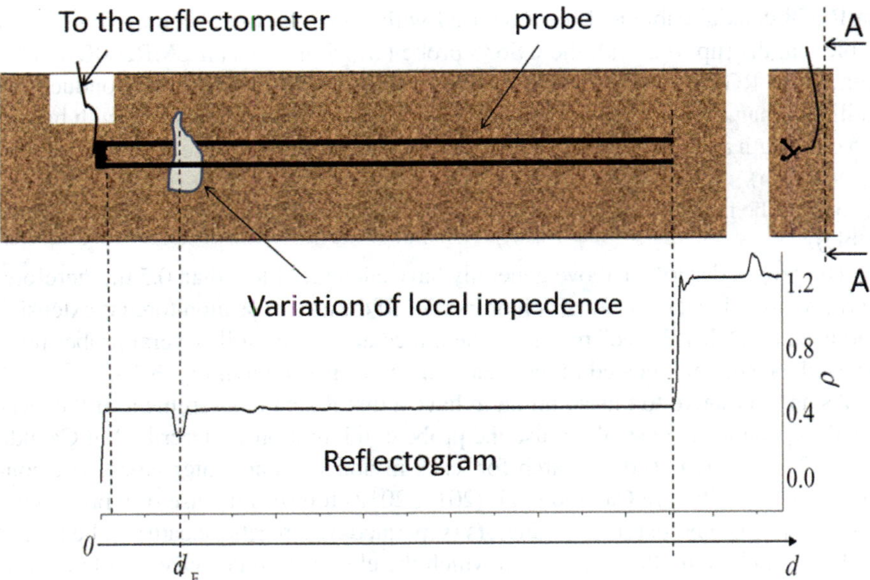

Fig. 3.12 Scheme of the TDR measuring apparatus with a distributed two-wire probe (drawing not to scale). The schematisation of the corresponding reflectogram is also reported

directly provides the punctual profile of the electrical impedance of the ground in the area considered. Thanks to the fact that the distance covered by the EM signal is reported on the abscissa of the reflectogram, by using this distributed probe it is possible to quickly identify the position in which the measured impedance variation is located. The spatial resolution of the measured impedance profile depends on the characteristics of the TDR instrument used. For the HL1500 it is possible to have a spatial resolution of the order of less than a few centimetres, thus obtaining what can be considered a "continuous" profile of electrical impedance of the ground. This solution directly provides the soil impedance profile along the entire length of the sensitive element. This type of two-wire probes can have a length that exceeds 200 m; with obvious time savings in the execution of the measurements.

The probe can be buried at any depth, possibly also with micro tunnelling. As can be seen from the figure, to carry out the measurement, it is sufficient to let a coaxial cable emerge from the ground for connection to the TDR instrument. This type of probe is an extremely low cost, and can also remain permanently buried, always already "ready to use" and the operator should only connect the reflectometer to the coaxial cable and proceed with the acquisition of the reflectogram.

The crucial aspect that has been taken into account is the possibility of guaranteeing the presence of several measurement channels, in order to be able to simultaneously manage multiple probes (corresponding to as many measuring points). Through a suitable multiplexing system, it is possible to simultaneously manage up to 512 independent probes with a single reflectometer. In particular, with the HL1500 instrument identified in the previous sections, it is possible to use Campbell Scientific's SDMX50 series multiplexers. These are multiplexers of type 8:1; that is, up to eight probes can be connected to each multiplexer, as shown in Fig. 3.13. Like the HL1500, this multiplexer requires a 12 V continuous supply.

Among the multiplexers of the series mentioned above, the SDMX50SP (Fig. 3.14) consists of an electronic multiplexer board mounted inside a metal protection box and support to reduce the strain on the cables. The multiplexer is portable and has very small dimensions (the size of the metal box is 22.3 × 12.2 × 2.5 cm, while those of the support for the cables are 20.3 × 4.3 × 1.5 cm).

The measurement system can employ three levels of multiplexing, as shown in Fig. 3.15. It should be noted that the first level includes an HL1500 and a multiplexer. Eight coaxial cables can be connected to each multiplexer. Each of the cables can be connected to a probe or another multiplexer of the next level, arriving (in total) to control a maximum of 512 probes. In this way, it is possible to achieve a substantial reduction in costs.

The software: The data acquired through the TDR measurement system must be processed to trace the physical parameters of interest. Processing must be adapted according to the parameter to be obtained. Once the processing algorithm has been developed, it is possible to have the processing done in real-time and automatically. By choosing a TDR instrument with an open communication protocol (such as the HL1500), it is possible to implement the processing algorithm in the development environment that best suits the specific needs. A development environment particularly suited to the management of TDR instruments and the processing of acquired measurements is LabVIEW, an advanced development environment that offers hardware integration and wide-ranging compatibility with a wide range of tools. LabVIEW combines the flexibility of a programming language with the power of advanced engineering tools for the rapid development of any application.

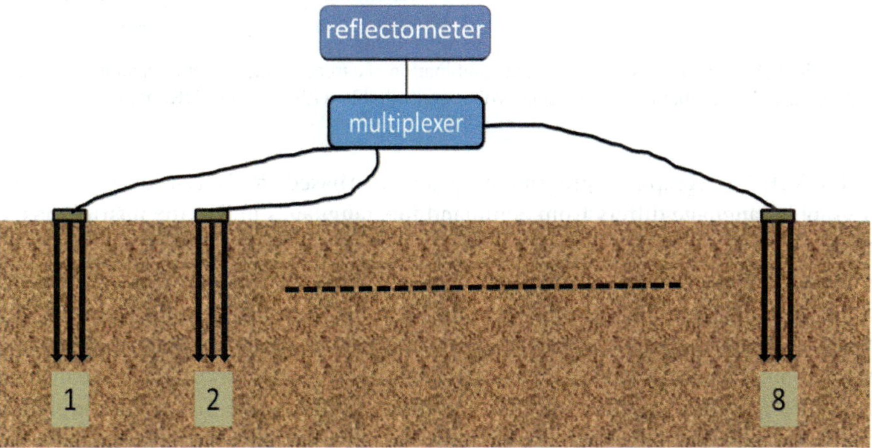

Fig. 3.13 Scheme of the management of eight probes through a multiplexer unit and a single reflectometer

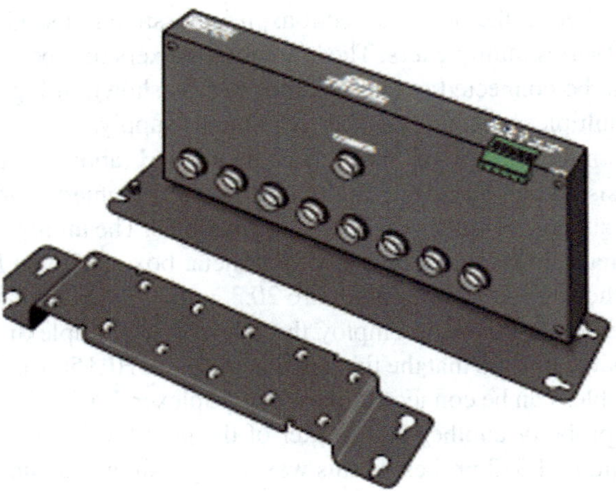

Fig. 3.14 Image of the SDMX50SP multiplexer

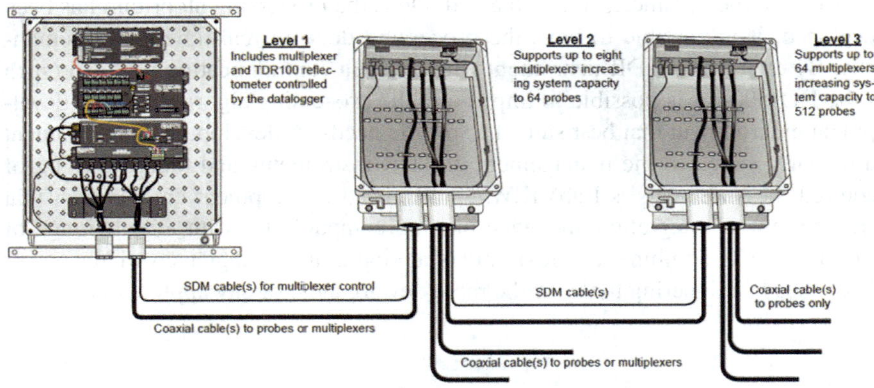

Fig. 3.15 Schematic representation of the combination of cascade multiplexers (on multiple levels), to maximise the number of independent probes controllable with a single TDR instrument

LabVIEW is a graphical programming language based on a Data-Flow philosophy. A graphic language differs from command line languages in that the instructions to be executed are not made explicit using a precise and complicated syntax, but by icons and connection lines, which are appropriately linked together, constitute the code. Each icon contains all the necessary for the execution of a particular function, making the management of the program more intuitive and manageable, even if more complicated to modify. Philosophy Data-Flow, on the other hand, means that the data flow determines the order in which the code is executed: while in command-line languages it is the order in which the instructions are written that dictate the execution, in LabVIEW none instruction is executed if all the inputs are not inputted to the function before. Being dedicated to the creation of interfaces with measurement tools, LabVIEW provides two spreadsheets that allow the visualisation of the two faces of the application being created:

- the control panel: allows viewing the interface with the user, with the consequent set of buttons, label, etc.
- the block diagram: containing the actual code of the application, i.e. the graphic source code.

The design criteria that motivate the choice of LabVIEW as a software development environment mainly lie in the possibility of creating a virtual instrument whose appearance and operations emulate real physical tools, thus making the software easy to understand and extremely user-friendly.

A very important aspect of using LabVIEW is that this development environment gives the possibility to create program executables, thus overcoming the problem related to the acquisition of licenses. As an example, in Fig. 3.16, a screenshot of software developed in LabView for the search for leaks in underground water pipes is reported.

Some examples: The physical parameter conductivity (σ) can be associated with various properties of the soil (salinity, solute transport, etc.). The evaluation of this parameter is intrinsically affected by high measurement uncertainty, especially in the case of granular materials (Corwin and Lesch 2005) in the traditional methods of measurement. Consequently, there is a growing interest in alternative solutions, such as those based on reflectometric methods, which guarantee, in addition to good accuracy, versatility and easy portability of the instrumentation. At present, these measurements are carried out using multi-row probes, inserted in the material

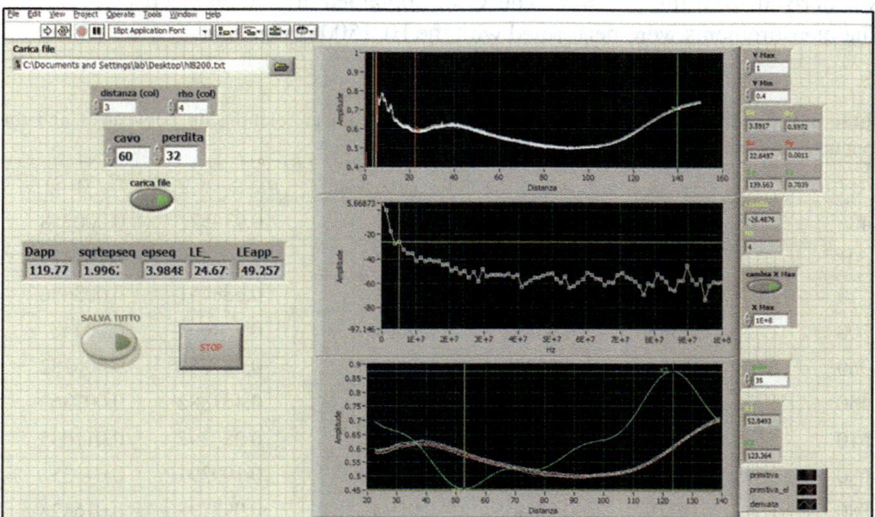

Fig. 3.16 Example of the graphical interface of an application developed in a LabView environment dedicated to locating leaks in underground water pipes through a TDR measurement system

under analysis, using improved versions of the Giese-Tiemann method, such as, for example, the Castiglione-Shouse method (Castiglione and Shouse 2003). All these methods link the σ_0 to the reflection coefficient ρ assessed by the reflectogram over very long times (corresponding to a static condition) and exploiting a proportionality coefficient k_p, whose value is to be determined through multiple preliminary calibration measurements carried out on materials of reference. Starting from the value assumed by the reflection coefficient at very long apparent distances, ρ_∞ (evaluated, as already mentioned at very long apparent distances) is possible to extrapolate the value of the static electric conductivity. The k_p value is calculated based on the linear relationship between σ [μS/cm] and the conductance Gs [μS] of the medium:

$$\sigma = k_p G_s \tag{3.6}$$

where

$$G_s = \frac{1}{Z_{TDR}} \frac{1 - \rho_{\infty,corrected}}{1 + \rho_{\infty,corrected}} \text{ with } \rho_{\infty,corrected} = 2\frac{\rho_\infty - \rho_{\infty,air}}{\rho_{\infty,air} + \rho_{\infty,shortcircuit}} + 1 \tag{3.7}$$

Z_{TDR} is the characteristic impedance of the measuring apparatus, $\rho_{\infty,air}$ is the reflection coefficient measured with the probe in the air, $\rho_{\infty,short\,circuit}$ is the reflection coefficient with the probe short-circuited at the end. In order to calculate the k_p value, one must perform measurements on electrolytic solutions of known conductivity (Cataldo et al. 2010a, b). For each solution, a reflectometric measurement was carried out at long distances (i.e., with d_{app} up to 2000 m) and the corresponding G_s was derived from time to time. The values obtained are summarised in Table 3.2. The measurements were performed with the HL1500 instrument.

Table 3.2 Experimental data for the evaluation of the linear relationship between σ and the conductance Gs of the medium

σ (μS/cm)	ρ_∞	$\rho_{\infty,air}$	$\rho_{\infty,short\,circuit}$	$\rho_{\infty,corrected}$	G_s (μS)
40	0.9097	0.9339	−0.9694	0.97457	0.000258
60	0.803	0.9339	−0.9694	0.862449	0.001477
170	0.6377	0.9339	−0.9694	0.688751	0.003686
200	0.5856	0.9339	−0.9694	0.634004	0.00448
360	0.4845	0.9339	−0.9694	0.527768	0.006182
450	0.4218	0.9339	−0.9694	0.461882	0.007362
510	0.2548	0.9339	−0.9694	0.286397	0.011095
680	0.172	0.9339	−0.9694	0.199391	0.01335
880	0.0757	0.9339	−0.9694	0.098198	0.016423
1080	−0.0515	0.9339	−0.9694	−0.03546	0.021471
1150	−0.1067	0.9339	−0.9694	−0.09347	0.024124

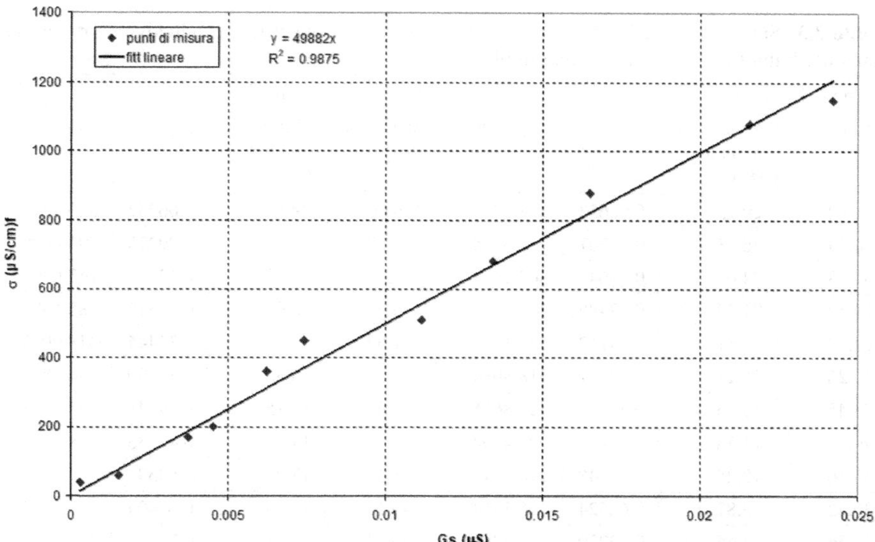

Fig. 3.17 Linear relationship between σ and the conductance Gs of the medium, the slope gives us the value of k_p

Thus, the graph G_s graph was plotted, as shown in Fig. 3.17.

From the slope of the regression line obtained, k_p is determined to be used for estimating the conductivity of soils at various levels of humidity. Then, conductivity measurements were made on red soil samples and white soil samples, moistened to known humidity values. Tables 3.3 and 3.4 shows the experimental summary data on the evaluation of electrical conductivity for various reference humidity levels for the two different types of soil considered.

Figure 3.18 shows, for the four soil samples, considered, the measurement of electrical conductivity at the different humidity values.

Traditionally, electrical conductivity measurements using the TDR techniques are characterised by a particularly long and extremely delicate calibration phase in terms of uncertainty from which the final conductivity measurements will be affected. Starting from these considerations, recently, two innovative approaches have been developed for the TDR measurement of the static conductivity of granular materials, characterised by a faster calibration procedure compared to traditional TDR methods (Cataldo et al. 2010a, b). The first approach considers a transmission line model and requires that the probe is inserted in air and bi-distilled water. The second method uses an LCR meter that measures the inductance (L), the resistance (R) and the capacitance (C) and allows the evaluation of k_p.

Starting from the full value of the TDR measurements (Fig. 3.19), the approach based on the transmission line model (TLM) allows to evaluates σ_0.

The measurement system (TDR unit, connection cable, tri-wire probe) is modelled through a three-section transmission line, including some elements with concentrated

Table 3.3 Summary experimental data on the evaluation of electrical conductivity for various reference humidity levels for red soil samples

Sample 1: red soil				Sample 2: red soil			
Saturation (%)	Volumetric humidity (%)	ρ_∞	σ (μS/cm)	Saturation (%)	Volumetric humidity (%)	ρ_∞	σ (μS/cm)
100.00	29.15	0.06653	838.7772	100.00	29.15	0.06952	833.4921
90.34	26.33	0.10707	769.6175	89.17	25.99	0.09275	793.4439
82.33	24.00	0.08644	804.1479	79.19	23.08	0.12313	743.6392
74.50	21.72	0.12488	740.8545	70.98	20.69	0.31312	485.6895
64.52	18.81	0.19717	633.1234	61.69	17.98	0.37444	418.0045
55.26	16.11	0.28511	518.8084	55.97	16.32	0.31399	484.6839
45.45	13.25	0.47017	323.8655	51.97	15.15	0.39645	395.1921
40.42	11.78	0.50414	293.4004	49.31	14.38	0.43333	358.5723
35.16	10.25	0.60548	210.3249	45.37	13.23	0.5459	257.8208
30.42	8.87	0.62524	195.3564	40.79	11.89	0.67404	159.9317
25.56	7.45	0.73279	119.9789	30.82	8.98	0.74167	114.1788
21.64	6.31	0.732	120.4978	27.30	7.96	0.7059	137.9176
0.00	0.00	0.9366	0.471079	0.00	0.00	0.9333	2.20218

Table 3.4 Summary experimental data on the evaluation of electrical conductivity for various reference humidity levels for white soil samples

Sample 1: white soil				Sample 2: white soil			
Saturation (%)	Volumetric humidity (%)	ρ_∞	σ (μS/cm)	Saturation (%)	Volumetric humidity (%)	ρ_∞	σ (μS/cm)
100.00	33.18	0.13001	732.7424	100.00	33.33	0.18057	656.6638
91.79	30.45	0.24298	571.5048	89.53	29.84	0.29256	509.8561
78.65	26.09	0.33251	463.5975	82.15	27.38	0.2637	545.1381
67.57	22.42	0.36344	429.6879	76.93	25.64	0.34179	453.2555
59.93	19.88	0.46267	330.7865	65.58	21.86	0.4152	376.3306
54.73	18.16	0.4839	311.3807	58.90	19.63	0.47866	316.1175
50.89	16.88	0.57981	230.3405	54.40	18.13	0.52033	279.3697
47.12	15.63	0.66203	168.4534	50.25	16.75	0.62413	196.1874
41.77	13.86	0.56063	245.7341	47.53	15.84	0.64963	177.3844
37.23	12.35	0.67826	156.967	41.80	13.93	0.67717	157.7313
32.28	10.71	0.78529	86.53995	37.55	12.52	0.73021	121.6754
29.09	9.65	0.7498	108.921	32.60	10.87	0.81324	69.5411
0.00	0.00	0.9328	2.464992	29.20	9.73	0.7661	98.52809
				0.00	0.00	0.9319	2.938402

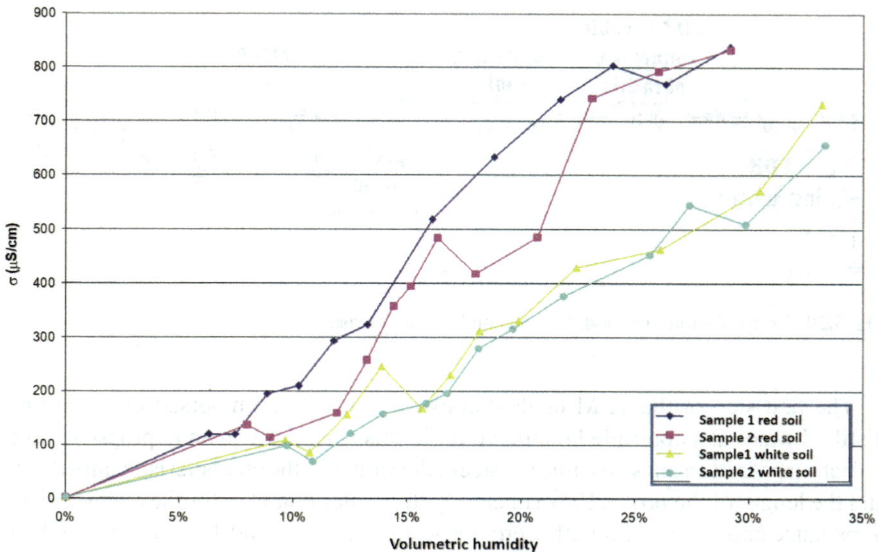

Fig. 3.18 Conductivity variation as a function of humidity expressed as a volumetric percentage

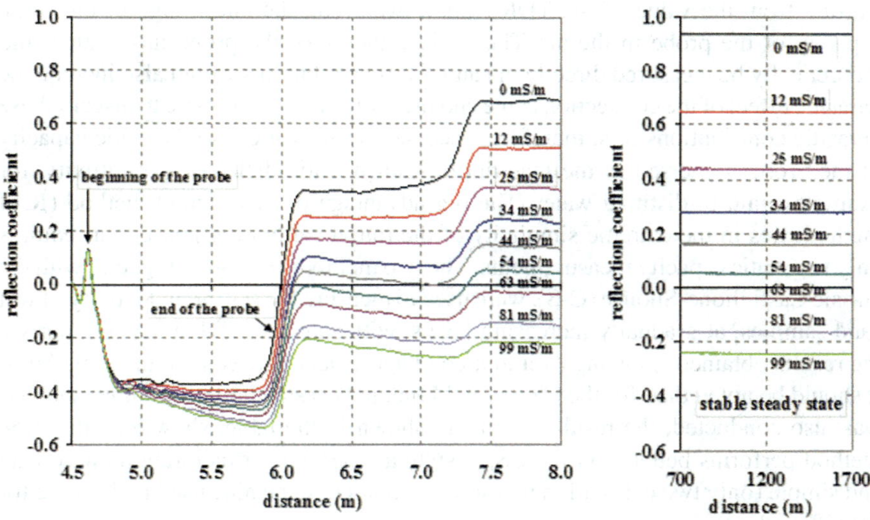

Fig. 3.19 TDR reflectograms acquired on nine electrolytic solutions with different electrical conductivity (left). Zoom of the portion at "very long distances" of the reflectograms (right)

constants able to take into account the parasitic effects due to the connections of the cables and the probe head (Fig. 3.20).

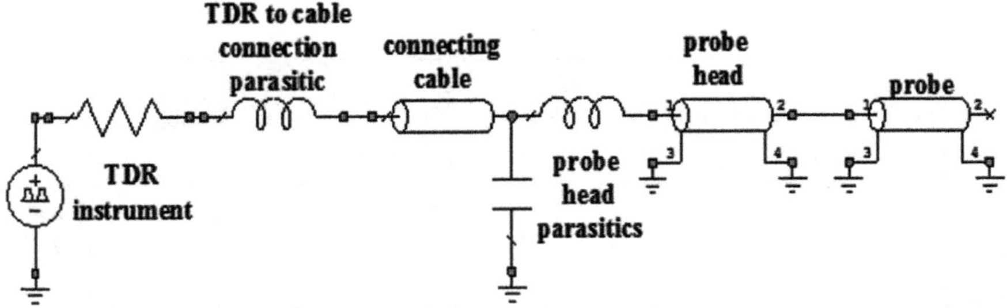

Fig. 3.20 Transmission line model of the conductivity measuring system

The first step of the TLM method considers the probe immersed in air and bi-distilled water at a controlled temperature. In this way, through an appropriate minimisation procedure, it is possible to esteem the values of the characteristic impedance and the length of the probe. Experimentally the values obtained for the characteristic impedance and the probe length were, respectively, 172 Ω and 15.2 cm, from which we obtain $k_p = 3.1$ m^{-1}.

The second approach proposed directly links the static conductivity σ_0 of the material under examination to the static conductance, measured (also in this case) starting from the value of the TDR signal, exploiting the knowledge of the static capacity of the probe in the air. The static capacity of the probe in the air could theoretically be measured directly via an LCR meter, but this would also include the parasitic effect of the connection cable and the probe head. To be able to discern these parasitic contributions it is, therefore, necessary to measure also the static capacity of the probe immersed in the material of known static dielectric permittivity, for example again bi-distilled water. The big advantage of this second method (ICM method) lies in the extreme simplicity of the initial calibration phase that requires only two static capacity measurements. The two methods proposed, together with the classic Castiglione-Shouse (C-S), were used to measure the static conductivity of wet sand samples, at gradually increasing levels, with saltwater. Table 3.5 summarises the results obtained, showing overall a good agreement between the three methods. It should be noted that, for the CS and ICM methods, a standard uncertainty analysis was also conducted, the results of which, shown in the table, show how the ICM method performs better than the CS, despite a much longer calibration phase lean and simple (only two calibration measures compared to the nine used in this case for the CS method).

Another very important application of the TDR technique is the monitoring of the water content of granular materials. In this case study, the calibration curves (ε-θ) relating to two different types of soil were obtained: of the white type and red type soils. Table 3.6 summarises the main chemical and physical characteristics.

Figure 3.21 shows the prepared soil samples, while in Fig. 3.22 the positioning of the probe for TDR measurements is shown. The measurements were performed with the HL1500 instrument and using a three-wire probe. The samples were taken to known humidity values, and the reflectogram corresponding to each humidity value

Table 3.5 Static conductivity measured by three different methods on samples of humidified sand with saltwater

Humidity (%)	$\sigma_{0,C\text{-}S}$ (mS/m)	$\sigma_{0,TLM}$ (mS/m)	$\sigma_{0,ICM}$ (mS/m)
0.0	–	0.0	–
2.0	0.8 ± 0.1	1.0	1.0 ± 0.1
4.0	2.5 ± 0.2	2.8	2.7 ± 0.1
6.0	4.1 ± 0.3	4.6	4.5 ± 0.1
9.1	6.3 ± 0.5	7.0	6.9 ± 0.2
12.1	9.9 ± 0.8	11.0	10.8 ± 0.3
15.1	14.3 ± 1.1	15.8	15.6 ± 0.4
18.1	20.2 ± 1.5	22.3	21.8 ± 0.5
21.1	30.7 ± 2.3	33.8	33.0 ± 0.8
24.5	41.5 ± 3.1	45.6	44.6 ± 1.1

Table 3.6 Chemical-physical characteristics of the considered grounds

Parameters	Red soil	White soil	Unit of measure
pH	8.21	8.35	–
Conducibilità	154.3	146.9	μS/cm
Sostanze organiche	17,613.75 (1.76%)	24,593.88 (2.46%)	mg/kg
Azoto totale	0.784	1.008	g/kg
Sodio	4.11	2.77	meq/100 g
Calcio	100.798	133.03	meq/100 g
Magnesio	12.86	16.45	meq/100 g
Potassio	2.65	3.78	meq/100 g
Fosforo assim.	203.99 (P_2O_5)	195.72 (P_2O_5)	meq/100 g
Calcare totale	4	8	%

reached was acquired. The value of the corresponding dielectric constant of the soil was calculated from each reflectogram. In this way, it was possible to obtain the calibration curve that associates the value of the dielectric constant of the soil (ε) with the corresponding humidity value (θ).

As an example, Figs. 3.23 and 3.24 show the reflectograms relating to the sample 1 of red soil and sample 1 of white soil. In particular, the reflectograms are indicated with "m", while the first derivatives of the reflectograms (used for the evaluation of the dielectric constant) are indicated with "dm". As expected, in the reported measures, as the humidity level decreases, the value of the minimums of the individual waveforms tends to increase and, at the same time, the apparent distance tends to decrease, due to the progressive decrease of the water content. As a direct consequence, the value of the second maximum of the derivatives of the signal TDR, corresponding to the probe endpoint, tends to shift to the left. By evaluating the

Fig. 3.21 Preparation of samples of red and white soil

Fig. 3.22 Placement of the three-wire probe in the sample (left). TDR measurements were performed with the probe fully inserted into the sample (right)

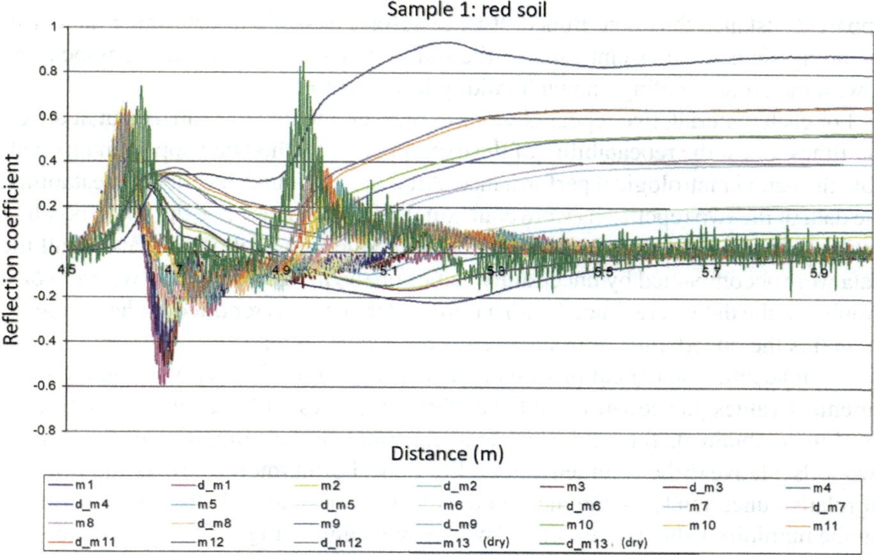

Fig. 3.23 TDR measurements performed on red soil for different reference humidity levels

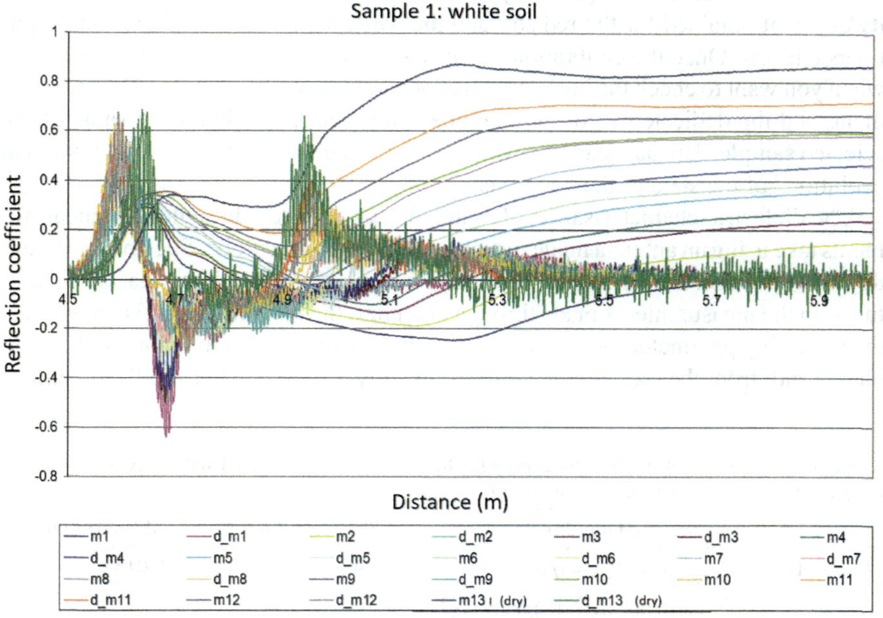

Fig. 3.24 TDR measurements performed on white soil for different reference humidity levels

apparent distance thus determined, it is, therefore, possible to determine the corresponding dielectric constant value of the soil sample under examination, associating it with the corresponding known humidity level (reference).

For each material, two repeated series of measurements were carried out at different times to test the repeatability of the experimental method developed and to evaluate the actual metrological performance. After having found excellent repeatability, the data of the two repetitions were contextually used to determine the corresponding calibration curve. For the final evaluation of the metrological performances, all the data were accompanied by uncertainty values on the measures of ε and reported on a graph ε-θ, the data were fenced with a third-order curve, as reported in the literature, is on this the uncertainty values on θ have been calculated.

From accurate statistical processing and propagation of uncertainty on the measurement values, the equations of the calibration curves with the relative confidence level were obtained. Table 3.7 shows the equations of the calibration curves for the two soils relative to the volumetric humidity value. Furthermore, the maximum values of relative uncertainty on the measured dielectric constant and absolute uncertainty on the humidity value are reported. By way of example, Fig. 3.25 shows the results relating to the two series of repeated measurements on the white ground. Excellent repeatability can be noted.

Figures 3.26 and 3.27 respectively show the calibration curves of dielectric humidity/constant obtained for the red soil and the white soil, with the relative 95% confidence levels. Once the calibration curves have been obtained for a given material, when you want to check the humidity level of a sample, make a single TDR measurement, and the dedicated algorithm automatically returns the humidity value of the selected sample. The data can be saved in appropriate databases to have a "historical evolution" of the state of the materials.

The dielectric characterisation of materials can be made through TDR measurements take using an antenna made in planar technology as a probe. Such probe is non-destructive and non-invasive, and therefore suitable for analysis on masonry structures. In the measurements made through antennas, the magnitude that is observed is the scattering parameter in the reflection of the antenna, S_{11} (defined in module and phase) and, from the measurement of this quantity, is possible to trace the quantities

Table 3.7 Equations of calibration curves for the two soils investigated with maximum relative uncertainty on ε and absolute uncertainty of θ

	Volumetric humidity	Max error on ε %	Max error on ε
Red soil	$y = -0.11764 + 0.049351x$ $- 0.00246x^2 + 0.000474x^3$	3.4	0.015
	$R^2 = 0.9637$		
White soil	$y = -0.1479 + 0.063572*x$ $- 0.00349x^2 + 0.0000698x^3$	4	0.02
	$R^2 = 0.9491$		

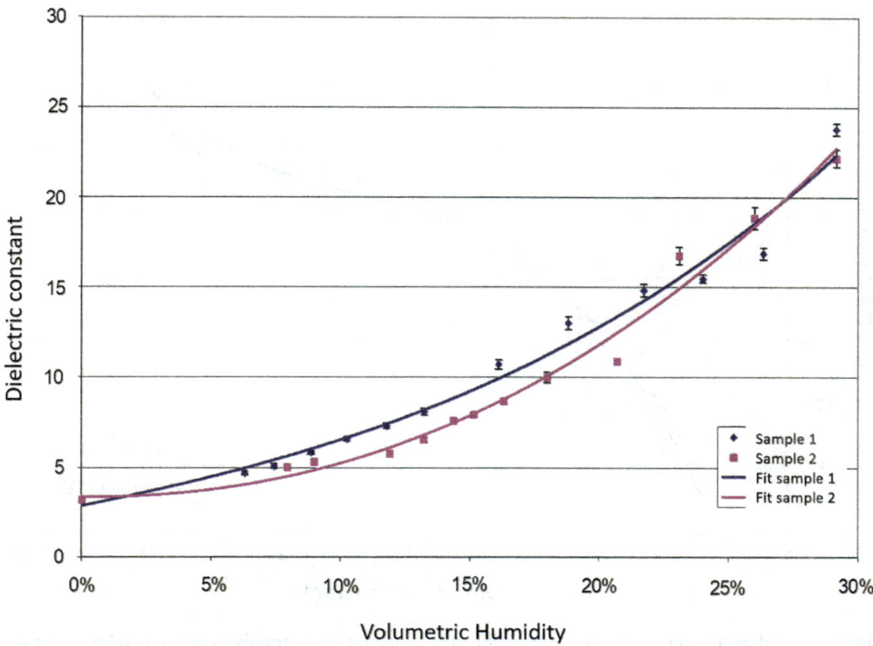

Fig. 3.25 White soil, 1st and 2nd repetition with bars of uncertainty on the dielectric constant when the volumetric humidity varies

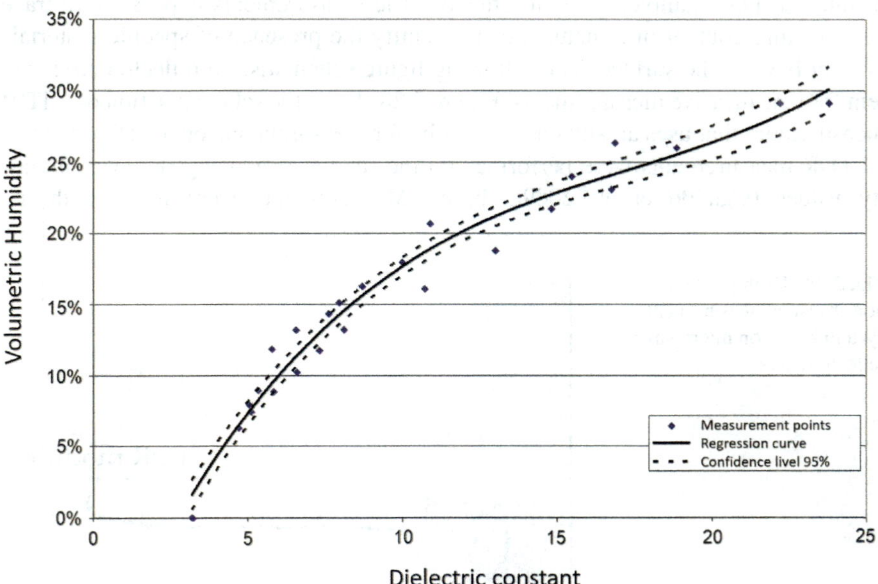

Fig. 3.26 Calibration curve for humidity/dielectric constant (with relative confidence level) for the red soil

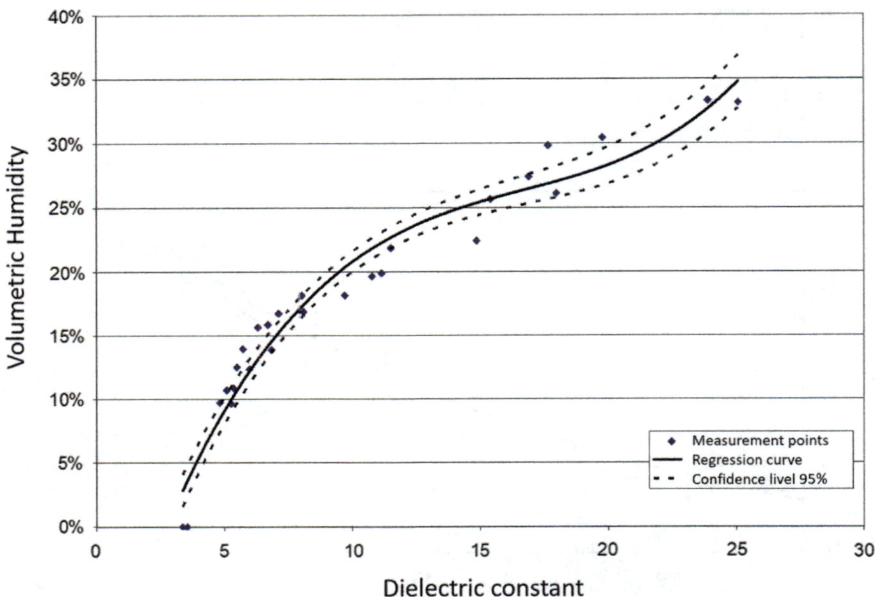

Fig. 3.27 Calibration curve for humidity/dielectric constant (with relative confidence level) for the white soil

of interest. For example, through reflectometric measurements is possible to trace the moisture content of structures or to identify the presence of specific materials present beyond the surface. The following figure schematises a reflectometric system for non-invasive measurements. Figure 3.28 shows the schematisation of a TDR measurement that uses an antenna as a probe for measurements on a wall structure.

TDR measurements were performed on the sand at increasing (known) humidity values (Cataldo et al. 2009a, b, c). Measurements were made with the

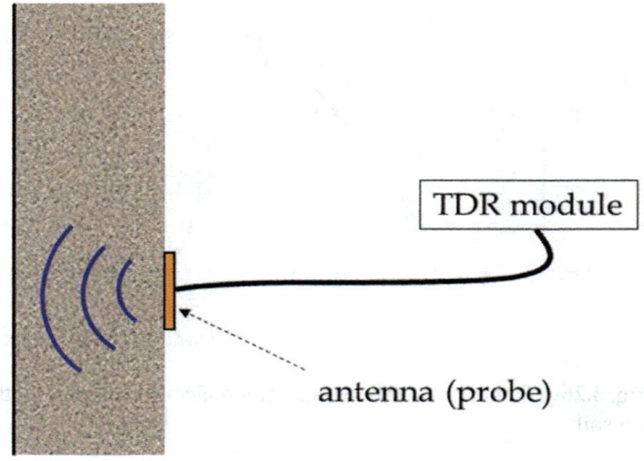

Fig. 3.28 Example of a non-invasive measurement system based on microwave reflectometers

DSA8200/TDR80E04 instrument. The diagram of the antenna designed in (Cataldo et al. 2009a, b, c) for this application is shown in Fig. 3.29.

For each humidity level, the corresponding reflectogram was acquired. Therefore, through an algorithm based on the Fourier transform, each reflectogram was automatically processed to obtain the corresponding value of the scattering parameter in reflection (S_{11} (f)), a function of frequency. Figure 3.30 shows the progression of the module of the scattering parameter in reflection for increasing values of humidity of

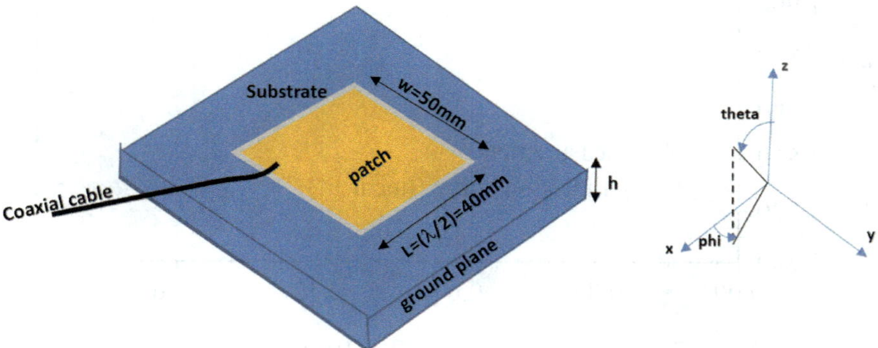

Fig. 3.29 Configuration of the microstrip antenna

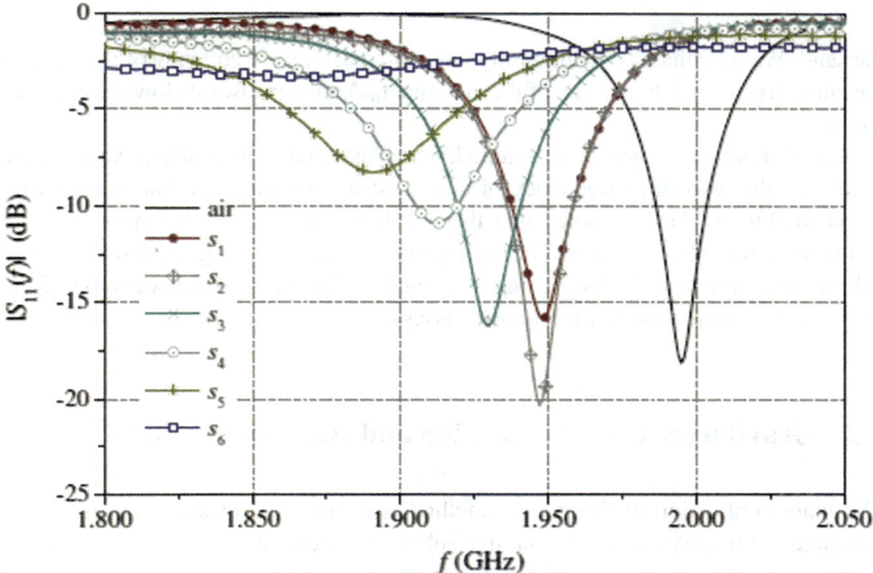

Fig. 3.30 Module of the scattering parameter S_{11} (f) of the antenna when it is placed in contact with sand at humidity levels gradually crescent (s1 refers to dry sand, air refers to the antenna that radiates in the air)

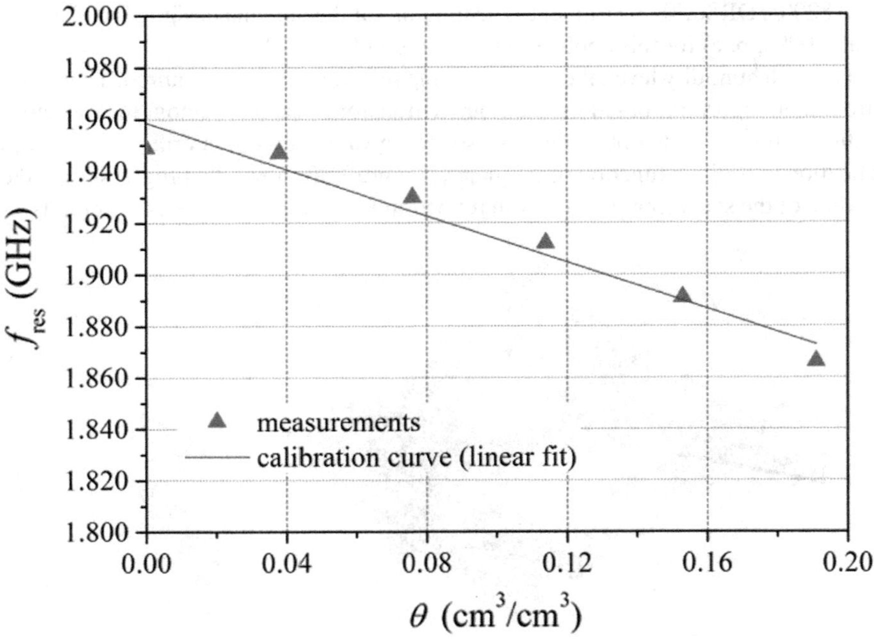

Fig. 3.31 F_{res}–θ calibration curve, related to the considered sand, obtained through TDR measurements (Cataldo et al. 2009a, b, c)

the sand. We note that, as the humidity value increases, the frequency of the peak of the curve (resonance frequency of the antenna, f_{res}), moves towards lower frequency values.

A calibration curve was then obtained, which associates the moisture value of the sand with the resonance frequency of a microstrip antenna placed in contact with the sand (Fig. 3.31). It is noted that the link between fres and the moisture level of the sand was almost linear. The straight line is shown in Fig. 3.31 represents a calibration curve (similar to the one described in the previous subsections) of the TDR measurement system for humidity assessment.

3.2 Gravimetry Data Acquisition and Instrumentation

The main application of the gravity method is to map subsurface geology and to calculate ore reserves for some massive sulfide orebodies directly. In the last years, a modest increase in the use of gravity techniques in specialised investigations for shallow targets was noted by an analysis of international literature. Shallow targets investigation is related also to forensic investigations (especially in finding cavities, hiding places, etc.).

In gravity measurements, many factors must be taken into account: (i) variation with latitude; (ii) variation with elevation; (iii) variation with the time; (iv) variation with geology.

All these considerations make very time expensive a gravimetric survey.

Variation with latitude:
As seen in Chap. 2, the gravitational acceleration g increase with latitude θ, based on the most recently accepted spheroid approximation, this increase is given by (Moritz 1984).

$$g = 978.0327\left(1 + 0.0053024\,sen^2\theta - 0.0000058\,sen^2 2\theta\right)$$

With g measured in Gal. This equation includes both the Newtonian attraction of the Earth as a spheroid and the centrifugal force caused by its rotation about its axis.

Variation with elevation:
The variation with elevation take into account

(i)   the free-air effect

$$\Delta g_{FA} = \frac{-2GM_E\Delta Z}{R_E^2} = -0.3086 \text{ mGal/m}$$

(ii) the Bougher effect

$$\Delta g_B = 0.04192d \text{ mGal}$$

d is the mean density of the slab in g/cm^3.
The Bouguer effect is negative in the presence of void spaces.
Free-Air and Bouguer effects may be combined as

$$\Delta g_E = -(0.3086 - 0.0419d) \text{ mGal/m}$$

Local irregularities in the topography around a gravity station may give rise to significant effects. Figure 3.32 illustrates the source of the Free-Air, Bouguer and terrain effects at a station P.

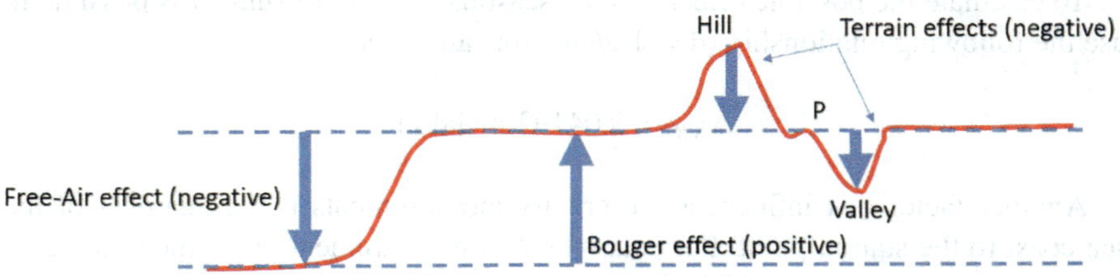

Fig. 3.32 Free air, bouguer and terrain effects

Variation with time:

Gravity force is defined as the vectorial sum of the two force, time-independent, the Newtonian attraction and centrifugal force. Unfortunately, Sun and Moon have a sufficiently large gravitational attraction (about 0.3 mGal) that cause serious time-varying changes in the measured gravity values on the surface of the Earth. Is it possible to calculate these tidal effects? For this purpose, one must consider the latitude, the longitude of the station and the Universal Coordinated Time (UTC).

The correction for Sun and Moon is

$$\Delta g = \frac{3 \cdot G \cdot m \cdot R}{d_0^3} \cdot \left(\cos^2 \psi - \frac{1}{3} \right)$$

where d_0 is the distance between the centre of the Earth and the centre of the Moon/Sun, ψ is the zenithal distance of the Moon/Sun (referring to the centre of the Earth). Numerically

$$\frac{3 \cdot G \cdot m \cdot R}{d_0^3} = 0.165 \text{ mgal for the Moon}$$

$$\frac{3 \cdot G \cdot m \cdot R}{d_0^3} = 0.075 \text{ mgal for the Sun}$$

Since the gravity meter is inserted in a fluid (the air), it applies the principle of Archimedes that makes the gravity measurements depending on the changes in atmospheric pressure. These changes imply changes in the mass of the air column above the gravity point of measurement.

Was evaluated that an increase in atmospheric pressure causes a decrease in the observed gravity and vice versa. The approximate relationship between these changes is given, experimentally by Merriam (1992):

$$\Delta g_p = -0.36 \ \mu \text{ Gal/mbar}$$

Gravity measurements can be affected by Rainfall. Rain can increase the moisture content and groundwater level in porous soils and rocks and can change the level of lakes and rivers in the near vicinity of the station (Lambert and Beaumont 1977; Dragert et al. 1981).

To calculate the possible effect of such seasonal rainfall or run-off is possible to use the following relationship ($d = 1$ g/cm^3 for rainwater):

$$\Delta g_R = 0.04192 \text{ mGal/m}$$

Another factor that influences the gravity measurements is the nearness of the sea coast to the station point. For example, in a measurement performed on a cliff adjacent to deep water, the tidal effect (Δg_T) could be:

$$\Delta g_T = 0.02 \text{ mGal/m}$$

Table 3.8 Rock and minerals types densities

Rock Type	Range (g/cm^3)
Sediments(wet)	
Overburden	
Soil	1.2 – 2.4
Clay	1.63 – 2.6
Gravel	1.70 – 2.40
Sand	1.70 – 2.30
Sandstone	1.61 – 2.76
Shale	1.77 – 3.20
Limestone	1.93 – 2.90
Dolomite	2.28 – 2.90
Sedimentary rocks (av.)	
Igneous rocks	
Rhyolite	2.35 – 2.70
Andesite	2.40 – 2.80
Granite	2.50 – 2.81
Granodiorite	2.67 – 2.79
Porphyry	2.60 – 2.89
Quartzdiorite	2.62 – 2.96
Diorite	2.72 – 2.99
Lavas	2.80 – 3.00
Diabase	2.50 – 3.20
Basalt	2.70 – 3.30
Gabbro	2.70 – 3.50
Peridotite	2.78 – 3.37
Acid igneous	2.30 – 3.11
Basic igneous	2.09 – 3.17

Mineral	Range (g/cm^3)
Metallic minerals	
Oxides, carbonates	
Bauxite	2.3 – 2.55
Limonite	3.5 – 4.0
Siderite	3.7 – 3.9
Rutile	4.18 – 4.3
Manganite	4.2 – 4.4
Chromite	4.3 – 4.6
Ilmenite	4.3 – 5.0
Pyrolusite	4.7 – 5.0
Magnetite	4.9 – 5.2
Franklinite	5.0 – 5.22
Hematite	4.9 – 5.3
Cuprite	5.7 – 6.15
Cassiterite	6.8 – 7.1
Wolframite	7.1 – 7.5
Sulfides, arsenides	
Sphalerite	3.5 – 4.0
Malachite	3.9 – 4.03
Chalcopyrite	4.1 – 4.3
Stannite	4.3 – 4.52
Stibnite	4.5 – 4.6
Pyrrhotite	4.5 – 4.8
Molybdenite	4.4 – 4.8
Marcasite	4.7 – 4.9
Pyrite	4.9 – 5.2
Bornite	4.9 – 5.4
Chalcocite	5.5 – 5.8

Rock Type	Range (g/cm^3)
Metamorphic rocks	
Quartzite	2.50 – 2.70
Schists	2.39 – 2.90
Graywacke	2.60 – 2.70
Marble	2.60 – 2.90
Serpentine	2.40 – 3.10
Slate	2.70 – 2.90
Gneiss	2.59 – 3.00
Amphibolite	2.90 – 3.04
Eclogite	3.20 – 3.54
Metamorphic	2.40 – 3.10

Mineral	Range (g/cm^3)
Cobaltite	5.8 – 6.3
Arsenopyrite	5.9 – 6.2
Bismuththinite	6.5 – 6.7
Galena	7.4 – 7.6
Cinnabar	8.0 – 8.2
Non – metallic minerals	
Petroleum	0.6 – 0.9
Ice	0.88 – 0.92
SeaWater	1.01 – 1.05
Lignite	1.1 – 1.25
Softcoal	1.2 – 1.5
Anthracite	1.34 – 1.8
Chalk	1.53 – 2.6
Graphite	1.9 – 2.3
Rocksalt	2.1 – 2.6
Gypsum	2.2 – 2.6
Kaolinite	2.2 – 2.63
Orthoclase	2.5 – 2.6
Quartz	2.5 – 2.7
Calcite	2.6 – 2.7
Anhydrite	2.29 – 3.0
Biotite	2.7 – 3.2
Magnesite	2.9 – 3.12
Fluorite	3.01 – 3.25
Barite	4.3 – 4.7

Variation with geology:

The bulk density is a characteristic of the various rock types and individual minerals which constitute the Earth. Table 3.8 provides a list of typical rock types and minerals, with their density range.

Therefore the results of a gravimetric survey may be interpreted in terms of the subsurface geology with certain assumptions and limitations.

Instruments for gravity measurements:

There are two base types of gravimeters, the absolute gravimeters that measure the free fall of a body in a vacuum, using lasers and optical interferometry and relative gravimeters that measure differences in gravity from station to station. The absolute gravimeters allow obtaining accuracies of better than 0.01 mGals under favourable conditions (i.e. Niebauer et al. 1986; Torge 1989).

The relative gravimeters rely on the gravity changes to the elongation of a spring which supports a proof mass. The astatic or unstable, and stable are two types of basic field-portable relative gravimeters with different spring balance configurations. The astatic use physic concept related to the equilibrium between the moment of forces in a state close to unstable equilibrium, which gives them great mechanical sensitivity. The LaCoste—Romberg meters operate on this principle. The stable gravimeters are simpler in mechanical principles but require much higher precision of sensing of the position of the proof mass. This gravimeter (Fig. 3.33) has a spring called initial zero-length connected to its mobile arm. On balance, the gravimeter arm is in a horizontal position. Any variation in gravity causes an elongation to the spring which can be cancelled manually by acting on a knob connected, through gears, to a measuring screw. The screw moves a system of levers which, acting on the upper

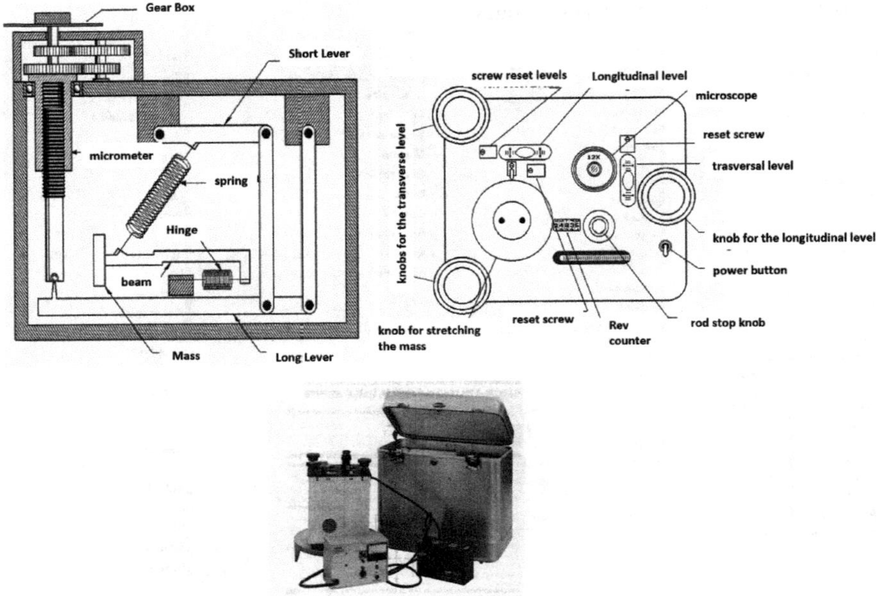

Fig. 3.33 Scheme of LaCoste-Romberg gravimetry

part of the spring, brings the arm of the gravimeter back into a horizontal position. A galvanometer connected to the gravimeter detects the perfect horizontal position of the arm.

When carrying out gravity measurements, first and foremost, a correct setting of the instrument is required, i.e. the horizontality is required for the balance of the arm of the mobile unit and the perfect verticality of the plane of oscillation of the same. For this purpose, the gravimeter is equipped with two levels, one transverse and one longitudinal, located on the base of the upper part. The transverse level forms an angle of 90° with the direction of the arm of the gravimeter and serves to ensure that the plane of oscillation of the mobile equipment is vertical. The longitudinal level, placed in a direction perpendicular to the previous one, guarantees that at the time of measurement, the arm of the mobile unit is perfectly horizontal in the plane of oscillation. The gravimeter has three conductive plates inside, two fixed and one attached to the mobile end of the arm. The movable plate is located between the two fixed plates to constitute two capacitors, connected in a Wheatstone bridge. When the arm is balanced, the capacitance of the two capacitors is the same and the galvanometer, at central zero, does not indicate current flow. The La Coste—Romberg gravimetric model D-168 is equipped with an accessory part that allows keeping the arm in the horizontal position using an impulsive electrostatic force applied to the condensers described above. With the help of an acquirer, gravity measurements can be automatically detected and recorded continuously. The model D allows a reading resolution of 1 μgal over a range of 200 mgal.

Before attributing the differences in gravity observed in the same station at different times to physical reasons, it is necessary to be sure that the instrument has worked to the best of its ability and that the calculation of the tidal effect and the consequent reduction in values of gravity has been performed correctly.

The causes of error are:

(1) Commissioning: particular care is required in setting up the gravimeter; a levelling error can produce noticeable reading errors (quantitatively a division of the bubble level can correspond to a variation in the gravity value ranging from 5 to 50 μGal); (2) The reading: the perception of the reading line is a subjective fact: by performing with the same gravimeter, repeated measurements of differences in gravity between two points are obtained even significantly different values (a few tens of gal). Only the operator's experience and his knowledge of the gravimeter can significantly reduce reading errors; (3) Cancellation procedure (balancing): make sure that the cancellation knob is always moved in the same direction (clockwise or counterclockwise) when returning the arm of the gravimeter to the equilibrium position. Approaching the reading line from opposite directions, there were errors of 8–10 μGal; (4) Instrumental drift: the mechanism of a gravimeter LaCoste—Romberg consists of a system of extremely sensitive levers and springs. Even if maintained at constant values of temperature, pressure and humidity, the characteristics of the springs, with time, may be subject to variations since a molecular settlement occurs. However, the instrumental drift is a characteristic of each instrument and normally decreases with the ageing of the instrument itself. The drift is approximate of the order of 4 μgal/day; (5) Thermal instability: the mechanical characteristics of gravimeters depend on temperature, pressure, magnetic field and humidity. For this reason, the LaCoste—Romberg gravimeter is equipped with a thermostat and a pressure compensator; and also the spring container and lever systems are sealed and insulated. However, it has been proved that the thermostat does not respond quickly to changes in the temperature of the external environment. It has been found that when the gravimeter is removed from its case and is taken to a cold environment, its internal temperature is reduced by approximately 0.3 °C.

This effect produces an apparent change in gravity, estimated as an instrumental drift.

Additional causes of error in the determination of g are: (1) The effect of the tide: on the earth's surface there is a small but measurable gravitational effect due to the presence of the Moon and the Sun. It is of the order of 240 μgal and varies over time with a maximum gradient of about 50 μgal/h. The effect due to other celestial bodies, <1 μgal, is below the resolution limit for a gravimeter normally used in field surveys and is therefore neglected. To obtain precision measurements, the effect due to the terrestrial tide must be carefully removed. Since the laws governing the relative motions of the Earth, the Moon and the Sun are known, the gravitational effect at any point on the earth's surface and at any time can be easily calculated. In the tide, diurnal and semi-diurnal components exist as is known and therefore, the assumption of linear drift with time for one-day data is grossly inaccurate. Therefore in the microgravimetric profiles, where the gravity differences measured to have very often the same order of magnitude as the tidal effect, it is essential to make the reduction. In terms of error, the one introduced by an inexact tidal reduction is the most critical, since the tidal effect determines variations in the gravity of the order of 1 μgal/min. The tide value obtained using standard forecasting tables or calculated with calculation programs is usually of sufficient accuracy even if the ideal procedure would consist in performing, on the area to be investigated, continuous recordings of gravity. (2) Noise: station points are often, for reasons of need, located on roads or near buildings. Many urban noises, characterised by a very high frequency, do not cause disturbances to the gravimeter; while some lower frequency noises (<50 Hz) such as traffic, can produce visible deflections on the arm. Errors due to noise can be of the order of several tens of gal; using particular devices they can be reduced to some algal. Even the wind is the cause of error because it can tilt the gravimeter taking it out of the bubble or inducing vibrations to the arm. (3) Jumps and micro jumps: in the gravimeter readings sudden and irreversible jumps can occur, that is sudden increases or decreases in the value of gravity read (of a few algal or several μgal) that affect all subsequent readings. Due to the jumps in the case of repeated readings on the same station the readings, even after the reduction due to the tidal effect, are significantly different from each other. The jumps that can have any size and sign are linked to causes of any kind very often external, for example, occasional blows during transport, and are however difficult to identify.

Survey Procedures: The gravimetric surveys using relative gravimeters must follow the steps as follows:

1. Set up your gravimeter and ensure that it is operating stably. This is particularly important if the gravimeter has been subjected to rough transport, or has been off-power before the start of the survey. If the setting of its levels has not been checked for some time, do so, following the manufacturer's instructions. Charge up your complement of batteries. On software-controlled gravimeters, also check and adjust the longterm drift correction and the temperature compensation, following the manufacturer's instructions. Also, conform to the other software set-up procedures appropriate to your gravimeter.

2. Establish your grid of proposed gravity stations, keeping in mind the recommendations for stability and freedom from nearby topographic irregularities. Mark each station. Select a number of these stations, which extend across the proposed survey grid, for use as base stations. These stations should be relatively easy for access, e.g. along roads or cleared lines, etc.

3. make gravity measurements at your proposed base stations, starting at one of these stations and returning to it at the end. Repeat the loop of these base stations at least once, so that each of these is read at least twice. If your base stations are reasonably in a line, you will read each twice on each loop (going and coming). Make and distribute the appropriate drift corrections for each loop and station.

4. In the case of microgravimetric surveys, in times of major weather frontal movements, take the barometric pressure readings, preferably at the same time and place as the gravity readings, and make the appropriate correction for atmospheric pressure to each gravity measurement.

5. Take the mean of the two or more sets of base station gravity values, corrected for drift, instrumental level, tides and barometric pressure changes, etc., as being the correct values for each base station. All such values will then have been corrected to a common time of measurement, so far as barometric pressure and instrumental drift is concerned. Their gravity values may be considered to be equally valid and accurate, for use as base stations for other survey stations in their vicinity. Therefore, it will not be necessary to use the same base station each day.

6. Complete the gravimetric survey by measuring at all stations systematically by sections. For each section, use one of the stations which were established under Step 4. as a base station. Through the use of these base stations, at the beginning and end of each day, or portion thereof, you will automatically correct all the newly established station values for cumulative instrumental drift and barometric level changes from the time you read your first station on the survey;

7. Establish the elevation and coordinates of each gravimeter station on the grid, by means appropriate to the accuracy required of the survey. This may be done before, at the same time as, or after the making of the gravity measurements themselves, depending on expediency and the type of elevation and positional control employed. In any event, the final gravity value determined for a station, when corrected for elevations, etc., shall refer to the station elevation as determined by levelling, if the latter differs from the instrumental level. Make the appropriate correction for elevation, as per Variation with Elevation;

8. Tie to national gravity grid. To determine the absolute level of your gravity stations, to allow your survey data to be incorporated into a regional or national database, and to be able to utilise the results from other, independent, gravimetric surveys, it is recommended that you tie one or more of your set of gravity base stations to the nearest station of the national gravity network. A double loop is recommended for greater precision of such a tie.

3.3 Magnetic Instruments and Data Acquisition

The magnetic instruments are classified into two main types: the mechanical instruments that measure the direction or a component of its direction of the magnetic field; and the magnetometers that are instruments capable of measuring the amplitude and/or a component of the magnetic field. The first advances in designing these instruments were made during World War II when Fluxgate Magnetometers were developed for use in submarine detection.

For mechanical instruments, a typical example is a magnetic needle that consists of a small test magnet that is free to rotate in the horizontal plane. The test magnet will align itself along the horizontal direction of the Earth's magnetic field. This is clearly because the positive pole is attracted to the Earth's negative magnetic pole, and the negative is attracted to the Earth's positive magnetic pole. In this way, the declination of the magnetic field measurements is possible. The magnetic needle was invented by the Chinese at least two thousand years ago.

Other devices mechanical devices that measure other components of the magnetic field are the dip needle and the torsion magnetometer. The dip needle measures the inclination of the magnetic field. The torsion magnetometer, through mechanical means, the strength of the vertical (or horizontal) component of the magnetic field.

For the magnetometers, the fluxgate magnetometer is the more used instrument in the geophysical measurements of the magnetic field. It was originally designed and developed during World War II for low-flying aircraft as a submarine detection device. Today it is used in several fields of applications such as archaeology, environment, forensic, etc. A scheme of the fluxgate magnetometer sensors is shown in Fig. 3.34.

Each bar is wound with a primary coil with an inverted direction. As affirm the Ampere's law the alternating current (AC) that passed through the primary coils cause a large, and varying magnetic field in each coil. Successively, as to affirm the induction principle, the magnetic field induced another magnetic field in the two cores that have the same strengths but opposite orientations, at any given time during the current cycle. Since the magnetic fields induced in the cores by the primary coil produce a voltage potential in the secondary coil. In the presence of an external magnetic field component, the behaviour in the two cores differs, by an amount which depends on the external field. Thus, by simply reorienting the instrument so that the cores are parallel to the desired component, the fluxgate magnetometer is capable of measuring the strength of any component of the Earth's magnetic field. The sensibility of the fluxgate magnetometers is of about 0.5–1.0 nT.

Another type of magnetometer is the proton magnetometer. It is currently the most common type. It is cheaper than the optically pumped magnetometer and also has a lower sensitivity (typically about 0.1 nT). They are based on the precession

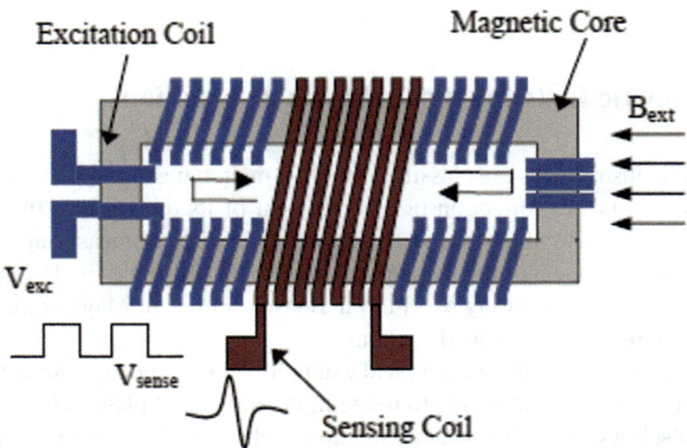

Fig. 3.34 Scheme of fluxgate magnetometer sensor

of protons in the magnetic field. The proton nucleus of the hydrogen atom has a magnetic moment which aligns with a direction of an external magnetic field. The nucleus is spinning and when the external magnetic field changes the nucleus aligns with the new direction. Since it is spinning it does not align instantly, but it twists around its centre, similar to a gyroscope. This effect is called precession. The angular velocity (or frequency) of precession is proportional to the magnetic field. The proton magnetometer has a container filled with a hydrogen-rich liquid (e.g. water or an alcohol). A coil is wounded around the container. When an electric current is passed into the coil, the magnetic field is generated (about 5–10 mT), and the protons are aligned with this field. Then the current is switched off, and protons start to align with the Earth's magnetic field, precessing. In the coil, the electric current is induced by electromagnetic induction. The frequency is measured and strength of the magnetic field computed.

There are two requirements for successful readings. First, the coil has to be roughly aligned so as its field is in a large angle with the direction of the measured field. And second, the field to be measured should be uniform throughout the container. Otherwise, protons in different parts of the container would precess with different frequencies and the readings would be wrong. A small, strongly magnetic bodies (e.g. a piece of iron) could cause non-uniformity of the magnetic field in the container. The measurement of one sample with this type of magnetometer takes several seconds. Figure 3.35 show the scheme of proton magnetometer.

In the few last years, a new way to measure magnetic anomalies was introduced. The instrument that allows this is the magnetic gradiometer.

The geometrical spreading (attenuation) affirm that the strength of the magnetic field decreases with a square of a distance. Therefore changing the height of the magnetometer's sensor, one can decrease or increase the measured values of the magnetic field. With the sensor closer to the ground, the effect of the small near-surface magnetic object will be emphasised.

Conversely, if one wont to increase the height of the sensor, the response of small near-surface bodies will decrease (Fig. 3.36).

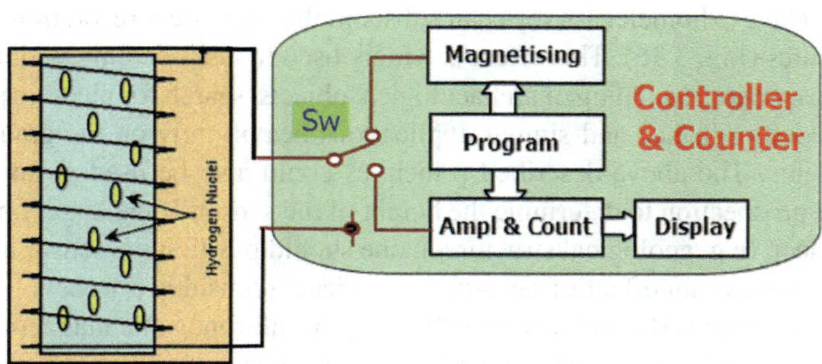

Fig. 3.35 Scheme of a proton magnetometer

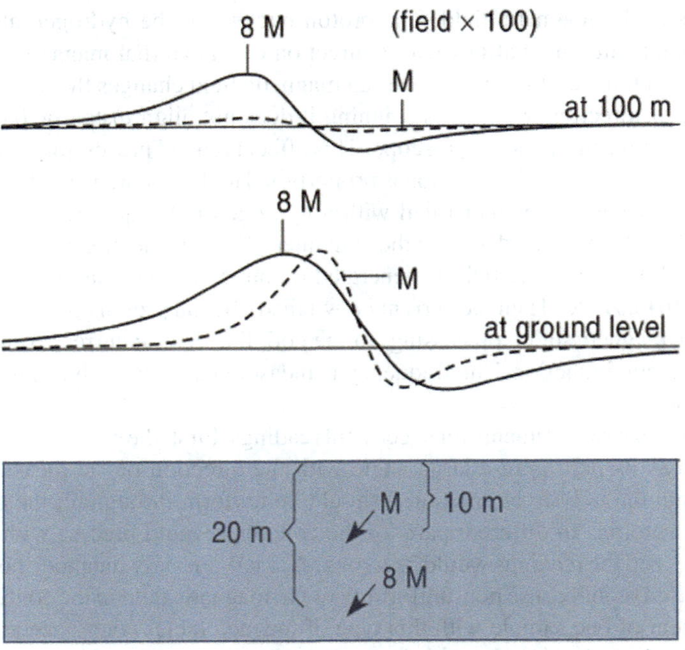

Fig. 3.36 Scheme Anomalies at different heights of two dipoles (Musset and Khan 2000). Note the rapid decrease of the amplitude of the shallower dipole

From the same figure it is also clear that if we subtract these two measurements, one will get the high difference for bodies close to the surface, whereas anomalies caused by deep bodies will (almost) cancel. This principle is often used in field studies. The magnetometer is equipped with two sensors in different heights. The height difference usually varies between 0.5 and 1.5 m. The magnetic field is measured by both sensors simultaneously, hence, in this case, one does not need to correct data for the diurnal variations (the Earth's magnetic field cancels by subtracting the two readings). The gradiometer surveys can substantially increase a resolution for near-surface bodies (Fig. 3.36). This effect is widely used in tasks dealing with the near-surface prospection. Mapping of archaeological objects, search for metallic pipelines or unexploded ordnance and similar applications heavily rely on the gradiometric measurements. The above-described principles could also be used in an ordinary one sensor prospection to determine the height of the sensor. If the target objects are large and deep (e.g. geological structures), one should position the sensor as high as possible to remove an effect of near-surface objects (considered to be a "noise" in this case). The near-surface objects are often highly anthropogenic magnetic objects, pieces of metal (parts of cars, agricultural equipment, cans, etc.). In contrast, in search for small objects, like the archaeological ones, one should position the sensor near the ground (e.g. height of 0.5 m is often used).

The Magnetic survey: three specific criteria must be considered carrying out a magnetic survey for forensic application related for example to environmental crimes: (i) the nature (buried drums, steel pipes or sheet metal) and the depth of the target; (ii) the target's dimensions and therefore the require surveys precision and accuracy; (iii) the target orientation (for objects having a linear surface expression, such as pipes and sheets of metal).

The anomalies created from objects within the first ten metres (usually these are the depths to which a man hides objects such as metal drums, landfill bodies in general, human bodies) can produce relatively intense and narrow anomaly profiles. The deeper the target, the broader the anomaly will be. Further, if the magnetised body is at greater depths, the anomaly will be less intense; more intensely magnetised bodies will have larger amplitudes and larger targets will have broader anomalies. The choice of station and line spacing, i.e. the density of the survey grid are influenced by the relationships of body size, depth and magnetisation. To understand consider that if data points are taken only every five metres in a square grid pattern, and the anomalous peak is only two metres in width or length, the peak will be missed altogether. A minimum of two data points must have to detect an anomaly. For this, the station spacing should be less than half the expected width of the target. To determine the strike length of a body, the same holds for the line spacing it should be less than half the expected length of the target, to have at least two survey lines crossing the target. This *detectability* threshold of twice the sample spacing is also referred to as the *Nyquist frequency*. Figure 3.37 illustrates these points.

Once determined the line and station spacing, both natural or human-made noise's sources must be considered.

The natural noise is related to the time-dependence of the magnetic field, the so-called time-based (diurnal) variations. They are fluctuations with a period lasting of several hours to one day. They can cause a variation of the order of 50 nT per hour (Fig. 3.38) and are not predictive.

The natural noise can be removed using tie-line corrections or base-station corrections (repeated measurements of the magnetic field at the same point over time), or by measuring the vertical gradient of the magnetic field.

The human-made sources of noise are electromagnetic and electrical fields. These sources can seriously influence negatively any magnetometer survey. Therefore, surveying directly under power lines can be problematic. Other sources of human-made noise are the many buried ferrous objects near the surface. But in some environmental crime cases, these objects, (tin cans, bed springs, appliances, etc.) are easily detectable because they can introduce magnetic spikes in the overall results.

The Doppler noise (caused by the rotation of the sensor in the Earth's field while walking) is a source of noise, affecting only proton-precession magnetometers.

There are several modes to perform a magnetic survey. They depend on the degree to which need the noise removed from the data.

Walking mode: In this data acquisition mode, the operator can take continuous readings (at suggested sampling rates of up to every 0.1 s). For a walking velocity of about 3 km/h, data will be collected at approximately every 10 cm. For the very shallow targets that are usually encountered in forensic applications, this tight spacing

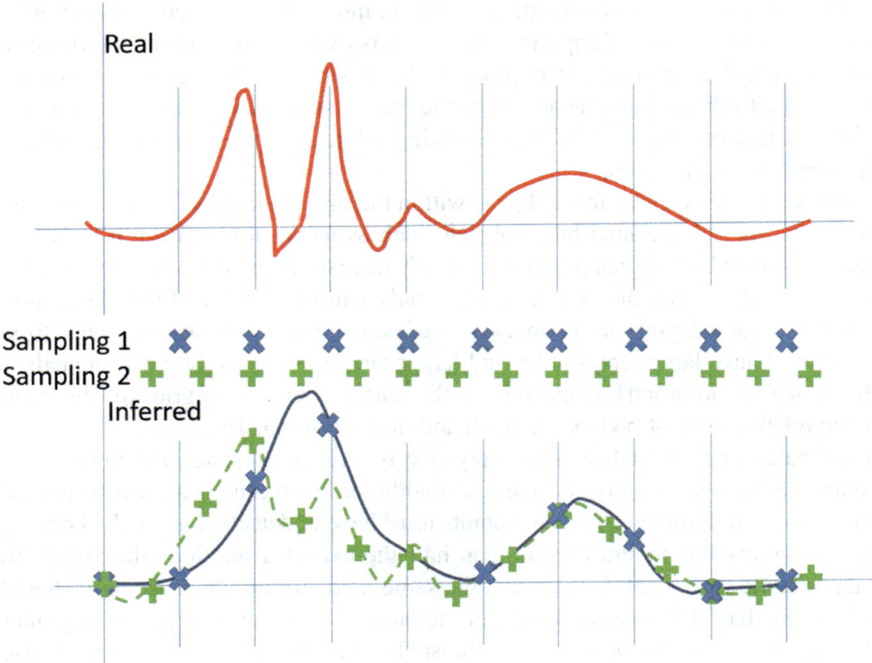

Fig. 3.37 Sampling interval and anomaly resolution

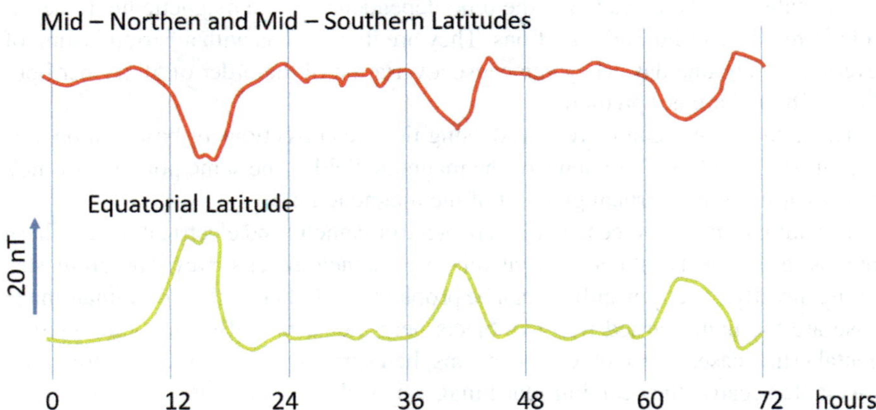

Fig. 3.38 Diurnal variation

of data is necessary. At the same walking velocity of 3 km/h, at 0.5 s sampling rate is equivalent to readings approximately every 0.5 m. The sampling rates depend on the magnetometer that allows choosing several sample rate, but this also depends by the operator. Figure 3.39 illustrates the set-up for a walking mode survey.

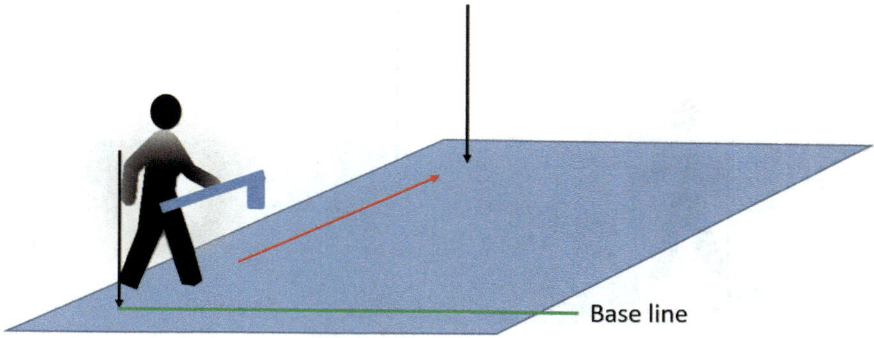

Fig. 3.39 Typical walking mode survey designed

Stop and go mode: This acquisition mode is advisable when one search for larger and deeper targets. The stop-and-go mode with the automatic station, typically at a 15 m spacing, incrementing can be used to evidence targets at depths below 30 m.

At least two points are needed to define an anomaly, which is why the station spacing is half that of the expected depth. This is illustrated in Fig. 3.40.

Gradiometer mode: Using either two vertically or horizontally spaced sensors, one can be performed the gradiometer mode survey. Each sensor read the magnetic and the difference between the two readings is divided by the distance. The measured value of a gradiometer survey is expressed as nanoTeslas per meter (nT/m). The advantages are that the magnetic field measurements are independent of time-based variations since it measures a difference in magnetic fields. Typically the vertical gradiometer sensor spacing is one meter. Gradiometer measurements are sensitive to near-surface objects. Figure 3.41 show the set-up for a gradiometer survey.

Note: The direction of the targets is most often random. Consequently, most environmental grids are laid out in a square pattern.

Survey grid: A baseline and one or several tie lines constitute a survey grid. The baseline is a zero reference line for the grid, and the tie lines serve to correct the

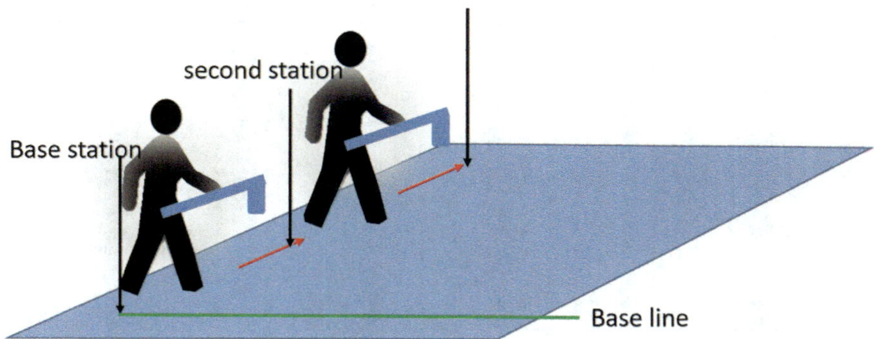

Fig. 3.40 Typical stop and go mode

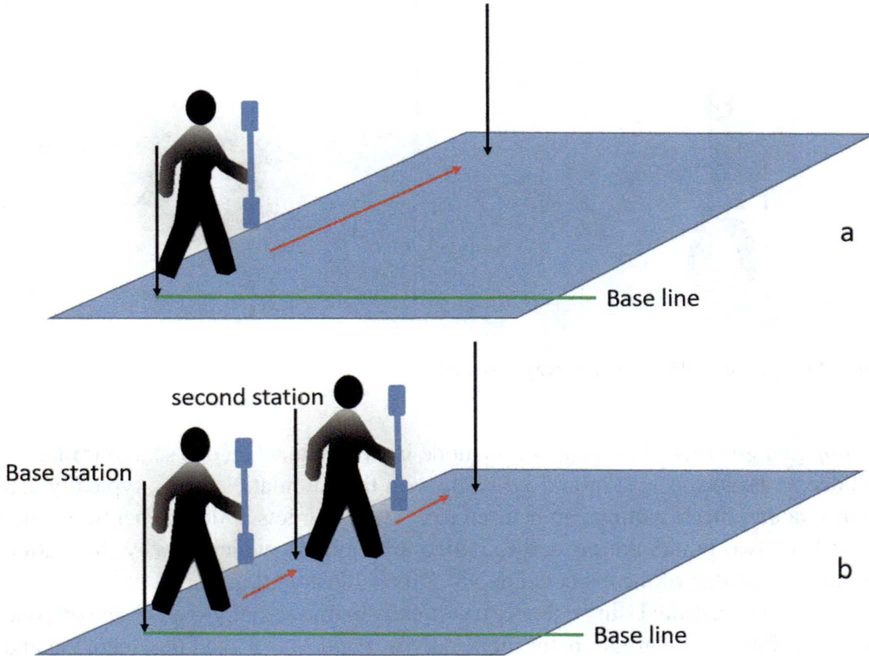

Fig. 3.41 Typical gradiometer survey: **a** walking mode; **b** stop and go mode

skewness of the survey lines. In a square survey grid, the station separation on each line is identical to the line separation and lines spaced are every meter or two, with data points every meter. Figure 3.42 show a typical survey grid, with baselines at 0 and 40 and survey lines every two metres.

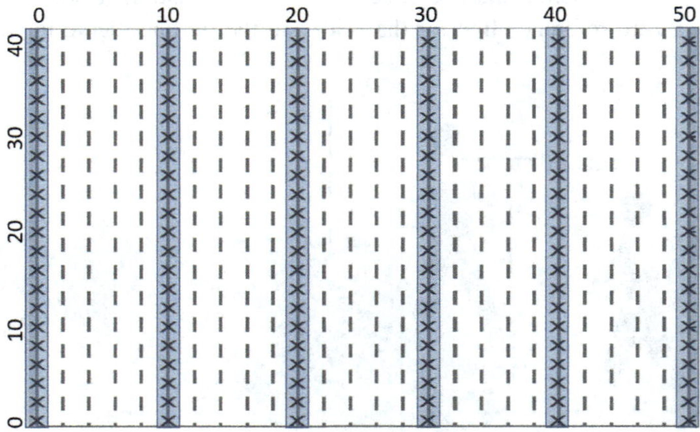

Fig. 3.42 Typical grid used in the magnetometric survey

Gradiometer on the drone: Despite the multiplicity of fields of application of magnetic method, there are currently no portable magnetometric instruments that can be used, without the direct presence of the operator in the field, in environments that are little or not at all accessible, such as—for example—those in which the presence of strong of obstacles (vegetation, walls). The magnetic survey of superficial anomalies related to the presence of hidden structures, human burials, or, for example, of illegal dumps, needs to be carried out at a relatively small distance compared to the earth's surface. This has so far prevented the development of this application, due to the impossibility of having aircraft, which could perform flights at very low altitude. The solution to the problem is offered today by the spread of small, remotely piloted aircraft, which can fly at low altitude and carry small technical instruments on board.

On the other hand, traditional geophysical sensors, if suitably miniaturised, can be reduced to such dimensions that they can be mounted and used on drones. In this way it becomes possible to exploit the technology related to the use of drones, which has reached a level of maturity and ease of use such as to make it suitable for the most varied mapping applications with considerable time savings, to overcome the problems related to the aero-magnetic survey, currently limited by the above mentioned need for low-level flights, which are influenced by the particular conditions and the morphology of the terrain. Thanks to a drone, the data can be collected at a minimum and constant distance from the ground. Through a drone equipped with a system of magnetometric sensors, it thus becomes possible to obtain rapid geophysical measurements at relatively low altitudes, even in sites that are difficult to access, directly in situ, acquiring significant information on presence and actual speed confirmation of any anomalies present in the subsoil, attributable to hidden structures.

As part of the present book, it is possible to develop a modular system that includes the following elements (Fig. 3.43)

  (i)   drone;
  (ii) Sensors;
  (iii) Acquisition system;
  (iv) Control and management software;
  (v) Connection cables;

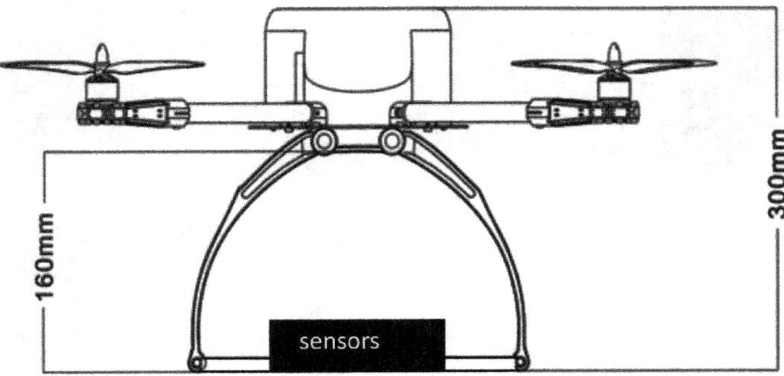

Fig. 3.43 Scheme of the acquisition system

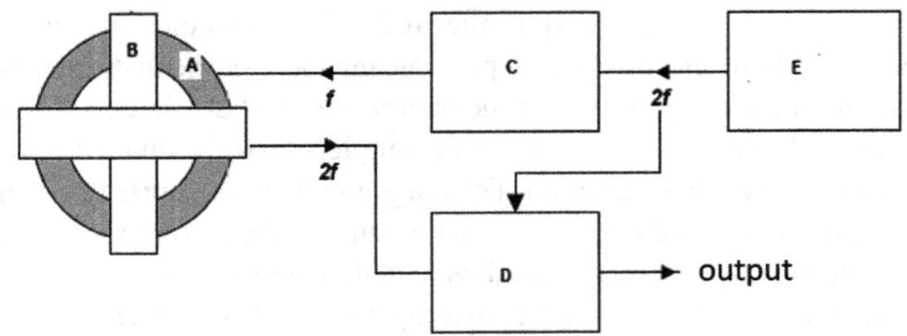

Fig. 3.44 General scheme of the sensor: Coils (A and B); frequency divider (C); phase detector (D); Frequency generator (E)

(vi)   Remote management and control unit via GSM/GPRS (or satellite network) which can be interfaced with the internet.

The data acquisition unit consists of the sensors capable of recording the total field at different heights and consequently, the vertical gradient of the field itself. In a nutshell, the components on which the construction of the sensor will be articulated, as shown in Fig. 3.44: coils (A, B), a frequency divider (C), a phase detector (D), a frequency generator (E) and an integrator circuit (D). The operation of the sensors will be based on the magnetic non-linearity of a core of ferromagnetic material.

The electronics circuits are necessary for recording the magnetic signals based on a theremino master circuit, an open-source system that unlike similar systems (e.g. Arduino), works immediately at power on and does not require firmware programming. The Theremino Master module is not a card with a programmable microcontroller, but an Input-Output device, like a Mouse. In fact, in a very simple way, the same user, with simple programming rudiments, can configure the modules to measure physical quantities of all types: temperatures, radiation, magnetic fields, etc. (Fig. 3.45).

The heart of the electronic circuit (Fig. 3.45) is formed by a PIC24FJ64GB002 (integrated circuit), this device is built-in CMOS technology, and its function is to

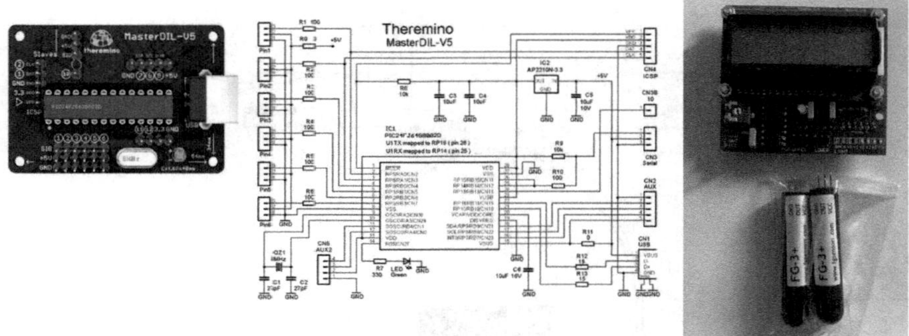

Fig. 3.45 General scheme of the module Master Theremino DIL-V5

read the sensor data. This is a signal that can vary from 30 to 120 kHz, but in this case, the point of greatest linearity of the sensor is used, which is found around the frequencies between 60 and 70 kHz corresponding to the zero-field, i.e. around the axis geomagnetic East-West; in this context, the PIC works in frequency counter mode. After obtaining the required data, the frequency values acquired by the sensors will be converted, through a mathematical algorithm, into nano Tesla, a unit officially used for reading the geomagnetic field and, finally, recorded on a micro SD card as a LOG file. Once downloaded on a common PC, through a specific program, it will be possible to visualise on the screen the temporal trends, in graphic colour format.

3.4 Electrical-Resistivity Tomography Field Data Acquisition

The "best practices" for the ERT data acquisition are presented. As underlined in Chap. 2, ERT hardware technology and software have had a rapid evolution related to data acquisition and analysis. Expert utilisers of ERT methods have benefited, in data acquisition and analysis, of numerous possibilities to select several parameters related to the acquisition mode and inversion processing. In some cases, these possibilities can substantially affect the results. As a result, it is now possible for them to know: (i) how to design robust survey geometries; (ii) how select the acquisition and inversion parameters; and (iii) how it is possible to obtain good images of the investigated medium to facilitate the interpretation of the results.

ERT Survey-design: The survey design must be allowed to retrieve high-resolution information while covering the largest possible area. Since the design is constrained by the number of electrodes that can be used in a single survey, these two criteria are in contradiction. For instance, to retrieve information with higher resolution, more electrodes are needed, but in turn, the covered area is smaller. Therefore, a trade-off between these two criteria, resolution and area, is to be found. To do so, several design alternatives are evaluated in terms of the two objective criteria, namely resolution and covered area.

In ERT data collection, Wenner, Schlumberger, or dipole-dipole arrays were historically used. The use of these fixed geometries was because scientists moved the two current and two potential electrodes by hand, and data processing could be performed without complex inversion. Such work was highly labour-intensive and time-consuming. The advent of multi-node cables and multi-channel instrumentation enables much faster data collection and data processing can be performed in minutes on a low-end PC. Whereas selection of an ideal geometry has been the subject of past research (Furman et al. 2003, 2007; Stummer et al. 2004), the ability to resolve properties in the subsurface is dependent on the electrical conductivity of the subsurface, which is unknown (Day-Lewis et al. 2005). Is important to select an appropriate number of quadrupoles and this is related to geometric factor and accuracy of forwarding modelling. Other factors are related to the instrument used, how quickly it can acquire data, the decision whether or not to collect time-lapse data, and to the limit on the spacing that is governed by the resistivity of the earth materials and the capability of the equipment to inject current.

Therefore the question is which of them is most suitable for forensic purpose prospection. To answer this important question, we must define the: (i) spatial resolution; (ii) strength of the response, and (iii) depth of distinction. The spatial resolution is related to the dimension of the buried features. The rule should be that the spatial extent of the resistivity anomalies should not be larger than the features that cause them (Schmidt 2013). The strength of the response is related to the resistivity contrast between the features of interest and the hosting medium. Anomalies related to feature with weak-resistivity contrast should be enhanced to be as strong as possible. The depth of distinction describes how well features can be distinguished when they are buried at various depths (Schmidt 2013).

The Wenner array is technically the Wenner alpha array. This type of array is sensitive to vertical changes in the resistivity of the investigated medium (Loke 2001; Leucci 2015). However, it is less sensitive to horizontal changes in the resistivity within the investigated medium.

Therefore, the Wenner array is a good one to resolve vertical changes and therefore, horizontal structures, and it is not considered for detection of horizontal changes (vertical structures). The depth of investigation for the Wenner array is approximately $0.5a$, where a is the spacing between the electrodes pairs. The array has a strong signal strength, and this is important in surveys where the investigated medium presents high background noise. A disadvantage is the relatively poor horizontal coverage.

Increasing the number of levels (n) increases the depth of investigation but decreases the horizontal coverage (Leucci 2019).

The dipole-dipole array: This array is sensitive to horizontal changes in resistivity; it is relatively insensitive to vertical changes in resistivity (Loke 2001; Leucci 2015). It is good to map vertical structures. The investigation depth depends on and n factors. Compared to the Wenner array, the dipole-dipole array has a shallower depth of investigation but has greater horizontal coverage. To use this array, the multichannel instrument should have a very good resistance contact between the electrodes and the investigated medium surface, high sensitivity and very good noise-rejection circuit.

The Wenner-Schlumberger array is a hybrid between the Wenner and Schlumberger arrays (Loke 2001; Leucci 2015, 2019). It is moderately sensitive to both horizontal and vertical structures, and therefore this array is a good compromise between the Wenner and dipole-dipole array. Compared to the Wenner array it has slightly better horizontal coverage while compared to the dipole-dipole array, it has a slightly worse horizontal coverage.

Considering that the buried structures are 3D, a fully 3D resistivity survey should, in theory, give the most accurate results. Generally, 3D data sets are constructed from several parallel 2D survey lines. Ideally, there should be a set of survey lines with measurements in the x-direction, followed by another series of lines in the y-direction. The use of measurements in two perpendicular directions helps to reduce any directional bias in the data (for more see Leucci 2019).

There are field situations in which a typical grid of electrode lines is limited by physical conditions (the presence of obstacle or the necessity to investigate the around the buildings, etc.). In this case, it is possible to use a "non-conventional" array that consists both of multiple L-shaped arrays for a grid of electrodes by surrounding an area with a square of electrode lines and circular array.

Circular array: The fields of application of the circular acquisitions are manifold: think of the characterisations of columns for architectural purposes or the study of tree trunks to assess their state of health. The circular array based on the esteem of geometrical factor "K". Weidelt and Weller (1997) proposed the calculation of the geometric factor for the study of cylindrical samples, such as drilling cores. This analytical geometric factor was calculated considering the dipole-dipole array. Given the electrodes A and B of current and M and N of potential, positioned on the circumference of a cylinder having length L and radius a, have been defined:

α as the angle subtended by the single dipoles;

β as the angular separation between the centres of the dipoles (Fig. 3.46).

The geometric factor is thus defined:

$$K = \frac{a}{f(\beta - \alpha) - 2f(\beta) + f(\beta + \alpha)}$$

If $L/a < 1$ the function for each angle (ε) is given by:

$$f\left(\varepsilon; \frac{L}{a}\right) = \frac{a}{\pi L} \log \frac{1}{\mathrm{sen}\left(\frac{\varepsilon}{2}\right)}$$

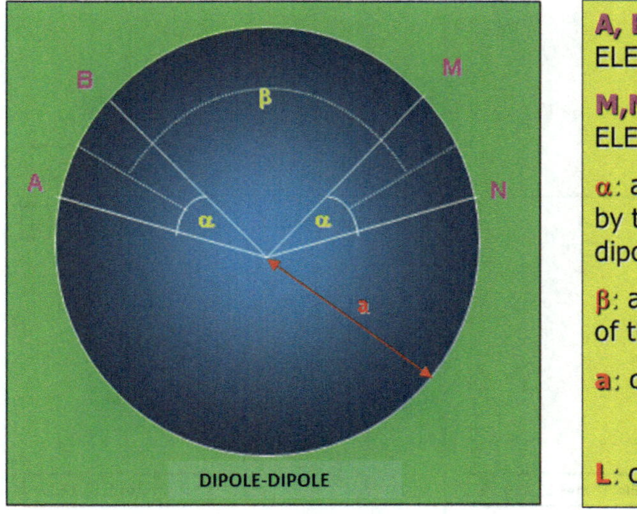

Fig. 3.46 Diagram of a cylinder section and scheme of the position and distance of the electrodes

If $L/a > 1$, the values that Weidelt and Weller have collected in a table and made available can be used for the function f. However, this geometric analytical factor has limitations related both to the difficulty of calculating the K and to its difficult application if one wants to take resistivity measurements on cylindrical surfaces that are not perfectly regular, such as the surface of a tree trunk.

As an alternative to the analytical geometrical factor k, a graphical method is proposed to calculate the geometric factor k obtained, in this case, from the measurement of the actual path taken by the current flow to reach the potential electrodes. The various phases that led to the identification of the method are described in detail.

Take, for example, a circumference (Fig. 3.47) with a diameter of 0.30 m on which 24 equidistant electrodes with a distance of 0.04 m are arranged. The circumference with the position of all the electrodes on it is designed. At this point the distances AM, AN, BM and BN to be replaced then in the equation that describes the standard geometrical factor for dipole-dipole array (see Chap. 2) are measured for each quadripole. In Fig. 3.47b, for example, the current electrodes are the 1st and 2nd while those of potential are the 8th and 9th. It can be seen that the values of graphically measured distances deviate considerably from those measured if the same electrodes were arranged on a linear profile having the same values of a and n.

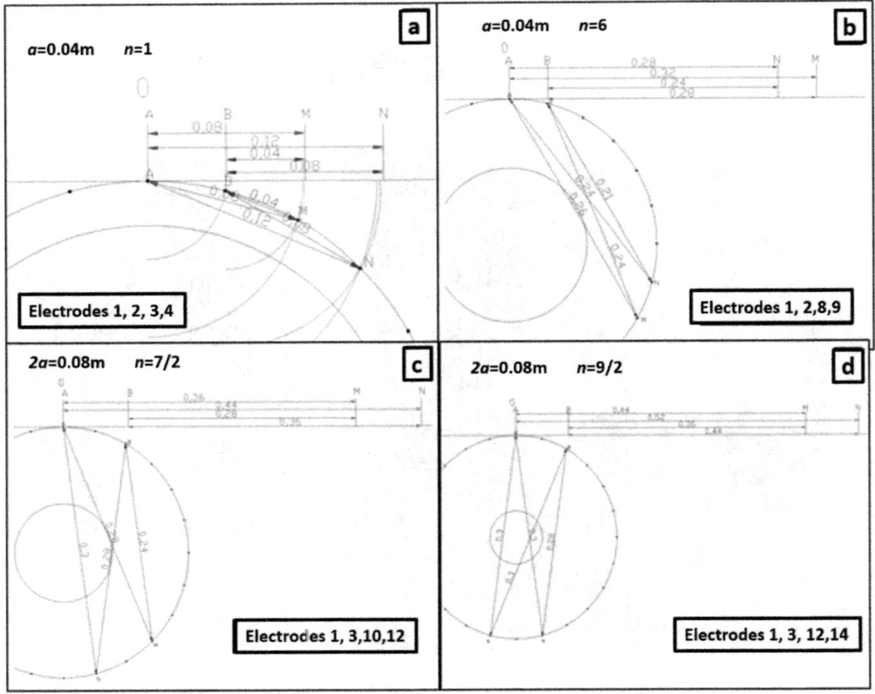

Fig. 3.47 Method of calculation of the K graph; a difference of the distances values calculated along the circumference and along with linear profiles having the same characteristics as a and n

Table 3.9 Comparison of distances for each quadripole

Quadripole	AM		AN		BM		BN	
	Linear	Circumf	Linear	Circumf	Linear	Circumf	Linear	Circumf
1-2-3-4	0.08	0.08	0.12	0.12	0.04	0.04	0.08	0.08
1-2-8-9	0.28	0.24	0.32	0.26	0.24	0.21	0.28	0.24
1-3-10-12	0.36	0.28	0.44	0.30	0.28	0.24	0.36	0.28
1-3-12-14	0.44	0.30	0.52	0.30	0.36	0.28	0.44	0.30

In Table 3.9 for each quadruple, the values of the distances calculated are compared if the electrodes were placed both along a line and along a circumference.

The real distance between the electrodes arranged on a circular profile is less than the distance of the same electrodes arranged on a linear profile. It can be deduced that the value of the apparent resistivity calculated with the classical methods is different from that obtained using the distance obtained between the electrodes and the one obtained with the graphic method. Figure 3.48 shows the graph and the relative table comparing the values of the linear K, of the analytical K and the K graph relating to a dipole-dipole sequence prepared for the study of a 0.40 m high cylinder with a diameter of 0.30 m. In the ordinate, the values of K are reported, while in abscissa those of the angle beta (angle formed by the axes of the segments identified respectively by the current dipole and by the potential one).

The analytical and the linear K have an almost similar course but diverge as the distance between the dipoles ($n * a$) and therefore, the investigated depth increases. The graphical K has an irregular course, occasionally presenting some points in common with those of the analytical K. To calculate the analytical K: inserting: the radius value (r) of the circumference, the number of electrodes and the interelectrode distance, it is possible to automatically calculate the distance between two points P and Q by the formula:

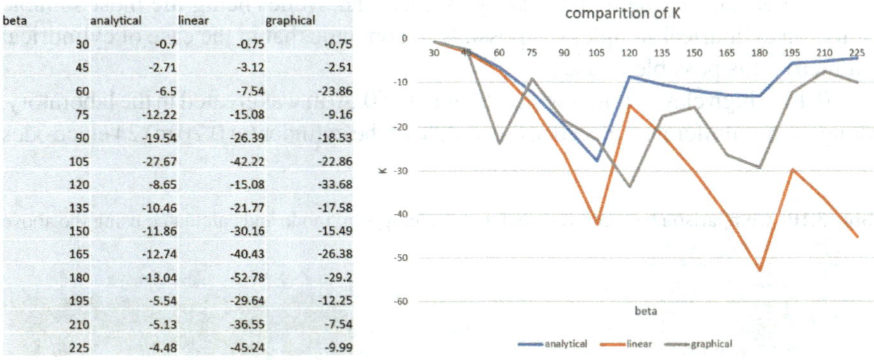

beta	analytical	linear	graphical
30	-0.7	-0.75	-0.75
45	-2.71	-3.12	-2.51
60	-6.5	-7.54	-23.86
75	-12.22	-15.08	-9.16
90	-19.54	-26.39	-18.53
105	-27.67	-42.22	-22.86
120	-8.65	-15.08	-33.68
135	-10.46	-21.77	-17.58
150	-11.86	-30.16	-15.49
165	-12.74	-40.43	-26.38
180	-13.04	-52.78	-29.2
195	-5.54	-29.64	-12.25
210	-5.13	-36.55	-7.54
225	-4.48	-45.24	-9.99

Fig. 3.48 Table and comparison chart of the values of the analytical, linear, and graphical geometrical factor K

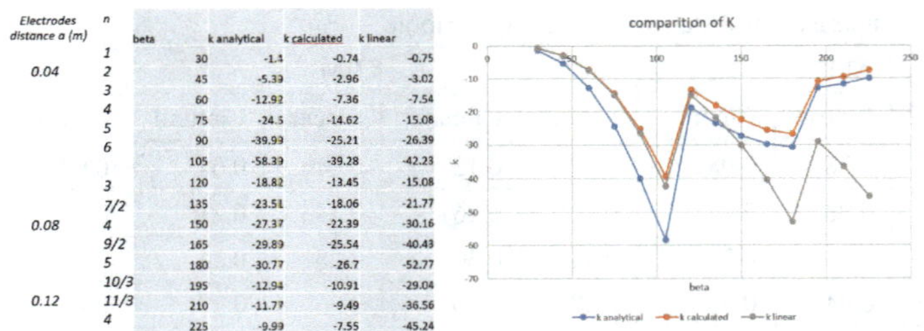

Electrodes distance a (m)	n	beta	k analytical	k calculated	k linear
0.04	1	30	-1.4	-0.74	-0.75
	2	45	-5.39	-2.96	-3.02
	3	60	-12.92	-7.36	-7.54
	4	75	-24.5	-14.62	-15.08
	5	90	-39.93	-25.21	-26.39
	6	105	-58.39	-39.28	-42.23
0.08	3	120	-18.82	-13.45	-15.08
	7/2	135	-23.51	-18.06	-21.77
	4	150	-27.37	-22.39	-30.16
	9/2	165	-29.89	-25.54	-40.43
	5	180	-30.77	-26.7	-52.77
0.12	10/3	195	-12.94	-10.91	-29.04
	11/3	210	-11.77	-9.49	-36.56
	4	225	-9.99	-7.55	-45.24

Fig. 3.49 Table and comparison chart of the values of linear, analytical and calculated K

$$\overline{PQ} = 2r * sen\left(\frac{\varphi}{2}\right)$$

where φ is the central angle subtended by the PQ arc.

The calculation speed, the ease of application whatever the size of the cylinder studied, as well as the precision with which the distances are calculated using the above formula, allow recalculating the values of K previously determined graphically. The surprising result is shown in Fig. 3.49: the values of this new k are significantly different from those of the graphically calculated K.

This difference is attributable to the degree of precision with which the distances have been estimated. In the graphic method, for convenience, each distance was measured with centimetre accuracy. A numerical example can clarify how the different degree of approximation gives different results of the value of k. Table 3.10 shows the distances AM, AN, BM and BN, their reciprocal and the values of k relative to $a = 0.04$ m and $n = 6$.

The latter calculated K has a trend very similar to that of the analytical K, and in particular, to the major depths, where the linear K diverges from the analytical one, the two graphs are parallel (Fig. 3.50).

Since the values of k calculated with the above method are similar to those of the analytical K (already demonstrated by Weidelt and Weller being the most suitable for use on cylindrical samples), is possible to conclude that in the case of cylindrical symmetric it is possible to use it.

A 0.40 m high clay cylinder with a diameter of 0.30 m was created in the laboratory. Along the circumference placed at the middle of the cylinder (at 0.20 m) 24 electrodes

Table 3.10 Comparison between K calculated in a graphical mode and calculated using the above formula

		Graphical K		Calculated K	
AM	1/AM	0.24	4.167	0.238	4.202
AN	1/AN	0.26	3.846	0.260	3.849
BM	1/BM	0.21	4.762	0.212	4.714
BN	1/BN	0.24	4.167	0.238	4.202
K=2*3.14/(1/AM-1/AN-1/BM+1/BN)		-22.86		-39.28	

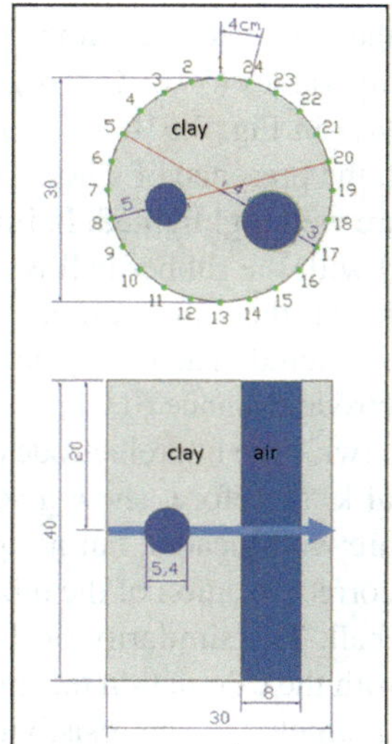
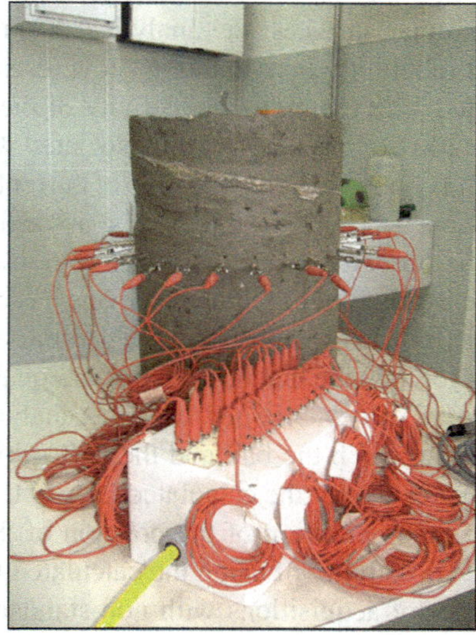

Fig. 3.50 Scheme of the clay cylinder, the ball-shaped body, the PVC pipe and the location of the electrodes

were placed at an interelectrode distance of 0.04 m. These electrodes are copper nails with a 0.002 m section and 0.04 m length. On the median section of the cylinder, at the 5th electrode, an empty rubber ball having a 0.054 m diameter was inserted and the centre of which is 0.077 m from the electrode. Furthermore, at the 17th electrode, an empty PVC tube with a diameter of 0.08 m (Fig. 3.50) has been inserted for the entire height of the clay cylinder, whose axis is 0.07 m from the electrode itself.

The dipole-dipole array was used.

Results were represented on a circular section. In Fig. 3.51a there are the resistivity data obtained with the use of the linear K, in Fig. 3.51b of the analytical K and in Fig. 3.51c of the calculated k. Already from the visualisation of the data, it was deduced that the linear K was not to be taken into consideration in the measurements

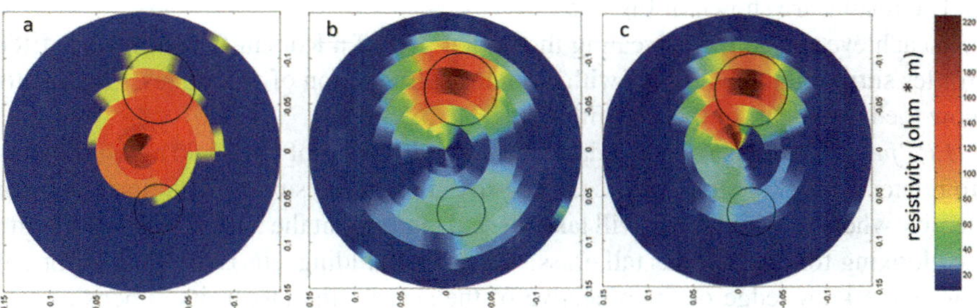

Fig. 3.51 Circular sections relating to the resistivity values calculated with the linear K (**a**), the analytical K (**b**) and the calculated K (**c**). The black circles on each section indicate the position and size of the ball (down) and the PVC tube (up)

made on circular surfaces; the analytical K and the calculated K, on the contrary, show good reliability. In a context of low resistivity values (5–15 Ω * m), that can be associated with the high conductivity of the clays in Fig. 3.51b, c areas of high resistivity (200 Ω * m) that can be associated with the presence of a body, the PVC pipe, which is opposed to the passage of current, are well highlighted. In Fig. 3.51b, a high resistive (35 Ω * m) that can be associated with the rubber ball is also well located. The observed values of resistivity, lower than the real ones, are probably attributable to the difficulty that the device has encountered in acquiring an object of dimensions (0.05 m) comparable with the interelectrode distance (0.04 m): this does not happen for the tube that has a diameter equal to twice the interelectrode distance. The resistivity values obtained with the analytical k, therefore, show a section of the cylinder in which the ball and the PVC pipe are well located, but not perfectly defined. The results for the calculated k show the correct location of the tube but not an equally good definition of the position of the ball. The similarity of the results suggests the possibility of using the calculated k with the calculation technique also in other types of acquisitions with non-standard geometries. The resistivity model relative to the linear k differs totally from reality, which is why its use in this kind of acquisitions is considered inadequate.

Linear non-standard array: This array considers a non-standard arrangement of the electrodes on the soil surface. The survey is conducted along each perpendicular line or transect. In the next step, the current electrodes remain at the end of one line, while the potentials are moved along the line. Then the current electrodes are moved one electrode position, and the potential electrodes are moved as previously described. The process is repeated until the current and potential electrodes cover the L geometry. This sequence of observations produces a series of resistivity towards and beneath the central portion of the array. The blue coloured lines in Fig. 3.52 represent the array geometry, where the apparent resistivities are measured for the performed ERT array. This process is discussed in detail by Tejero-Andrade et al. (2015). Several L-arrays can be combined to surround a structure to build a 3D matrix of observations.

Figure 3.52 depicts the results obtained using a series of L lines acquired in Rudiae (Lecce, south Italy).

The results are shown in Fig. 3.53.

To achieve the goal of revealing the buried part of a Roman Amphitheatre, ERT profiles surround the buildings with a random distribution of electrodes. The results show the buried part of the amphitheatre (Fig. 3.53).

ERT field Procedures: The fundamental considerations of ERT field measurements are related to survey strategies regarding the aims of the survey. The first step is to identify where the field study will take place. Is important the accessibility of the site (i.e., looking for obstacles as tall grass, sinkholes, building, etc.). Another important item is the knowledge of the objective of the survey: the target, its dimensions its depth. Important item is also the local geology (subsoil highly or poorly conductive) this can be related to the capability of the instrument to inject enough current to reach a certain depth related to the objective of the survey. The interval between each

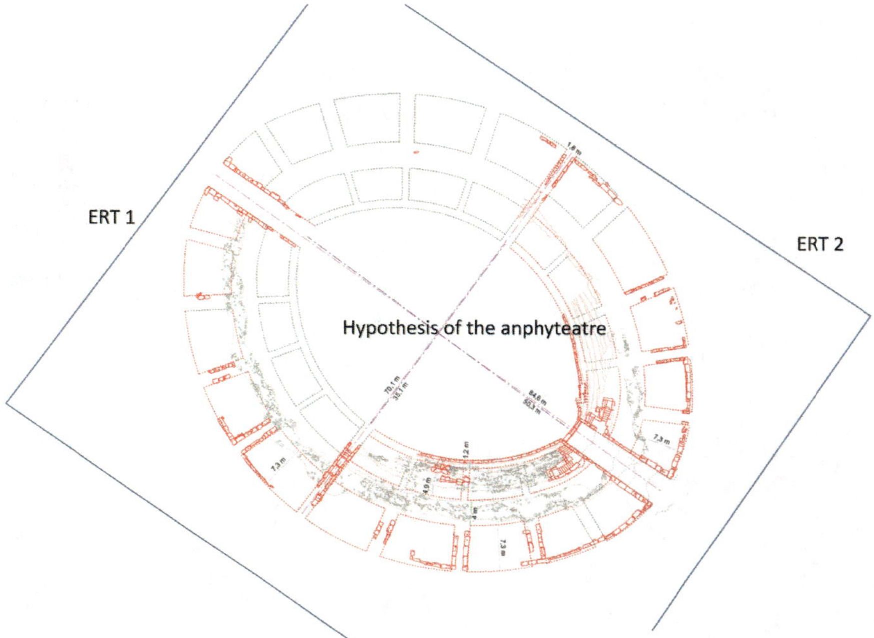

Fig. 3.52 The apparent resistivity measured lines in a non-conventional array

electrode is related to the maximum resolution and the depth of investigation into the ground depends on the length of the electrode spread.

For example, when using the dipole-dipole electrode array, the investigation depth will be about the 20% of the electrode spread length (a straight line over a 100-m distance allows a depth of 20 m). To esteem, the depth of investigation related to a particular array sees Table 2.3.

Another important rule that one must keep in mind is if its size of the target is less than a quarter of the depth it cannot be detected (a target of one meter cannot be seen deeper than four meters in depth). Also, you cannot expect to detect a target smaller than half the electrode spacing. If the electrode space is a two-meter interval, the smallest object to be detected would be one meter near the surface.

Another example is the follow: if the target is 6 ms deep, the total electrode spread length will be the expected depth divided by 0.2 (or 20%), which is 30 (this means that to spread the electrodes over at least 30 m to get to the depth of a 6 m).

In the field, the survey is important to verify the contact resistance between the ground and the electrodes. The instrument will issue warnings for electrodes which are not connected right or not planted firmly in the ground so that they can be checked before the actual survey starts (for more see Leucci 2019).

The results can visualise on the computer on a graphical colour plot, called a pseudo-section, in real-time using the software related to the instrument. Nowadays, the ERT instruments can visualise the results in real-time. This is important because

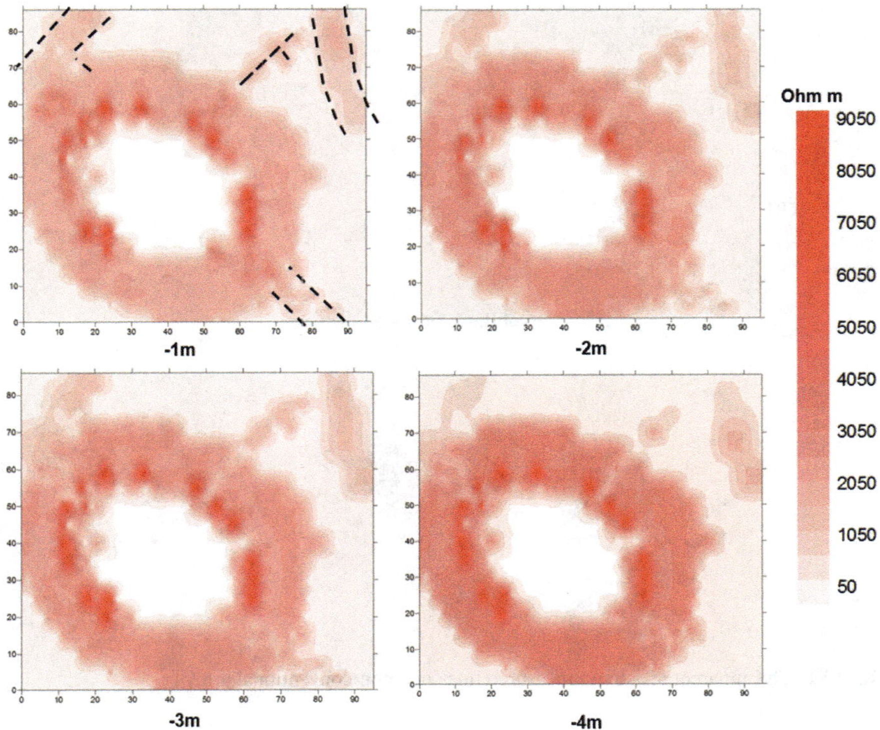

Fig. 3.53 3D model of resistivity distribution: depth of 1, 2, 3 and 4 m

one can control the data quality and assurance directly in the field. That way, you can change any setting or repair any issues (i.e. connect a disconnected electrode or change instrument settings if noisy data is recorded).

3.5   Self-potential Data Acquisition

The self-potential (SP) method is well known because it was successful used in the minerals exploration. Nowadays the SP method has been repurposed in several application fields such as archaeology (Leucci et al. 2014), in agriculture (Golovko and Pozdnyakov 2010), environmental and engineering (Corwin 1990), in geotechnical (Corwin and Butler 1989) in geothermal exploration (Corwin and Hoover 1979), etc.

The basic SP equipment is consisting of a pair of electrodes connected by wire to a digital multimeter with a high input impedance capable of reading accuracy of ±0.0001 volts (Fig. 3.54). There are, however, two restrictions on the electrodes and voltmeter which are most important. They are: (1) the measurement technique can introduce no spurious potentials, and (2) the reference or base electrode must be

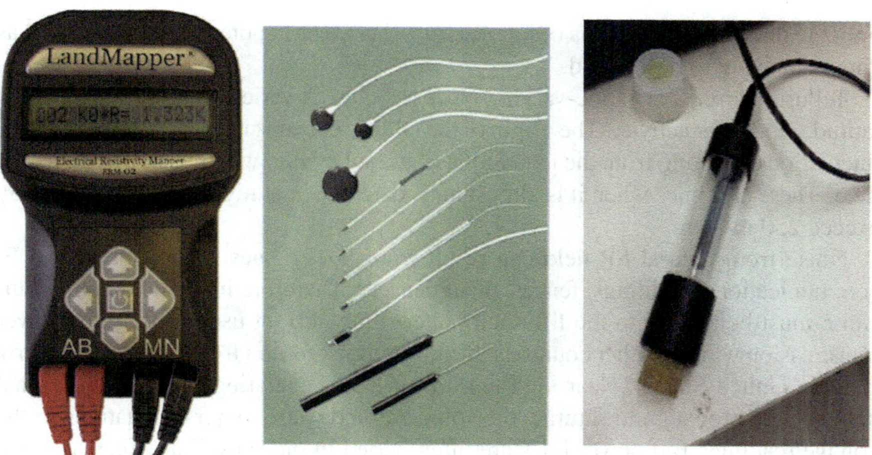

Fig. 3.54 Solid-state Ag-AgCl non-polarizing electrodes LandMapper ERM-02, hand-held device for measuring electrical resistivity, conductivity and self-potential

placed outside the system, above the water table, and not in a reducing environment such as a bog or swamp.

Figure 3.55 shows the method of conventional SP data acquisition schematically.

The base station electrode is always attached to the negative lead of the voltmeter while the roving electrode is placed in a shallow hole in the ground and attached to the positive lead of the voltmeter. The voltage between the base station electrode and the roving electrode is recorded with the defined sign convention. After making the voltage measurement, the roving electrode is picked up and moved to station 2. Here another voltage measurement is made. The procedure is repeated until the end of the wire is reached, at which point the wire is rewound, and the base station is moved to the end of the completed line, where a secondary base station is established. This process is repeated until the survey is complete. For every station must be taken some minutes (about 5–10) to obtain the stabilisation of the field self-potential.

SP data contains the superposition of several different electric fields, the direct current field and time-varying field (telluric currents and cultural noise). In SP surveys,

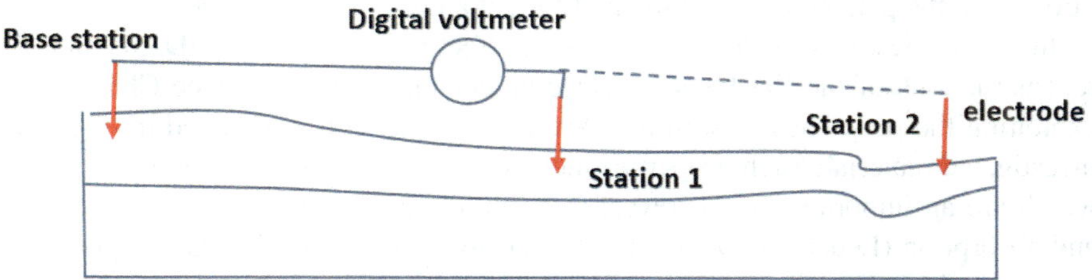

Fig. 3.55 Self-potential field-data acquisition scheme

only the direct current field is of interest; all other electric potentials are regarded as "noise" and are to be avoided.

Telluric currents are time-varying electric fields. It varies with the time of day, latitude, and solar activity. The origin of the telluric currents is in the magnetospheric fluctuations resulting from the interaction of the solar wind with the earth's magnetic field. There are times when it is almost nonexistent, e.g. at night; other times it may exceed 250 mV/km.

Noise from natural SP field can result from power lines, radio (mainly ULF) communications, buildings, fences, pipelines, etc. Therefore in the field data acquisition must be careful to the fifty-hertz alternating current used in Italy for power transmission while in other countries the power transmission frequency is sixty hertz (i.e. the United States). Near high tension lines or generating stations, this signal may be strong enough to saturate the voltmeter and cause major fluctuations in the voltage readings. But power lines are often buried in the subsoil and they alter the ground potential for several meters. No measurement should be made within 10 m of a ground wire to avoid any cultural direct current offset.

Cultural noise due to power transmission can be removed using a suitable notch or low pass filter (50–60 Hz). Also, metal pipes alter the potential electrical and therefore, if possible, measurements should not be made within 20 m of any metal object. Pipelines are often electrically charged to prevent corrosion. Measurements should not be made within 500 m of an electrically-protected pipeline.

Obviously must say that this is not a noise if the SP measurements are related to the analysis of the corrosion. For more see Leucci 2019.

3.6 Seismic Tomography Sonic and Ultrasonic Data Acquisition

The tomography has origin in medical research to produce images of tissue density (Hounsfield 1973). In this type of tomography, the object (the patient) is moved through a large doughnut-shaped machine, where an X-ray beam and a set of electronic X-ray detectors are located opposite each other. The source and detectors are rotated around the target region, collecting the amount of radiation being absorbed throughout the patient's body at many different angles (Goldman 2007).

In geophysics, seismic tomography is an effective technique for 2D, 3D and 4D reconstructions of the Earth's subsurface and/or other materials (see Chap. 2). It exploiting the properties of seismic wave energy after it has travelled through the investigated material. Such properties include travel time, ray paths and amplitude, which are an important key to reveal information about seismic velocity, density, and absorption (Leucci 2019). Currently, seismic tomography is widely applied to a variety of scales and geometries. The local scale, tomography is convenient for environmental or civil engineering or forensic investigations, economic exploration and archaeological research.

Depending on the input data, seismic tomographies fall into three main categories: transmission tomography using P or S first arrivals, i.e., direct, diving and refracted waves; reflection tomography using P or S reflection waves; diffraction tomography using P or S scattered waves (e.g., diffractions, reflections, and converted transmissions). Transmission tomography is an appropriate technique to define: horizontal layering structures; regions exhibiting low complexity velocity distributions.

Seismic tomography survey is typically performed in the following manner. An array of receivers, normally with an equal receiver spacing, is fixed in a given location. Data are then recorded for multiple source locations (Fig. 3.56).

These source locations are usually equally spaced along a survey line, and data are acquired sequentially along the survey line. A set of data acquired with a fixed receiver array and a sequential set of source locations is referred to as a sweep. Another sweep may then be acquired by placing the sources along a different survey line or by moving the receiver array.

Typical geometry (transmission, refraction or reflection tomography) has to be chosen according to the aim of the investigation, in agreement with the potential targets' testing problems, the cost-effectiveness of the specific investigation, the required depth of investigation, and the accessibility of the site. To acquire significant data, a minimum distance from the transmitter to receiver, related to the texture

Fig. 3.56 Experimental set-up for sonic and ultrasonic measurements

and unit dimension, must be considered. Lower distances are indicative of limited portions of structures and lead to local results. Since seismic data collection strategy varies dependent upon the desired information, technique, and the processing algorithms employed. The desired information is largely controlled by target depth, target dimensions, and expected sensitivity of the target to seismic energy (e.g., target to background contrasts in physical properties). Transect length and geophone spacing changed between experiments, but the general practices are constant. These general practices consisted of acquiring seismic records with shot points located perpendicularly adjacent to geophone locations along with a profile and on the opposite face in respect to the receiver position in transmission tomography (Fig. 3.57a) and shot points coincident with geophone locations in refraction and reflection tomography (Fig. 3.57b).

Structures more complex, (as the microcracking related to the stability of the structures), are more demanding of the seismic technique. As the search for a smaller object, improved seismic imaging of the surveyed materials is required. The more refined the approach to gathering data, the better the observed resulted image. In the field of optics, if an image is required to be clearer, the imaging equipment needs better

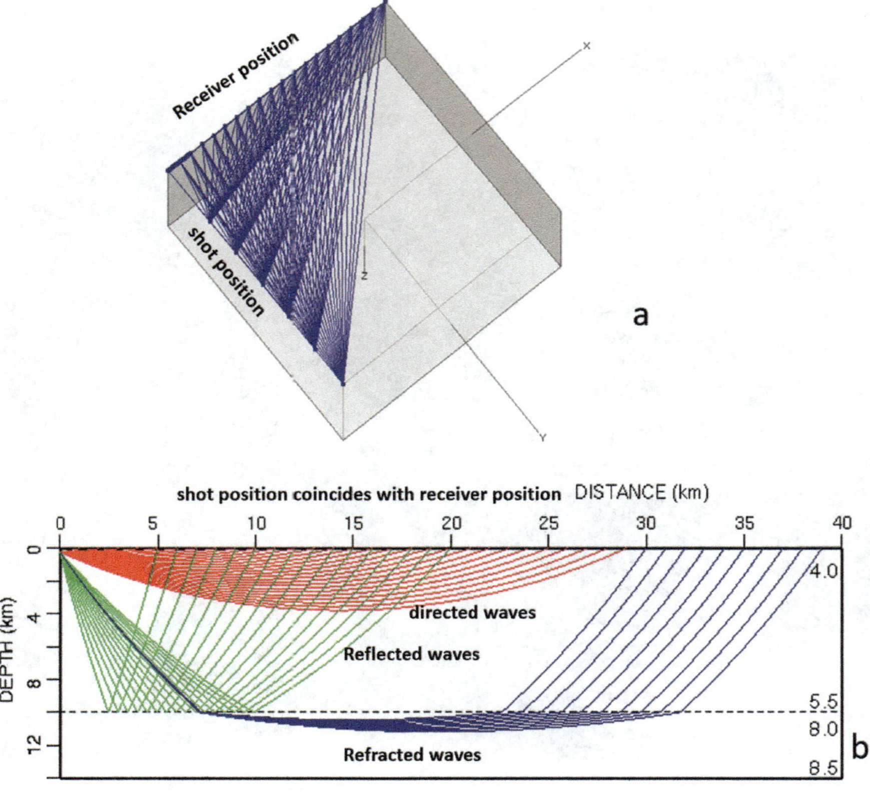

Fig. 3.57 Traveltime tomography: **a** transmission; **b** refraction and reflection

resolving power. Seismic tomography exploration has borrowed part of this term. Thus, more complex structures may require higher-resolution data. This includes both lateral and vertical resolution. The high lateral resolution demands close line and shot/receiver spacing (i.e., spatial sampling), whereas higher vertical resolution requires higher temporal sampling of the received data. Good field procedures will produce a better signal-to-noise ratio and higher resolution.

The frequency content of the received seismic signal limits vertical resolution. The wider the bandwidth and the higher the frequencies received, the greater the resolution of the final results and the greater the definition and imaging of studied material. Therefore, the survey recording parameters mustn't compromise the survey objectives by either temporally sampling the data too far apart or band-limiting the received frequencies. The seismic signal must be recorded adequately if it is to be reproduced in the computer centre at a later date.

The target objective determines the target depth. Acoustic impedance contrasts (i.e., a sonic velocity or density contrast) determine if the target at a particular depth will be detected. The following three issues are closely tied to the depth and nature of the target: The energy source must have adequate power at the desired frequencies to obtain the target visualisation. If the source is too strong, the dynamic range of the recording instruments may be saturated, ruining the fidelity of the data. If the source is too weak, the poor signal-to-noise ratio may result, and the target will be imaged poorly. There must be sufficient fold (number of shot and receiver) to maintain a target signal-to-noise ratio adequate for the geophysicist to make a correct interpretation. The source/receiver geometry must have an adequately long offset to optimise velocity calculations and multiple attenuations to resolve the target properly.

The best way to establish technical specifications is to perform tests in situ. Before any survey, all previous acquisition parameters and results must be reviewed to see if a change in parameters might improve on previous work. If acquisition parameters are changed, a previous line should be reshot to prove that the newer parameters are an improvement over the previous parameters.

Other elements are the channel numbers: These are limited to the number of channels the recording equipment can handle. Generally, the greater the number of channels, the shorter the station spacing which can be used in the field. The geophone array design that depends on the survey type see above.

Geophones are tuned to give a peak voltage amplitude output over a desired frequency spectrum of ground motion. They can be tuned to commence their peak output at an initial motion frequency value, from as low for sonic to greater for ultrasonic (see Chap. 2). This is their "natural" or "resonant" frequency. The energy source is related to the best signal-to-noise ratio.

Related to the seismic signal amplitude attenuation, there are two types of amplitude decay, namely spreading loss and absorption. As a seismic wave expands outward from a shot, the energy per unit area of the wavefront is attenuated as the inverse of the square of the distance from the source (Fig. 3.58). This phenomenon is called energy geometrical spreading loss.

The spreading loss per unit area is given by

Fig. 3.58 Spreading loss

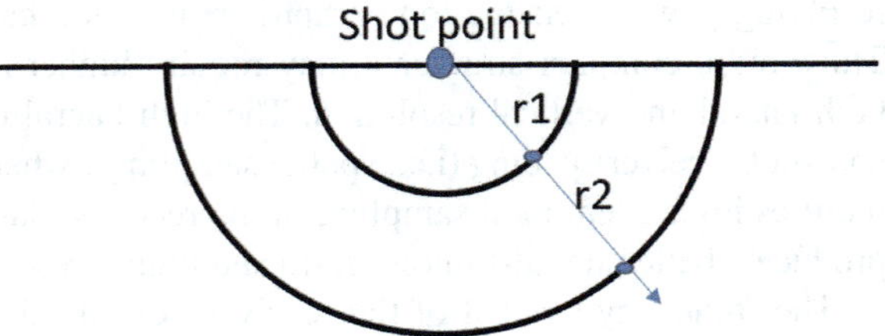

Amplitude Spreading Loss in dB $= 10\log r_2/r_1$

where r_1 and r_2 are two radial distances from a shot.

Another amplitude loss occurs as a wavefront passes through the material. It varies exponentially with distance

$$A_2 = A_1 e^{-\alpha(r_2-r_1)}$$

where A_1 is the amplitude at distance r_1, A_2 is the amplitude at distance r_2 and α is the absorption coefficient of the material. Therefore

$$absorpion\,loss\,in\,\mathrm{dB} = 10\log\!\left[e^{-\alpha(r_2-r_1)}\right] = 4.3^{\alpha(r_2-r_1)}$$

in amplitude terms (accounting for both types of losses)

$$A_2 = A_1\frac{r_1}{r_2}e^{-\alpha(r_2-r_1)}$$

Assuming a typical value of is 0.25 dB per wavelength λ, where r_1 and r_2 are the radial distances from the shot, v is the velocity of a seismic wave through the material, f is the wavefront's predominant frequency, and wavelength $\lambda = v/f$, then absorption loss will be,

$$absorpion\,loss\,in\,\mathrm{dB} = \frac{1.1f(r_2-r_1)}{\mathrm{v}}$$

According to the above equation, higher frequencies experience greater absorption loss than do lower frequencies. The combined effect of both losses explains why deeper events in shot records generally have lower frequency content.

The information contained in a signal can be characterised by three quantities: signal-to-noise ratio, bandwidth, and duration. The signal-to-noise ratio can have different meanings depending on the circumstances. For example, diffractions from out of the reflection plane are noise on 2-D data but are part of the signal in 3-D surveys. The recorded signal bandwidth is related to the used source and, such has seen above, the source frequency. Generally, it is lower than the signal source frequency. The materials behave like a low pass filter.

The duration of recorded signals depends on the nature of the source and target depth. It can be range from a few ten or hundred milliseconds.

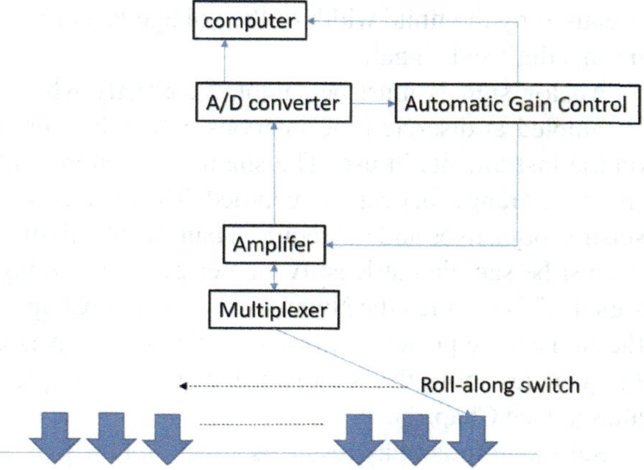

Fig. 3.59 Scheme of the recording system

In data recording, an analog seismic signal travelled from the geophones along electrical conductors (the cable) to a roll along with a switch in the recording truck, after which it is converted to a digital signal and recorded on the computer. The basic components of the recording are shown in Fig. 3.59.

The scheme shows a Roll-along switch. It allows the observer to record a selected subset of the geophones connected to the recording computer. The roll-along switch minimises the need to move the recording computer. The switch has many input seismic channels and outputs the seismic data to the recording system. The multiplexer. This is an electronic switch that timeshares data from multiple channels. It changes multiple parallel inputs to a serial output ready for amplification, digitising, and recording. The multiplexer cycles through all of the inputs during each digital sampling interval.

The amplifier. This amplifier receives all analog signals input to it and passes them on to the A/D converter with an amount of gain determined by the gain controller. The A/D converter. Analog signals are converted to digital signals with this device. It allows the analog stream of data to be recorded in digital form. The received incoming signal must be filtered to prevent aliasing before conversion to a digital form—the gain controller. The received signal includes reflections, refractions, transmission, and environmental noise, all of which may have amplitudes varying in a range from microvolts to volts.

The proposed scheme is general since each actual recording system has unique aspects that can be understood best by studying the manufacturer's documentation.

Another consideration is related to the instrument noise and sampling interval. Several types of electronic noise can occur in recording instruments. Coherent repetitive signals (50, 60, or 100 Hz) are caused by electrical interference or crossfeed between the instruments and nearby power supplies or cables. Modern electronic instruments can suffer from several types of incoherent noise that are inherent in the nature of the components. For example, the noise caused by the randomness of conducting electrons, whereas shot noise is caused by the discreteness of charge carriers within semiconductors. Quantisation noise is another type of incoherent noise that is caused by the finite width of the voltage level represented by the least significant bit in a digitised signal.

Analog signals must be sampled correctly when digitising. The seismic signal is sampled at discrete time intervals called the sample intervals, which depending on the instruments in use. The smaller the sample interval, the larger the potential frequency range that can be recorded. The choice of sample rate is dependent on the seismic objectives and the required bandwidth. To digitise an analog signal properly, it must be sampled at least twice per cycle of the highest frequency present in the signal. This is called the Nyquist criterion (see Chap. 1). For a given sample interval, the highest frequency that satisfies this criterion is called the Nyquist frequency. Frequencies above the Nyquist frequency are sampled incorrectly and are said to be aliased (see Chap. 1).

Seismic tomography results is a section in wiggle traces visualisation (Fig. 3.60). It shows on the abscissa the geophone position (in meters) and on the y-axis the seismic wave travel time. For more see Leucci (2019).

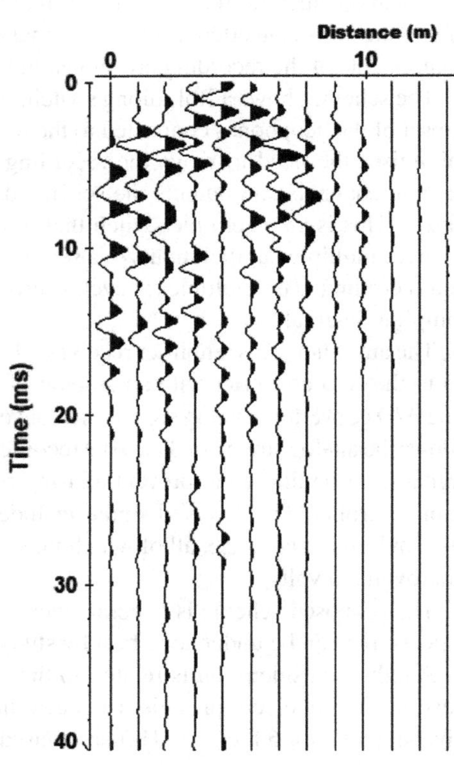

Fig. 3.60 Typical seismogram

3.7 GPR Data Acquisition

In Ground-penetrating radar surveys design is of primary importance the choice of some acquisition parameters to obtain the best results in resolution the objective target. To achieve the targeted depth and resolution, numerous parameters can be considered. For this purpose are important the following parameters: antenna frequency, time window, sample interval, samples per trace, step size, stacking, and transect spacing and orientation. The relative dielectric constant is the physical parameter related to the properties of the materials and therefore, cannot be controlled in GPR surveys.

As seen in Chap. 2, the ability of EM waves to effectively penetrate the ground to a particular depth is primarily dependent upon two factors: the frequency of the waves, and the characteristics of the ground (Conyers 2004). The chose of the correct antenna frequency for a GPR survey is critical (physically only wave frequency can be controlled). Lower frequency antennas with long wavelengths provide the deepest penetration, whereas high-frequency antennas with short wavelengths are only able to image shallow features (Conyers 2004; Neubauer et al. 2002). Conyers (2004) suggests as a general rule that 400–900 MHz antenna should be used to image features within 1 m of the surface, whereas images 1–3 m below the surface are best imaged with 250–500 MHz antenna. Goodman et al. (2009) recommend selecting a radar frequency (and time window length) that will collect information to a depth of at least 1.5–2 times that of the target area. Leucci (2008, 2019) suggests a way to esteem the maximum penetration depth as a function of the GPR system (performance figure) and antenna frequency. The results of are resumed in Table 3.11.

Table 3.11 Antenna parameters, determined in experimental mode, based on the values of attenuation calculated in the field survey (Leucci 2019)

Antenna Frequency (MHz)	Antenna Time range (ns)	a (m)	TPL (dB)
1000	20	2.84×10^{-4}	82.7
1000	15	2.29×10^{-4}	82.4
1000	8	1.43×10^{-4}	81.7
500	100	0.0011	74.8
200	300	0.0032	63.1
200	150	0.0019	62.1
100	500	0.0056	53.8
100	250	0.0034	53
35	500	0.0073	54.9
35	1000	0.0123	54

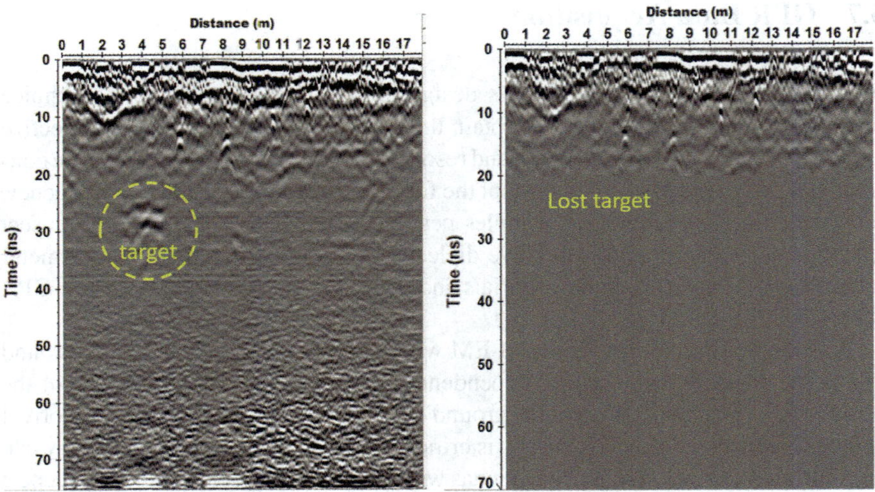

Fig. 3.61 Example of a shorter time window that losing the target

The parameter time window represents the amount of time for which the receiving antenna will record two-way travel time data. It is a function of the objective of the survey. Shorter time window risks losing the target of the survey (Fig. 3.61).

The equation (Conyers 2004):

$$W = 1.3\,(2D/v)$$

Can be used to esteem the time window. Here W is the length of the time window (in ns), d is the maximum depth to be resolved (in m), and v is the minimum electromagnetic-wave velocity in the surveyed material (in m/ns). Consider the uncertainty in velocity and depth variations the time window would be increased by 30% (Annan, 2005).

The sampling interval is related to the time between points collected for each recorded waveform. It can be estimated using the following equation

$$t = 1000/6f$$

where t is the sampling interval (in ns) and f is the centre frequency of the antenna (in MHz) (Sensors and Software 1999). Sampling interval can be taken to be 1.5 times the centre frequency of the antenna and generally does not exceed half of the period of the highest frequency.

Table 3.12 indicates the calculated sampling interval for a variety of antenna frequencies (Annan 2001). An increase in antenna frequency requires an increase in the sample rate due to the preferential attenuation at high frequencies. Increases in sample rate should be based on the Nyquist principle—the greatest vertical resolution that can be expected is one-quarter the wavelength (Jol and Bristow 2003).

Table 3.12 Suitable sampling intervals and corresponding antenna frequencies (Annan 2001)

Antenna Frequency (MHz)	Antenna Time range (ns)	a (m)	TPL (dB)
1000	20	2.84×10^{-4}	82.7
1000	15	2.29×10^{-4}	82.4
1000	8	1.43×10^{-4}	81.7
500	100	0.0011	74.8
200	300	0.0032	63.1
200	150	0.0019	62.1
100	500	0.0056	53.8
100	250	0.0034	53
35	500	0.0073	54.9
35	1000	0.0123	54

The spacing and orientation of the GPR profile is another important parameter. It can also affect the resolution of subsurface features. The effect of profiles spacing on the resolution is best demonstrated by several experiments in which line spacing was varied to assess the effect on the resolution. In general, Sensors and Software (1999) suggests transects should be separated by less than the long dimension of the footprint or first Fresnel zone (see Chap. 2). Based on the Nyquist rule, Leckebusch (2003) recommends a standard transect spacing of 25 cm for 400–500 MHz antenna.

An experiment conducted in surveys over buried tombs with profiles spacings of 0.25 m, 0.5 m, and 1.0 m show that the profiles spacings of 1.0 m did not adequately resolve the subsurface features, and a transect spacing of 0.5 m or less is highly recommended (Leucci et al. 2019). Here is demonstrates that in some cases, the Sensor and Software suggestions are a good chose to reduce the survey time acquisition.

Profiles orientation is also shown to affect resolution. A survey in the x and y directions with transect spacing 0.1 m was conducted inside a church to evidence the presence of a buried (Fig. 3.62).

The buried was not know in its spatial position. The survey determined that the buried were best resolved in perpendicular sections, and the best resolution overall combined data from both the x and y directions. Of course, it was also noted that single direction surveys with transect spacing 0.1 m had higher resolution (Figs. 3.63, 3.64 and 3.65).

Is important to note that the profiles acquired in y-direction evidenced the structures perpendicular to y-direction (Fig. 3.63) while profiles acquired in x-direction evidenced structures perpendicular to x-direction (Fig. 3.64). Of course, the xy direction profiles evidenced all structures (Fig. 3.65).

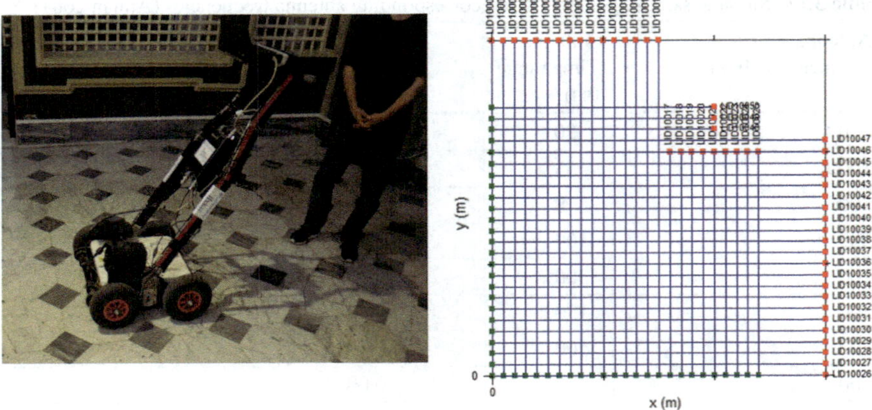

Fig. 3.62 GPR profile location

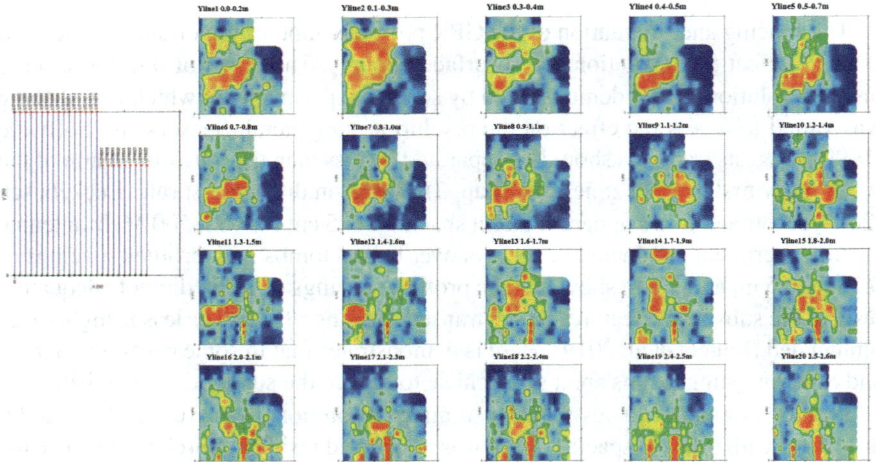

Fig. 3.63 GPR profiles in y-direction: the depth slice visualisation (see Chap. 4)

Figure 3.66 shows a comparison between the depth slices at 1.1–1.2 m depth in the y-direction (3.66a), x-direction (Fig. 3.66b), and xy direction (Fig. 3.66c). It shows that the hypothetical buried tomb was well resolved in xy direction results. The subsequent videoendoscopic analysis confirms the presence of the burial (Fig. 3.67).

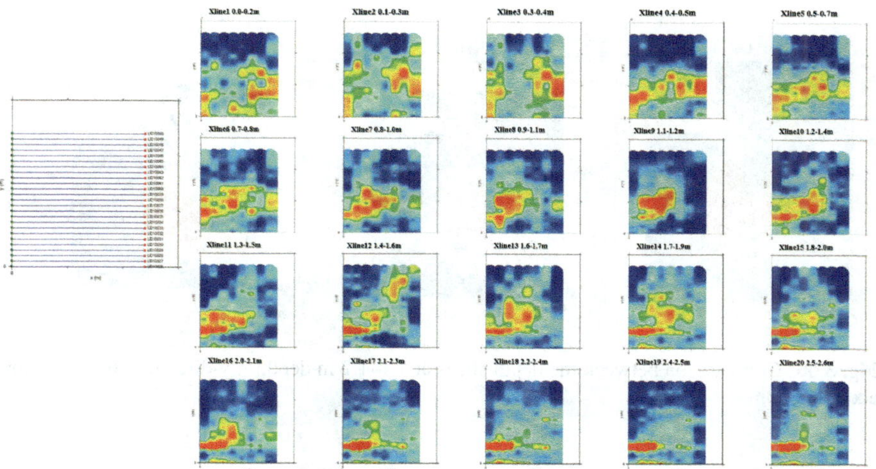

Fig. 3.64 GPR profiles in x-direction: the depth slice visualisation (see Chap. 4)

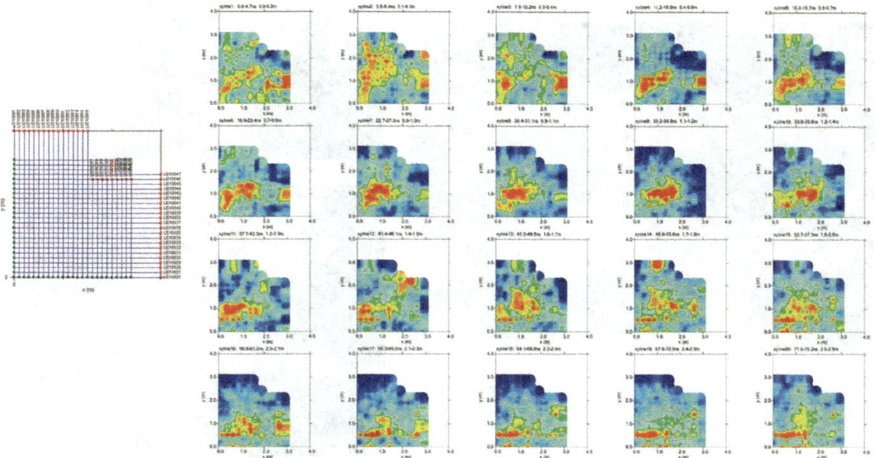

Fig. 3.65 GPR profiles in xy direction: the depth slice visualisation (see Chap. 4)

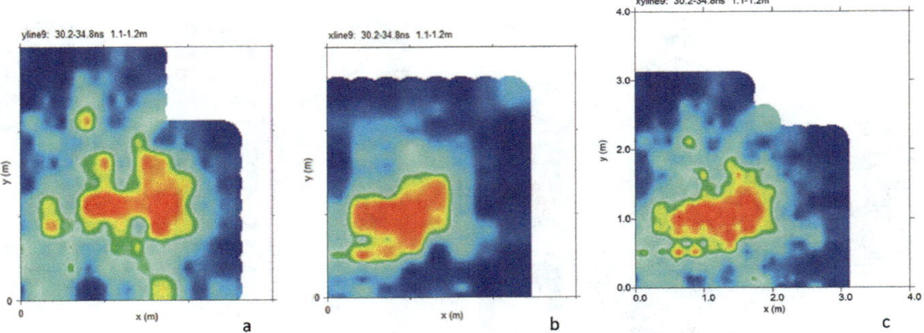

Fig. 3.66 Comparision between the depth slices at 1.1–1.2 m depth: **a** y-direction; **b** x-direction; **c** xy direction

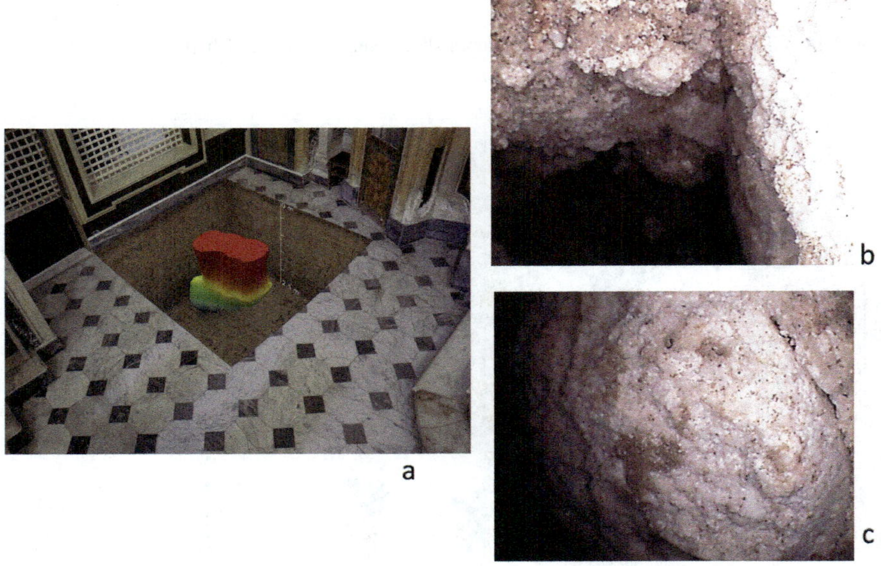

Fig. 3.67 **a** Virtual excavation with the iso-amplitude visualisation (see Chap. 4); **b** videoendoscopic analysis that shows the presence of cavity; **c** videoendoscopic analysis that shows the presence of probable calcification

References

Annan, A. P. (2001). *Ground-penetrating radar workshop notes*. Mississauga: Sensors and Software Inc.

Annan, A. P. (2005). Ground-penetrating radar. In D. K. Butler (Ed.) *Near-surface geophysics. Society of exploration geophysicists: Tulsa, Investigations in Geophysics* (Vol. 13, pp. 357–438).

Castiglione, P., & Shouse, P. J. (2003). The effect of ohmic cable losses on time domain reflectometry measurements of electrical conductivity. *Soil Science Society of America Journal, 67*(2), 414–424.

Cataldo, A., Cannazza, G., & De Benedetto, E. (2011). *Apparatus and method for detection and localization of leaks and faults in underground pipes*. European Patent Application EP 2 538 192 A1, 23 Giugno.

Cataldo, A., Cannazza, G., De Benedetto, E., & Giaquinto, N. (2012). A TDR-based system for the localization of leaks in newly-installed, underground pipes made of any material. *Measurement Science and Technology, 23*(10), 105010.

Cataldo, A., Cannazza, G., De Benedetto, E., Tarricone, L., & Cipressa, M. (2009a). Metrological assessment of TDR performance for moisture evaluation in granular materials. *Measurement, 42*(2), 254–263.

Cataldo, A., De Benedetto, E., Cannazza, G., Giaquinto, N., Savino, M., & Adamo, F. (2014). Leak detection through microwave reflectometry: From laboratory to practical implementation. *Measurement, 47*, 963–970.

Cataldo, A., Monti, G., De Benedetto, E., Cannazza, G., & Tarricone, L. (2009b). A noninvasive resonance-based method for moisture content evaluation through microstrip antennas. *IEEE Transactions on Instrumentation and Measurement, 58*(5), 1420–1426.

Cataldo, A., Monti, G., De Benedetto, E., Cannazza, G., Tarricone, L., & Catarinucci, L. (2009c). Assessment of a TD-based method for characterization of antennas. *IEEE Transactions on Instrumentation and Measurement, 58*(5), 1412–1419.

Cataldo, A., Piuzzi, E., Cannazza, G., & De Benedetto, E. (2010a). Improvement and metrological validation of TDR methods for the estimation of static electrical conductivity. *IEEE Transactions on Instrumentation and Measurement, 59*(5), 1207–1215.

Cataldo, A., Piuzzi, E., Cannazza, G., De Benedetto, E., & Tarricone, L. (2010b). Quality and anti-adulteration control of vegetable oils through microwave dielectric spectroscopy measurement. *Measurement, 43*(8), 1031–1039.

Conyers, L. B. (2004). *Ground-Penetrating Radar for Archaeology*. Walnut Creek, CA: Alta Mira Press.

Corwin, R. F. (1990). The self-potential method for environmental and engineering applications. In S. H. Ward (Ed.), *Geotechnical and environmental geophysics. Vol. I: Review and tutorial*. Society of Exploration Geophysics. P.O. Box 702740/Tulsa, OH 74170-2740.

Corwin, R. F., & Butler, D. K. (1989). *Geotechnical applications of the self-potential method; Report 3: Development of self-potential interpretation techniques for seepage detection*. Tech. Rep. REMR-GT-6, U.S. Army Corps of Engineers, Washington DC.

Corwin, R. F., & Hoover, D. B. (1979). The self-potential method in geothermal exploration. *Geophysics, 44*, 226–245.

Corwin, D. L., & Lesch, S. M. (2005). Apparent soil electrical conductivity measurements in agriculture. *Computers and Electronics in Agriculture, 46*(1–3), 11–43.

Day-Lewis, F. D., Singha, K., & Binley, A. M. (2005). Applying petrophysical models to radar travel time and electrical resistivity tomograms: Resolution dependent limitations. *Journal Geophysical Research, 110*, B08206. https://doi.org/10.1029/2004JB003569.

Dragert, H., Lambert, A., & Liard, J. O. (1981). Repeated precise gravity measurements on Vancouver Island, British Columbia. *Journal Geophysical Research, 86*, 6097–6106.

Friel, R., & Or, D. (1999). Frequency analysis of time-domain reflectometry (TDR) with application to dielectric spectroscopy of soil constituents. *Geophysics, 64*(3), 707–718.

Furman, A., Ferré, A. P., & Heath, G. L. (2007). Spatial focusing of electrical resistivity surveys considering geologic and hydrologic layering. *Geophysics, 72*(2), F65–F73.

Furman, A., Ferré, T., & Warrick, A. W. (2003). A sensitivity analysis of electrical resistivity tomography array types using analytical element modeling. *Vadose Zone Journal, 2*, 416–423.

Goldman, L. W. (2007). *Journal of Nuclear Medicine Technology, 35*(3), 115–128. https://doi.org/10.2967/jnmt.107.042978. http://tech.snmjournals.org/content/35/3/115.full.

Golovko, L., Pozdnyakov, A. I. (2010). Applications of self-potential method in agriculture. Moscow State University, Moscow, Russia Conference Paper January 2010. https://doi.org/10.4133/1.3445435.

Goodman, D., Piro, S., Nishimura, Y. (2009). *GPR archaeometry, in ground penetrating radar: Theory and applications*. In H. M. Jol (Ed.). Amsterdam: Elsevier (pp. 479–508).

Hounsfield G. N. (1973). Computerized transverse axial scanning (tomography): Part I. Description of system. *British Journal of Radiology, 46*, 1016–1022.

Huisman, J. A., Lin, C. P., Weihermüller, L., & Vereecken, H. (2008). Accuracy of bulk electrical conductivity measurements with time-domain reflectometry. *Vadose Zone Journa, 7*(2), 426–433.

Jol, H. M., & Bristow, C. S. (2003). GPR in sediments: advice on data collection, basic processing and interpretation, a good practice guide. In C. S Bristow, & H. M. Jol (Eds.), *Ground-penetrating radar in sediments* (pp. 9–28), Special Publication 211. Geological Society, London.

Kane, W. F., Beck, T. J., & Hughes, J. (2010) Applications of time-domain reflectometry to landslide and slope monitoring. In *Proceedings of 2nd International Symposium Workshop on Time Domain Reflectometry for Innovative Geotechnical Applications* (pp. 305–314).

Kim, M. D., Kim, J. C. M., Kim, D., Kim, J., Choi, H., & Kim, C. Kim. (2010). Detection of inorganic chemicals in a sandy soil using TDR: Effect of probe geometry and water content. *Geosciences Journal, 14*(3), 321–326.

Lambert, A., & Beaumont, C. (1977). Nano variations in gravity due to seasonal groundwater movements: implications for the gravity detection of tectonic movements. *Journal Geophysical Research, 82*, 297–306.

Leckebusch, J. (2003). Ground-penetrating radar: a modern three-dimensional prospection method. *Archaeological Prospection, 10*, 213–241.

Leucci, G. (2008). Ground penetrating radar: the electromagnetic signal attenuation and maximum penetration depth. *Scholarly Research Exchange, 2008*, Article ID 926091. https://doi.org/10.3814/2008/926091.

Leucci, G. (2015). *Geofisica Applicata all'Archeologia e ai Beni Monumentali* (p. 368). Dario Flaccovio Editore, Palermo. ISBN: 9788857905068.

Leucci, G. (2019). *Nondestructive testing for archaeology and cultural heritage: A practical guide and new perspectives*. Cham: Springer International Publishing.

Leucci, G., De Giorgi, L., Ditaranto, I., Giuri, F., Ferrari, I., & Scardozzi, G. (2019). New data on the messapian necropolis of Monte D'Elia in Alezio (Apulia, Italy) from topographycal and geophysical surveys. *Sensors, 19*(16), 3494. https://doi.org/10.3390/s19163494.

Leucci, G., De Giorgi, L., & Scardozzi, G. (2014). Geophysical prospecting and remote sensing for the study of the San Rossore area in Pisa (Tuscany, Italy). *Journal of Archaeological Science, 52*, 256–276. https://doi.org/10.1016/j.jas.2014.08.028.

Loke, M. H. (2001). *Electrical imaging surveys for environmental and engineering studies. A practical guide to 2-D and 3-D surveys. RES2DINV Manual*. IRIS Instruments, www.iris-instruments.com.

Merriam, J. B. (1992). Atmospheric pressure and gravity. *Geophysical Journal International, 109*, 488–500.

Moritz, H. (1984). Geodetic reference system 1980. *Bulletin Géodésique, 58*, 388–398.

Musset, A. E., & Khan, M. A. (2000). *Looking into the earth: An introduction to geological geophysics* (pp.139–198). London: Cambridge University Press.

Nemarich, C. P. (2001). Time-domain reflectometry liquid levels sensors. *IEEE Instrumentation and Measurement Magazine, 4*(4), 40–44.

Neubauer, W., Eder-Hinterleitner, A., Seren, S., & Melichar, P. (2002). Georadar in the Roman civil town Carnuntum, Austria: An approach for archaeological interpretation of GPR data. *Archaeological Prospection, 9*(3), 135–156.

Niebauer, T. M., Hoskins, J. K., & Faller, J. E. (1986). Absolute gravity: A reconnaissance tool for studying vertical crustal motions. *JGR, 91*, 9145–9149.

O'Connor, K. M., & Dowding, C. H. (1999). *Geomeasurement by pulsing TDR cables and probes*. CRC Press.

Robinson, D. A., Jones, S. B., Wraith, J. M., Or, D., & Friedman, S. P. (2003). A review of advances in dielectric and electrical conductivity measurement in soils using time domain reflectometry. *Vadose Zone Journal, 2*, 444–475.

Scheuermann, A., & Huebner, C. (2009). On the feasibility of pressure profile measurements with time-domain reflectometry. *IEEE Transactions on Instrumentation and Measurement, 58*(2), 467–474.

Scheuermann, A., Huebner, C., Wienbroer, H., Rebstock, D., & Huber, G. (2010). Fast time domain reflectometry (TDR) measurement approach for investigating the liquefaction of soils. *Measurement Science and Technology, 21*(2), 025104.

Schmidt, A. (2013). *Earth Resistance for Archaeologists.* In L. B. Conyers, & K. L. Kvamme (Series Eds.). AltaMira Press. 195 pages. ISBN: 978-0-7591-1204-9.

Sensors & Software. (1999). *Ground penetrating radar survey design.* Mississauga: Sensors & Software.

Smith, P., Furse, C., & Gunther, J. (2005). Analysis of spread spectrum time domain reflectometry for wire fault location. *IEEE Sensors Journal, 5*(6), 1469–1478.

Stummer, P., Maurer, H., & Green, A. G. (2004). Experimental design: Electrical resistivity data sets that provide optimum subsurface information. *Geophysics, 69*, 120–139.

Tejero-Andrade, A., Cifuentes, G., Chavez, R. E., Lopez Gonzalez, A., & Delgado-Solorzano, C. (2015). "L" and "Corner" arrays for 3D electrical resistivity tomography: An alternative for urban zones. *Near Surface Geophysics, 13*, 1–13. https://doi.org/10.3997/1873-0604.2015015.

Torge, W. (1989). *Gravimetry.* Berlin: de Gruyter.

Weidelt, P., & Weller, A. (1997). Computation of geoelectrical configuration factors for cylindrical core sample. *Scientific Drilling, 6*, 27–34.

Zegelin, S. J., White, I. & Jenkins, D. R. (1989). Improved field probes for soil-water content and electrical conductivity measurement using time domain reflectometry. *Water Resources Research, 25*. https://doi.org/10.1029/89WR01417.

Chapter 4
Forensic Geophysical Data Processing and Interpretation

Abstract Processing and interpretation of geophysical data determine the success or failure of an investigation in the forensic sciences. Furthermore to help data processing and interpretation is advisable the integration with other data (i.e., data from one or more geophysical techniques, investigators data, geological data, archaeological data, structural data, etc.). In this chapter will be discussed the methodologies and associated mathematical and physical parameters related to the processing of the geophysical data that can help in the resolution of forensic problems such as to individuate the presence of hidden objects.

Keywords TDR · Gravimetric · Magnetic · ERT · SP · Seismic ultrasonic · GPR data processing and interpretation

4.1 TDR Data Processing and Interpretation

This paragraph presents available TDR analysis and interpretation methods. To well understand data analysis, an esteem of materials relative dielectric constant will be performed.

A large number of waveforms which can easily be obtained with the TDR system need automated data processing. The purpose is to determine the dielectric permittivity

$$\varepsilon_r = \left(\frac{cT_s}{2l} \right)^2$$

where c is the velocity of propagation of EM wave, l is the length of the TDR sensor (Topp et al. 1980), T_s is the travel time of the reflected EM signal in the studied materials. In the TDR device, the quantity reflection coefficient is used to sampled the reflected signal. The propagation time T_s can be calculated graphically plotting the reflection coefficient as a function of the time and determining the time when the signal enters and leaves the TDR probe:

$$T_s = T_P - T_0$$

© Springer Nature Switzerland AG 2020

G. Leucci, *Advances in Geophysical Methods Applied to Forensic Investigations*,
https://doi.org/10.1007/978-3-030-46242-0_4

in which $T_P = t_b - t_e$ (related to reflections caused by signal at the beginning and at the and the end of the TDR probe); while T_0 is a correcting time. It refers to the travel time difference between the reflection caused by the beginning of the probe and the point where the signal enters the investigated materials.

As shown in Fig. 4.1 calculating the intersection points of the regression lines fitting the base and the rising section of each reflection is possible to esteem the precise times of the reflections of the beginning and the end of the TDR probe.

Before installation, the value of T_0 is found by performing a single measurement in air. Is important to note that each sensor (or probe) installed in the analysed materials has its own unique reflection pattern.

A typical TDR waveform is shown in Fig. 4.2.

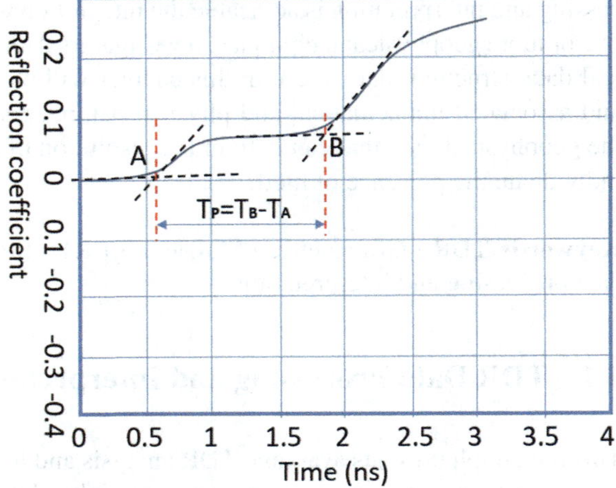

Fig. 4.1 Example showing the reflections of the beginning and end of the sensor

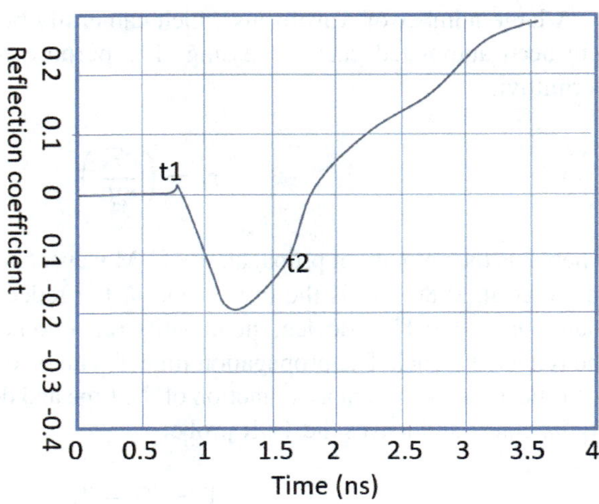

Fig. 4.2 Example of typical TDR waveform

The times that the step signal enters and leaves the sensor is the propagation time, and it is marked the time between the starting time t_1 and the end time t_2. Usually, t_1 is associated with a dramatic decrease of the waveform (spike and a change of slope of the signal), and t_2 is associated with a point where the waveform begins to increase. Since the propagation time can write as

$$t = t_2 - t_1$$

The TDR sensor is a major factor that influences the position of t_1 and therefore if the TDR sensor is provided, t_1 is a constant value for repeated measurements. There are two methods (labelled tangent line methods) commonly used to determine t_1. As shown in Fig. 4.3 t_1 can be determined by drawing lines tangent to the maximum increasing point before the spike and maximum decreasing point after the spike, and use the intersection of the two tangent lines to evaluate t_1 (Baker and Allmaras 1990).

To determine t_2 one can use one of the two methods the tangent line methods (Or et al. 2004; Evett 2000) and adaptive waveform interpretation with Gaussian filters (AWIGF) (Schwartz et al. 2013). Evett (2000) and Or et al. (2004) describes two variations of tangent line methods, flat line method and slope line method, with one numerical correction with linear regression. They claim that for the flat line method, one tangent line is taken at the local minimum point (t_{min}) after t_1, and the other tangent line is taken at the point with maximum first-order derivative after t_{min} (the second inflexion t_{Vmax2}). For the slope line method, one tangent line is taken on the anchor point t_a between t_1 and t_{min}; usually, the distance between t_a and t_1 is set twice as much as the distance between t_a and t_{min} as the default value. The other tangent line is taken on the second inflexion, t_{Vmax2}. Usually, the slope line method needs a correction with linear regression, which is another kind of analysis in TACQ (Evett 2000), where a tangent line is taken at the second inflexion, t_{Vmax2}, and a

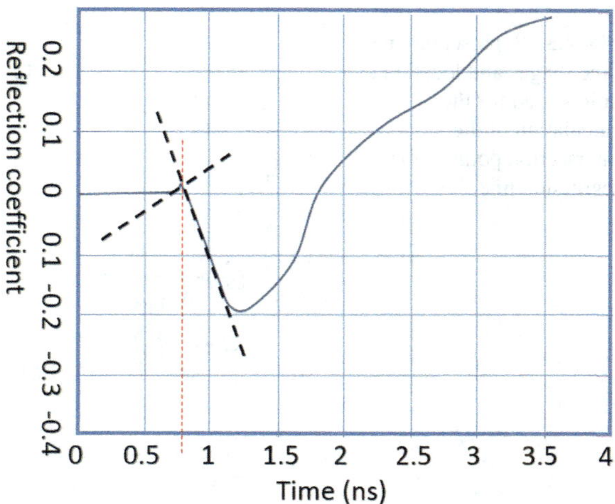

Fig. 4.3 Example of a typical method to determine t_1

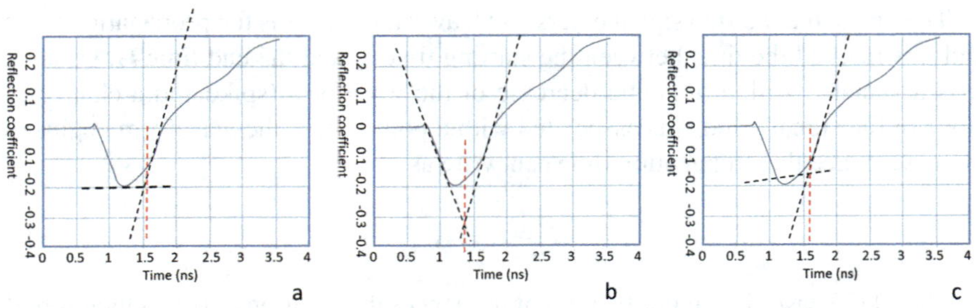

Fig. 4.4 Flat line method (**a**), slope line method (**b**), linear regression correction method (**c**)

regression line is taken with a swath of points between t_1 and $t_{V\mathrm{max2}}$. Figure 4.4 show the methods that graphically determine t_2.

Furthermore, it is important to underline that the analysed materials are heterogeneous; therefore, in the resulting waveform, additional reflections can occur. Is possible to note also additional noise introduced by extra connectors or other small discontinuity. Moreover, the reflections come less pronounced as the conductivity of the analysed materials increases (Dalton and van Genuchten 1986; Yanuka et al. 1988; Topp et al. 1988). To perform a reproducible analysis for each measurement done with the same sensor a number of time ranges are needed, which are shown in Fig. 4.5.

The values of T_S, ΔT_{min}, ΔT_{infl}, ΔT_{base}, have to be set for each probe separately. For the time being these values are obtained empirically from a careful interpretation of a single EM waveform, if necessary one under dry as well as one under wet conditions. When more experience with different kinds of materials has been obtained, an additional program will be made to handle this subject. The point A in Fig. 4.5 is related to waveforms obtained in the field and stored as 251 data points

Fig. 4.5 Representation of time ranges and location of points used for the calculation of the intersection point of the regression line

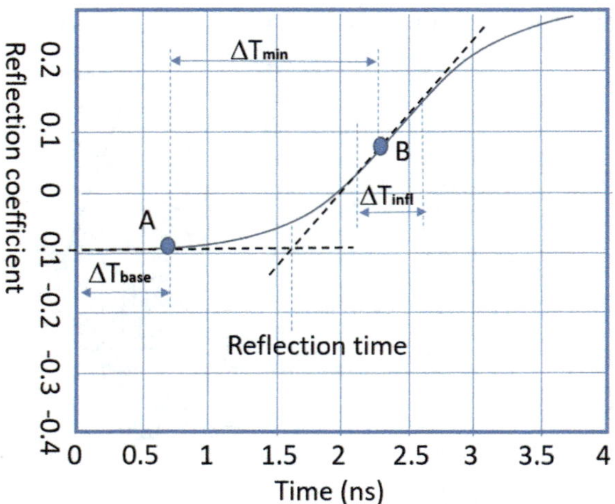

with equidistant time step T between two successive points. For each point A the corresponding time T is first calculated from the several single data and T. The reflections defining the beginning and the end of the sensor both show a definitive rise, caused by the change in transmission line geometry (cable to sensor) and the open end, respectively. Each reflection has a distinctive inflexion point (B) that is traced as the maximum in the first derivative curve. The first derivative D_i of all points is calculated with

$$D_i = (A_{i+1} - A_{i-1})/2T \text{ for } i = 1, \ldots, n \text{ (the number of measures)}$$

Because of the noise in the measurement, the curve $D_i(t)$ needs some smoothing. This is executed by calculating moving averages over a smoothing period (T_S):

$$D_i = \left(\sum_{j=i-m}^{i+m} D_j \right) / (2m\Delta t + 1) \quad m < i < n - m$$

where $2m\Delta t$ equals the smoothing period.

The line fitting the rising section of the reflection is drawn through the inflexion point, which is the maximum of the first derivative. The line is calculated with a linear regression over a range ΔT_{infl}. To find the range with which the regression for the baseline is calculated, the minimum value of A is determined within a range ΔT_{min} preceding the inflexion point B. The line, fitting the base section of the reflection, is calculated with a weighted linear regression ($A = at + b$) over a preset rage ΔT_{base}. More emphasis is s placed on points close to the reflection than to points further away.

$$a = \frac{\left[\sum (u_b - j + 1)t_j A_j - \sum (u_b - j + 1)t_j \sum (u_b - j + 1)\right] A_j/n}{\left\{ [\sum (u_b - j + 1)t_j^2] - [\sum (u_b - j + 1)t_j]^2 \right\}/n}$$

$$b = \left[\sum (u_b - j + 1)A_j - a \sum (u_b - j + 1)t_j \right]/n$$

where Σ stand for $\Sigma_{j=l_b}$ l_b and u_b stand for the lower and upper boundary, and n is the number of points between l_b and u_b. The upper boundary u_b is the index of the minimum value of A. The lower boundary l_b is calculated with $l_b = u_b - (\Delta T_{\text{base}}/T)$.

Calculating the intersection point of both regression lines the reflection time is obtained. By repeating this procedure for both reflections, at the beginning as well as at the end of the sensor, the difference between these times T_p can be entered in the equation to calculate T_s.

The interpretation of TDR data is a very difficult topic. In some cases can be performed using the construction of models that can simulate the expected results. But this topic can well understand in the next chapter where some TDR application will be proposed.

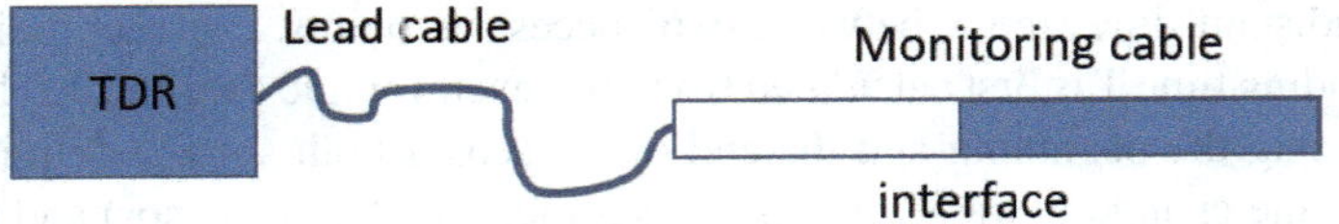

Fig. 4.6 Schematic of the laboratory setup for water level monitoring

At this time, an example can be explained one of the more interpretation that can be performed with the TDR data. The interpretation of TDR data depends on the application and the target.

In general, waveform inversion based on the TDR wave propagation model can simultaneously determine the interface (the position where dielectric property changes) and the dielectric property. As an example, a laboratory test for water level monitoring could be performed (this can be used for example, to monitoring the level of the leachate level in a landfill). The test is schematised in Fig. 4.6. Using the wave propagation model with known transmission line parameters is possible to calculate the precise water level.

A wave propagation model could be represented by the follows equation

$$F(0) = \frac{Z_{in}(0)}{Z_{in}(0) + Z_s} F_s = H F_s$$

where F(0) is the Fourier transform of the TDR waveform; F_s is the Fourier transform of the TDR step input; Z_s is the source impedance of the TDR instrument, typically $Z_s = 50\ \Omega$, $Z_{in}(0)$ is the input impedance at $z = 0$, and H is the system function.

For a given TDR measurement system, are known

the length Li;
the geometric impedance Zp,i;
the cable resistance parameter Ri;
the equivalent dielectric permittivity ε_i* of each uniform section of the nonuniform transmission line, the terminal impedances, Z_S and Z_L to predict the TDR waveform.

The following steps in the simulation of a TDR waveform must be considered: (i) determine the parameters Li, Zp,i, ε_i*, and Ri related to the model parameters of each uniform section; (ii) to avoid aliasing in discrete Fourier Transform an appropriate window size for frequency and time must be determined; (iii) Fast Fourier Transform must be applied to the source voltage in frequency domain; (iv) determine F(0) in frequency domain; (v) finally performs an Inverse Fast Fourier Transform.

Figure 4.7 showed a comparison between measured and simulated waveform related to the estimations of water level was the same as the water level set during the experiment.

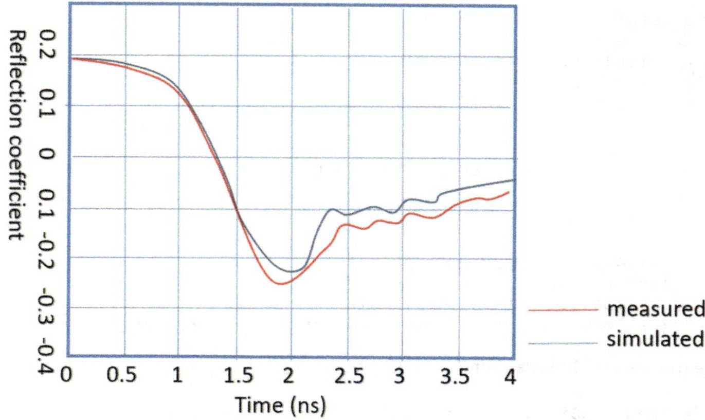

Fig. 4.7 Comparison of the measured and predicted TDR waveforms in water level monitoring

In the Matlab environment, some codes that involved simulation algorithms were implemented. Follow a code to TDR simulation

```
%% Init
clear, clc

%addpath('./lib')
addpath('c:/matlab701/work/tdr')

%% Setting up filenames and loading needed data
line_file = './lines/model.mat';
measurement_file = './measurements/lecroy/1_i2mtc_m_10pF.mat';


load(line_file)

%% Set stimulus signal
% This cell may be used to override the default stimulus provided by the
% datafile containing the line response. Uncomment ONLY the desired mode

%stimulusMode = 'datafile'; % Default stimulus or sythStimulus if none provided
in the line file
stimulusMode = 'synthetic'; % Synthetic stimulus with parameters given in the
synthStimulus struct
%stimulusMode = 'measured'; % Overrides the default stimulus with another
measured signal loaded from the stimulusOverride path

% Check a list of available signal shapes in the help of TLsourceSignal
% Delay and width factors are referred to the whole simulation time
synthStimulus = struct('shape', 'gauss_pulse', ...
    'delayfactor', 0.1, 'widthfactor', 0.01, ...
    'amplitude', 1);

%% Signal loading or synthesis
```

```matlab
switch stimulusMode
    case 'datafile'
        wave = load(sim_params.stimulusfile);
        tsig = wave.t;
        Vs = wave.v;

    case 'synthetic'
        tsig = freq_params.t; tsig = tsig - tsig(1);

        % Extracting relevant line parameters
        % TODO adapt to multiline
        line_velocity = sim_params.v_line;
        line_length = 0;
        for i = 1:length(network_topology)
            line_length = line_length + network_topology{i}.l;
        end
        t_expected = 2*line_length/line_velocity;

        Vs = TLsourceSignal(synthStimulus.shape, tsig, ...
            synthStimulus.delayfactor * t_expected, synthStimulus.widthfactor *
t_expected, ...
            synthStimulus.amplitude);
    case 'measured'
        wave = load(sim_params.stimulusfile);
        tsig = wave.t;
        Vs = wave.v;

    otherwise

end
%% Time domain response computation (FFT)
% implement Fourier analysis consistency validation

% Avoiding odd number of samples
if mod(length(tsig), 2)
    tsig = tsig(1:end-1);
    Vs = Vs(1:end-1);
end

f = freq_params.omega/(2*pi);
f_max = freq_params.fmax;

if length(tsig) ~= 2*(freq_params.Nh+1)
    % Resampling needed
    df_required = 1/range(tsig);
    Nh_required = round(f_max / df_required);
    f_required = df_required*(1:Nh_required)';

    Nt_required = 2*(Nh_required + 1);
    t_ = linspace(0, tsig(end), Nt_required); dt = mean(diff(t_)); % Time array
and dt
    Vs = interp1(tsig, Vs, t_); Vs = Vs';

    VrefH = interp1(f, VrefH, f_required, 'spline');
else
    Nt = 2*(freq_params.Nh+1);
end
```

```matlab
Vs = reshape(Vs, [length(Vs), 1]);

Vs_fft = fft(Vs)/Nt;

H = [VrefH; 0; conj(VrefH(end:-1:2))];
xRef_fft = H .* Vs_fft;
xRef = ifft(xRef_fft)*Nt;

%% Integrated time-domain response FIXME find a more orthodox way to remove the
baseline xRef = xRef - mean(xRef(1:50)); xRefInt = TLfftIntegral(tsig, xRef);
% % FIXME raw normalization steps, to be removed % xRefInt = xRefInt -
xRefInt(1);
% xRefInt = xRefInt / max(xRefInt); % Plot time domain response
meas_file = load(measurement_file);
vmeas = meas_file.v/2;
hTDR = TLplotTDR(tsig, xRef/max(xRef), 'fontsize', 12, 'linewidth', 2, 'xlim',
[0, 1.8e-6]);
%hTDR = TLplotTDR(tsig, xRef, 'measurements', vmeas, 'fontsize', 12,
'linewidth', 1, 'xlim', [-2, 60]*1e-9);
%hTDR = TLplotTDR(tsig, xRef, 'measurements', vmeas, 'fontsize', 12,
'linewidth', 2);

%% Ending program
```

The following Matlab code prepares the Transmission Line model for the
simulator.

```matlab
%% Init
clear, close all
clc
% File path and name for line model
savefile = './line_models/model.mat';
%% Define the model and save it % Define any number of chained lines and put
them in the line_model cell array
rg58 = struct( ...

'linemodel', 'coax', 'ri', 0.000255, 'ro', 0.00076, 't', 0.00027, 'l', 100, ...
% Line model and geometry
'epsilon_r', 2.26, 'esrtoggle', true, 'tan_d', 0.00031, '
debyetoggle', false, ...
% Dielectric properties
'sigma', 5.96*e^7, 'skintoggle', true, ...
% Conductor properties
'Lprofile', struct('shape', 'disabled'), ...
'Rprofile', struct('shape', 'gauss_rect', 'position', 0.5, '
width', 0.2, 'amplitude', 3, 'rise', 0.01), ...
'Gprofile', struct('shape', 'disabled'), ...
'Cprofile', struct('shape', 'disabled', 'position', 0.25, '
width', 0.02, 'amplitude', 100e-12) ...
);
biwire = struct( ...
'linemodel', 'biwire', 'd', 1e-3, 'D', 2e-3, 'l', 10, ... %
Line model and geometry
'epsilon_r', 2.1, 'esrtoggle', true, 'tan_d', 0.002, '
debyetoggle', false, ... % Dielectric properties
'sigma', 5.96*e^7, 'skintoggle', true, 'sigma_d', 1*e^-15, ... %
Conductor properties
'Lprofile', struct('shape', 'disabled'), ...

'Rprofile', struct('shape', 'disabled', 'position', 0.5, '
width', 0.2, 'amplitude', 1, 'rise', 0.01), ...
'Gprofile', struct('shape', 'disabled'), ...
'Cprofile', struct('shape', 'disabled', 'position', 0.5, '
amplitude', 1000e-12, 'width', 0.1) ...
);
line_model = {rg58, biwire};
save(savefile, 'line_model')
```

4.1.1 Gravimetric Data Processing and Interpretation

The processing that must be performed on the raw gravimetric data allows calculating the values of the Bouguer Anomaly. In Chap. 2, several factors (internal and external of the gravimeter) that affect gravimetry measurements were studied. These factors cause changes in the observed gravimeter readings with time. Earth tides, long term instrumental drift, changes in battery supply voltage and atmospheric pressure changes are the principal factors that must be considered in the gravity data processing. Following the steps of the gravity field data processing:

Tides Corrections—C_T:
As discussed in Chap. 3, the corrections for the tidal gravity effects of the Sun and the Moon are essential for all types of gravimetric surveys (Fig. 4.8), since they are a first step to make measurements time-independent. These corrections can contribute up to 0.3 mGal difference to the measurements. In micro-gravity surveys, where high precision is expected, these corrections are particularly significant since tidal rates of up to 30 μGals per hour can readily occur.

These corrections can be applied in real-time (directly on software-controlled instrumentation), or off-line, later.

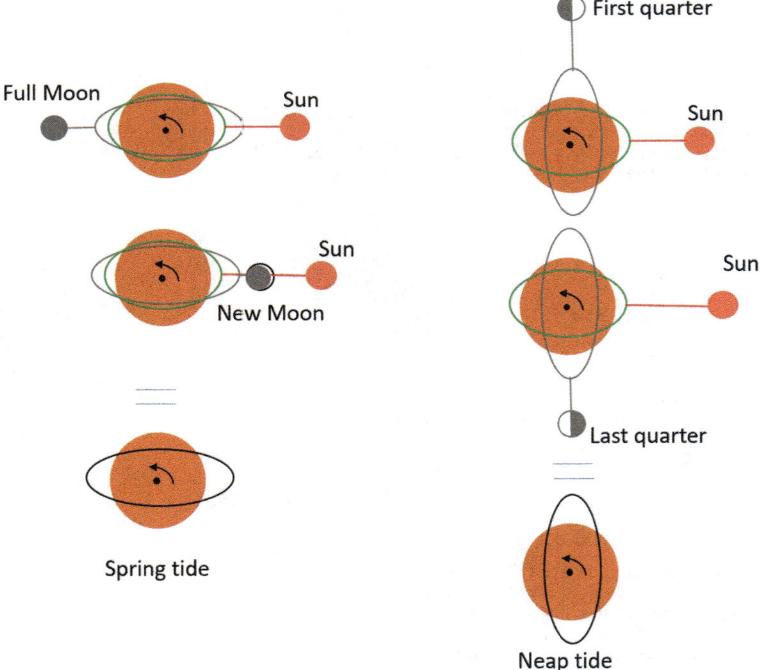

Fig. 4.8 Variation of the tide amplitude

The European Association of Exploration Geophysicists (EAEG) has published, for many years, tables of predicted tidal gravity corrections which could be used to estimate the tidal correction at any point at any time in the prospective year.

Depend on the software used for raw data correction the form of latitude, longitude and time values.

For example, using the formula of Longman (1959), the convention of the sign is as follows.

Latitude North is positive, and South is negative, Longitude West is positive, Est is negative. Latitude and Longitude values must be stated in degrees and decimals, not minutes and seconds. In the Longman formula, the time parameter is expressed in UTC (Universal Time Coordinated).

Instrumental Drift—C_D:
This correction allows making the gravity measurements time-independent. There are various factors which may contribute to the instrumental drift (see Chap. 3). To perform this correction (long term drift, battery supply voltage changes, and vibration, etc.) one must repeat measurements on the base station, for example, at the beginning and end of each measurements day, or more frequently (e.g. even in a cycling mode, if convenient). For a measurements line, once one defines the station position one can return to the first station hourly. Due to the instrumental drift, on the same point (base station), the gravity values change. To perform the correction, one must build the instrument drift curve represented by the measured gravity changes (on the base station) as a function of the time. The slope of the best linear fit to the resultant curve of gravity values with time will determine the residual long term drift rate of the gravimeter. This value is then used to adjust the drift correction factor.

Residual drifts during the survey day could be observed. These can be due to the battery supply voltage changes and vibration-induced short term drifts, as well as recent power-downs of the instrument. Also, in this case, residual drifts are determined using a repeat of measurements made at a base station.

If the following gravity values are observed at a specific base station, corrected for tidal effects and linear long-term drift.

P_0: base station gravity value, previously established and corrected
P_1: observed value at the start of the day, at time T_1 at the base station
P_2: observed value at the end of the day, at time T_2 at the base station
P_3: observed value at a new station, at time T_3 (between T_1 and T_2).

The residual correction C_D to be applied to the new station will be given by

$$C_D = [(P_0 - P_1) - (P_2 - P_1)] * [(T_3 - T_1)/(T_2 - T_1)]$$

The drift correction can be made graphically. On the measurements line one measure the gravity at the base station and can back in the base station hourly. In this way, one builds the instrumental drift curve (Fig. 4.9).

The drift correction which one can do at generic hour h is the corresponding Δg on the curve (Fig. 4.9).

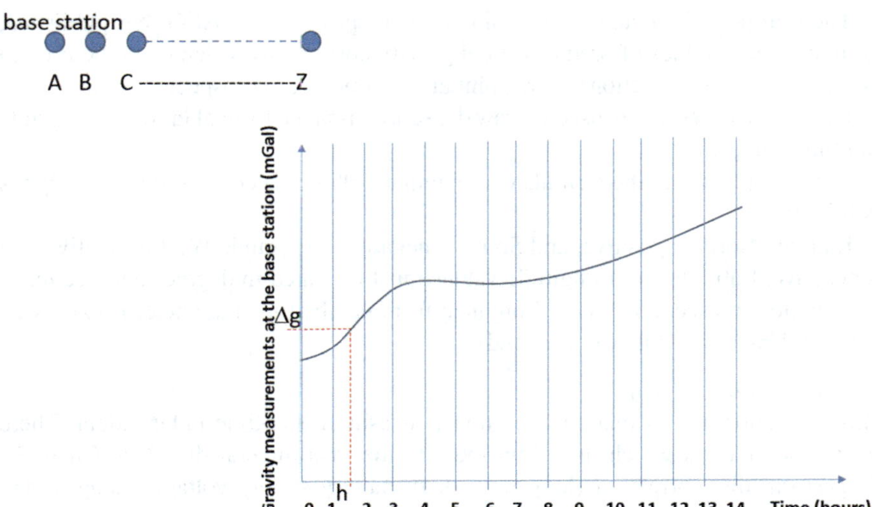

Fig. 4.9 Instrument drift curve

Atmospheric Pressure Changes—C_P:
The atmospheric consist of a fluid (the air) and the instrument is immersed in this fluid and responds to the Archimedes principle. Since the effect of atmospheric pressure changes on the observed gravity values. On days with a normal weather pattern, the pressure variation is in the range of 0.3–1 kPa (i.e. 1–3 μGals) per day. In a non normal conditions (e.g. a thunderstorm) the pressure can give rise changes totalling 5 kPa (i.e. 18 μGals) in amplitude, with temporal gradients of the order of 0.5 kPa/h (1.8 μGal/h) and spatial gradients of the order of 0.2 kPa (0.7 μGal) in 10 km distances.

Considering the use of a barometer in a gravity survey, the corrections for these effects can be made. The barometer must have a resolution of 0.1 kPa. In this way, a pressure read must be done at each gravity station.

The correction to the observed gravity values for changes in barometric pressure can be calculated using the following formula:

$$C_P = +3.6(P_1 - P_0) \text{ in } \mu\text{Gal}$$

where P_1 is the atmospheric pressure at the field station (or at the field station time) and P_0 is the atmospheric pressure at the base station (or base station time) at the start of the day, in kPa.

For example, on normal weather days, the barometric pressure will be constant, within 0.5 kPa over such an area; thus C_P so derived will be correct to within 2 μGals.

The atmospheric pressure change also with the elevation and this had consequent on gravity measurements. Also, these variations can be considered and therefore measured by the use of a barometer.

Changes in Groundwater and Surface Water Levels—C_{GW}:
Changes in groundwater levels depend on the Earth areas where one perform gravity measurements and are likely to be seasonal and will reflect the wet and dry seasons.

If Δw (m) of the groundwater level variation the correction in mGals will be given by

$$C_{GW} = -0.04192 \cdot \Delta w \cdot b$$

where b is the porosity of the soil.

Variation with Latitude—C_L:
As known gravity increased with latitude θ. The rate of increase of gravity with distance (north or south of the equator) is given by

$$\Delta g_L = 0.813 \cdot sen2\theta - 1.78 \times 10^{-3} sen4\theta \text{ in mGal/km}$$

Table 4.1 illustrate the gravity variation with latitude.

Table 4.1 Variation of the terrestrial gradient with latitude

θ (°)	Δg_L (mGal/km)
0	0
5	0.137
10	0.278
15	0.397
20	0.522
25	0.609
30	0.704
35	0.759
40	0.803
45	0.813
50	0.803
55	0.773
60	0.706
65	0.634
70	0.524
75	0.417
80	0.280
85	0.145
90	0

These corrections are negative with increasing latitude; north or south of the equator.

Variation with Elevation—C_E:
The correction for the combined Free-Air and Bouguer effects may be written as

$$C_E = (0.3086 - 0.0419 \cdot d) \cdot h \text{ in mGal}$$

where d is the mean density of the near-surface rocks, in g/cm^3; and h is the elevation of the gravimeter station in metres relative to a datum level. The value of d may be determined by measuring the density of rock samples, but another method, originally suggested by Nettleton (1976) is more useful in providing a macroscopic estimate of the value of d.

Figure 4.10 show the as to determine the value of the density d. It shows a hypothetical profile over the hill. Several profiles of the same data are shown, corrected for C_E based on different values of density. The corrected profile most nearly approximating a straight line across the topographic feature determines the optimum value for the density of the rocks in the vicinity, i.e. d_4 in this case.

Terrain Effects—C_{TE}:
Topographic features near the stations can influence gravimeter measurements. All such effects act to reduce the observed gravity results and therefore are negative. There are several ways to esteem the topographic effect and all are based on breaking the measurement stations surrounding areas into cells or prisms. To every cell or prism must be assigned an average height (relative to the level of the station) and density

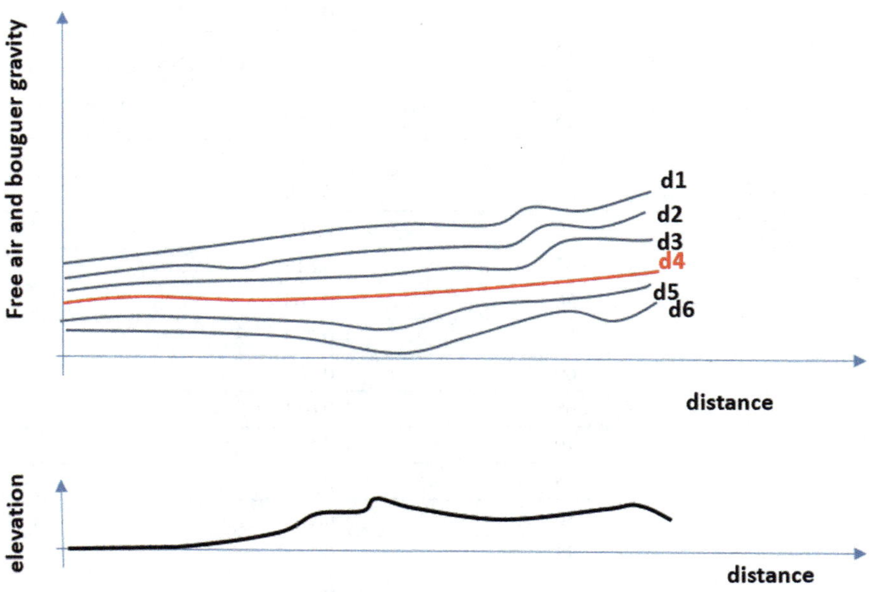

Fig. 4.10 Nettleton's Method for estimating the density of surface rocks

to each. Successively the gravity contribution of each cell or prism is calculated. Finally, all the contributions to determine the topographic effect (and correction) for that station must be summed. The traditional method of making topographic corrections is based on that of Hammer (1965) (Net of Hammer—Fig. 4.11). Hammer improved on the method of Hayford to simplify terrain corrections. His "Hammer net" was used for 70 year to make terrain corrections. The net is used with printed topographic maps. The net is developed with the centre posed at the measurement station. Successively radial lines from the gravity station and concentric circles are drawn at specific distances from the gravity station. An average elevation above or below the station elevation within each compartment is estimated.

The gravity values in the compartments are written as:

$$g_{comp} = Gd\,\Delta\theta \left[R_0 - R_i + \sqrt{R_i^2 + h^2} - \sqrt{R_0^2 + h^2} \right]$$

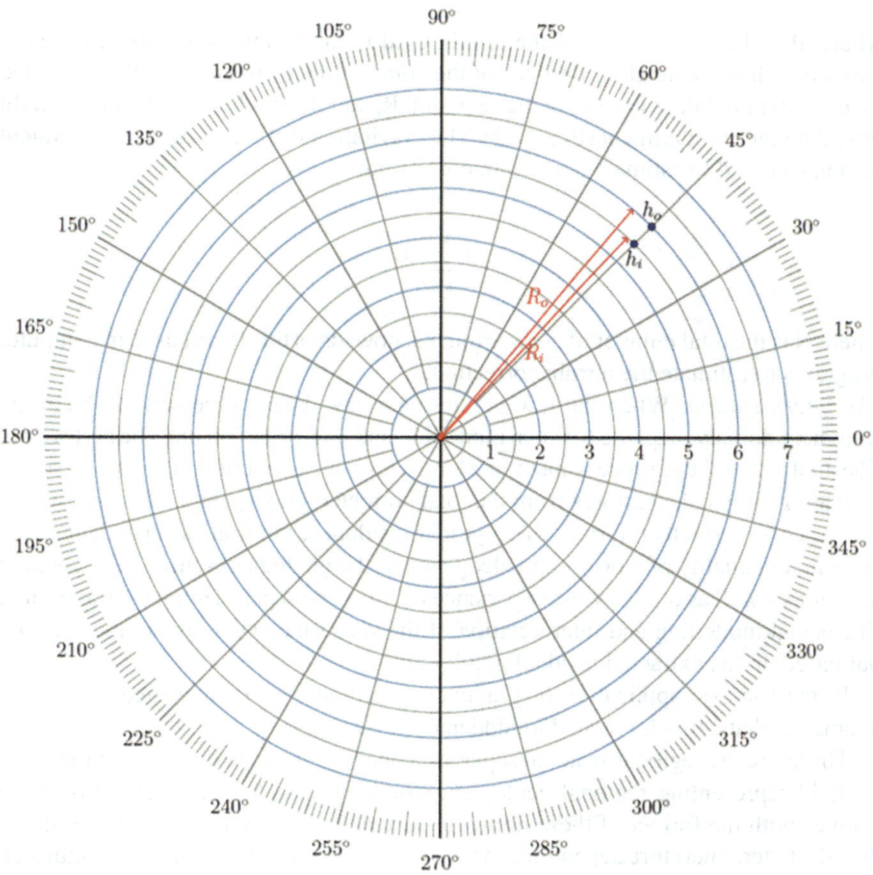

Fig. 4.11 The hammer's net

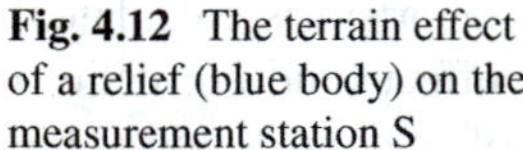

Fig. 4.12 The terrain effect
of a relief (blue body) on the
measurement station S

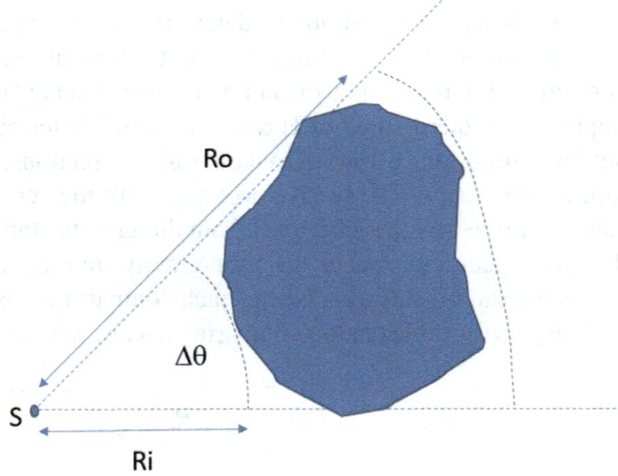

where: d is the bulk density of the terrain used in the simple or spherical Bouguer
correction, h is the height difference of the compartment, $\Delta\theta$ is the angle subtended
by the two radial lines bounding the segment, R_i and R_0 are the outer and inner radii
bounding the compartment (Fig. 4.12). The gravitational effect of each compartment
is then summed to estimate the terrain correction:

$$g_{ter} = \sum_{k=1}^{N} g_{comp}$$

where N is the total number of compartments, nowadays this is still used in computer
programs to estimate the terrain correction.

The interpretation: When the raw gravity data have been processed to obtain the
Bouguer anomaly map, the work on modelling and interpretation must be performed.
The first step of the process aims at identifying and extracting signals caused by
relevant geological structures from the total corrected gravity field. The outcome of
this process, normally referred to as regional-residual separation, is dependent on the
amount and quality of a priori knowledge about the geological setting of the area in
question as well as on skills and experiences of the person performing the separation.
The actual modelling and interpretation of the subsurface geology are then carried
out based on the extracted residual gravity field.

In the forensic application of gravimetry method, one is interested in shallow
anomalies that could be related to hidden objects.

Therefore the regional-residual separation needs to separate elements of the gravity field representing regional geological structures and elements caused by local
sources with the former of these normally assumed to be located at a greater depth
than the latter. Therefore depending on their depth, the anomalies can be distinguished
in:

- Regional anomalies are of great extension, attributable to deeper sources such as to influence all or most of the gravimetric survey. These anomalies are of interest for deep tectonic studies;
- local anomalies: they are of limited extension, attributable to less deep sources and such as to influence only some stations.

It is necessary to separate the local anomalies from the regional ones because depending on the study you want to do, they interest either one or the other. Since the current study is aimed at highlighting the presence of local rather than regional anomalies. In general, on an area that is not too large, the regional anomaly can be represented, as a first approximation, by a straight line or plane according to whether the anomalies are known along with a profile or a surface. The methods commonly used to isolate the regional field include graphical methods and analytical methods. In this case, the least-squares analytical method was used. The approximates regional anomaly can be represented by

$$R(x) = ax + b$$

If the total anomaly is indicated by $g(x)$, the local anomaly is defined by

$$A(x) = g(x) - R(x)$$

After the regional-residual separation, the subsurface distribution of density can be esteem. This value did information about the distribution of the subsurface hidden objects.

There are two approaches to interpret gravity data, the forward and inverse approaches.

In the forward approach, gravity anomalies in the region of measurement are calculated considered simple geometric models of geologic formations and structures. In this case, individual bulk densities are assigned.

In the inverse interpretation, one attempts to directly invert the gravity data to determine the distribution of subsurface densities. Must be underline that for the inverse approach, the observed gravitational field variations can be caused by many possible distributions of density.

In the forward approach, quantitative modelling of gravity anomalies are considered. It involves fitting of model responses to measured residual anomaly data by adjusting geometry and density of subsurface gravity source models. When satisfactory models have been found, they have to be interpreted into geological models. The process is critically dependent on the amount and quality of a priori information.

In any approach, a great help consists in the knowledge of the local geology. The so-called apriori information allows to better chose the may most likely approximate the subsurface body geometry. Once found the body or set of bodies that incorporate these models into the forward solution is possible to adjust their parameters (dimensions, shapes and densities), to obtain the closest desired fit to the observed gravity anomaly data. Successively the interpretation becomes an iterative process type trial

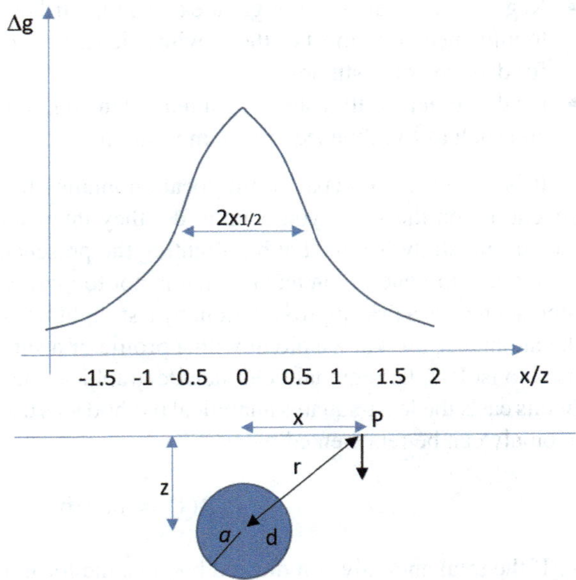

Fig. 4.13 Gravity effect of a sphere

and error. This process is linked to the possibility to calculate the gravity anomaly Δg due to a body of known geometrical shape.

Combined, a simple basic body is possible to build a complex body model.

A sphere is the most basic body and usually is used as a part of other models or could approximate symmetrical bodies. The gravity effect of a sphere at point P (Fig. 4.13) is:

$$\Delta g = \frac{GM}{r^2} = \frac{\frac{4\pi}{3}da^3 z}{\sqrt[3]{\left(x^2 + z^2\right)}}$$

where d is a density of the sphere, a is a radius of the sphere and z is the depth of the centre of the sphere.

The half-width of the anomaly at half of its value determines the depth of the sphere centre.

The gravity effect of the vertical cylinder is:

$$\Delta g = 2\pi Gd \left\{ L + \sqrt{z^2 + R^2} - \left[\sqrt{(z+L)^2 + R^2} \right] \right\}$$

where L is the vertical size (length) of the cylinder z is the depth of its top and R is its diameter. If

$$R \rightarrow \infty$$

One has an infinite horizontal slab.

Assuming a horizontal rod perpendicular to the x-axis at a depth z, the gravity effect is (Fig. 4.14):

$$\Delta g = \frac{Gm}{z\left(1 + \frac{x^2}{z^2}\right)}\left\{\frac{1}{\sqrt{1 + \frac{x^2+z^2}{(y+L)^2}}} - \frac{1}{\sqrt{1 + \frac{x^2+z^2}{(y-L)^2}}}\right\}$$

where m is the mass of the rod. If the rod is expanded into the cylinder with a dimension a then $m = \pi a^2 d$. When the length L of the rod is infinite (usually a good approximation when the L > 10z) then the above formula is simplified into the:

$$\Delta g = \frac{2Gm}{z\left(1 + \frac{x^2}{z^2}\right)}$$

The depth z of the rod could be estimated from the half-width of the anomaly z = $x_{1/2}$.

For a general 2D body (Fig. 4.15) a two-dimensional approximation, where the cross-section of the body S is assumed to be constant, and its strike extent is unlimited, can be used.

In this case, the vertical gravitational effect at point P of the horizontal prism is (Grant and West 1965):

$$\Delta g = 2G\Delta d \cos\theta d\theta dr$$

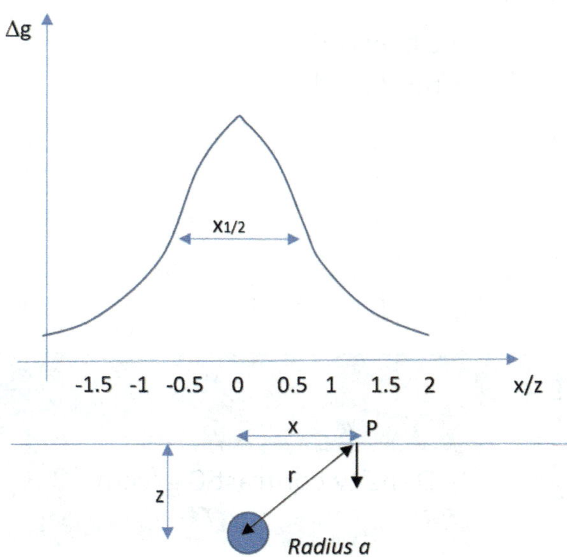

Fig. 4.14 Gravity effect of a cylinder

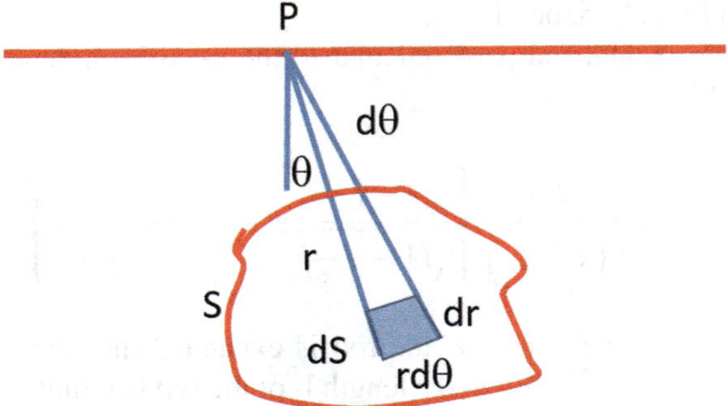

Fig. 4.15 Gravity computation basis for two-dimensional bodies

where θ is the angle of the prism relative to the vertical at P, and Δd is the density contrast of the prism. The gravity effect of the entire body is:

$$\Delta g = 2G\Delta d \sum \Delta \sin\theta \, \Delta r$$

Figure 4.16 show an example of the calculation of gravity contribution of 2D body and compared to experimental data.

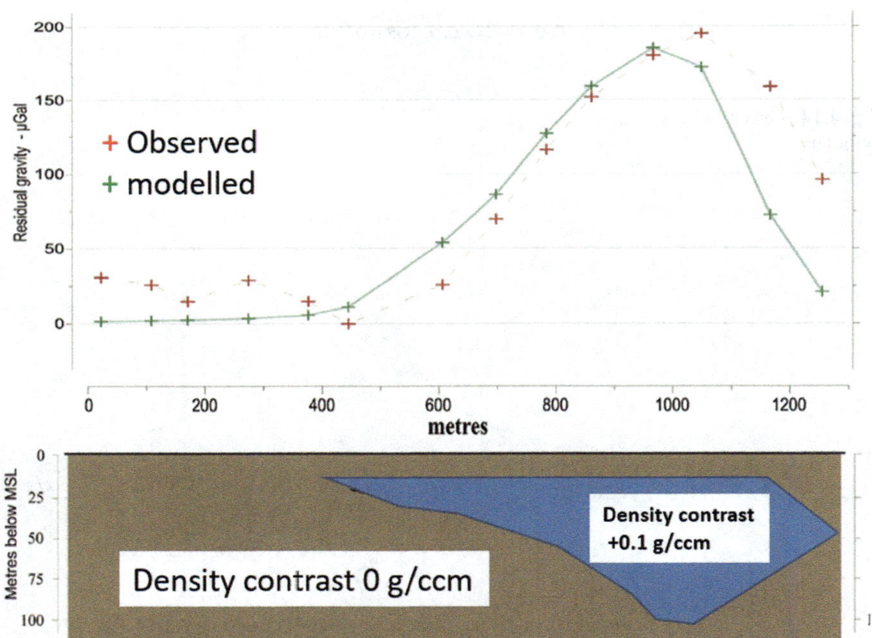

Fig. 4.16 Gravity computation of 2D model body

Following a matlab code for terrain correction

```matlab
function
[output]=corr(lat,lon,H_msl,dd_ref,month_ref,yr_ref,sec,min,hour,utc_diff,day,mo
nth,year)


%== GRS80 reference ellipsoid
%a=6.378137E8;   %  semi-major axis in [cm]!
%e_crt_2=0.006739496775544254; %Second eccentricity squared

% Hayford ellipsoid (after original Longman (1959):
a=6.378270E8;
e_crt_2=0.006738;
%======

%== Gravity & mass
G=6.673e-8;                              % Newton's gravitational constant, in
[cm^3g^-1s^-2]
M_m=7.3537E25;                           % Mass of the Moon, in [grams]
M_s=1.993E33;                            % Mass of the Sun, in [grams]
%======

%== orbit & distance parameters
m_s2m=0.074804;                          % Ratio of Sun's mean motion to that of
the Moon,

e_m=0.05490;                             % 0.054899720;    % Eccentricity of the
Moon's orbit, from
c_m=3.84402E10;                          % Mean distance between centers of the
Earth and the Moon, in [cm]
c_s=1.495E13;                            % Mean distance between centers of the
Earth and the Sun, in [cm],
omega=dms2rad([23 26 21.48]);            % Inclination of the Earth's equator to
the ecliptic, might be also: 23.452°.
i=deg2rad(5.145);                        % Inclination of the Moon's orbit to the
ecliptic
%======


%== Love numbers  and gravimetric factor:
Love_h2=0.612;                           %
http://physics.ucsd.edu/~tmurphy/apollo/doc/tides.pdf
Love_k2=0.303;                           %
http://physics.ucsd.edu/~tmurphy/apollo/doc/tides.pdf
beta=1+Love_h2-3/2*Love_k2;              % gravimetric factor
%======
%== auxiliary parameters
rev=360;                                 % 1 full revolution (rev.) equals 360°
rev_sec=rev*3600;                        % 1 full revolution in seconds
```

```
%% Input arguments check
if (size(lat,1)~=size(lon,1))
    error('Latitude and logitude vectors are not of equal size.');
elseif (size(lat,1)~=size(H_msl,1))
    error('Latitude and height vectors are not of equal size.');
elseif (size(lon,1)~=size(H_msl,1))
    error('Longitude and height vectors are not of equal size.');
elseif ((lat<-90.0) | (lat>90.0))
    error('Latitude outside of the range.');
elseif ((lon<-180.0) & (lon>180.0))
    error('Longitude outside of the range.');
elseif (H_msl<0.0) | (H_msl>=9000.0)
    error('MSL height outside of the range.');
elseif ((dd_ref<1) | (dd_ref>31))
    error('Reference day outside of the range.');
elseif ((month_ref<1) | (month_ref>12))
    error('Reference month outside of the range.');
elseif ((yr_ref<0) | (yr_ref>3000))
    error('Reference year outside of the range.');
elseif ((day<1) | (day>31))
    error('Day outside of the range.');
elseif ((month<1) | (month>12))
    error('Month outside of the range.');
elseif ((year<0) | (year>3000))
    error('Year outside of the range.');
elseif ((sec<0) | (sec>59))
    error('Seconds outside of the range.');
elseif ((min<0) | (min>59))
    error('Minutes outside of the range.');
elseif ((hour<0) | (hour>23))
    error('Hours outside of the range.');
end


%== computation_station (lat,lon,h)
lat_rad=deg2rad(lat);
lon_rad=deg2rad(lon);
H=H_msl*100; % H is now in [cm]

%=======


%== reference epoch for computation
UTC_hour=hour-utc_diff;
J_zero=Julian_day(yr_ref,month_ref,dd_ref)+12/24;
J_dn=Julian_day(year,month,day);
J_d=J_dn+(UTC_hour/24)+min/1440+sec/86400;
T=(J_d-J_zero)/36525; % number of Julian centuries from the reference epoch
%== Julian hour
t0 = (UTC_hour + min./ 60. + sec./ 3600.);
if (t0 < 0); t0 = t0+24; end
if (t0 >= 24);  t0 = t0-24; end
%=======


%== Mean longitude of Moon in its orbit reckoned from the referred equinox
%sm_rad=dms2rad([270 26
14.72])+deg2rad((1336*rev_sec+1108411.20)/3600)*T+deg2rad(9.09/3600)*T^2+deg2rad
(0.0068/3600)*T^3; % Longman (1959);
sm_rad=dms2rad([270 26
11.72])+deg2rad((1336*rev_sec+1108406.05)/3600)*T+deg2rad(7.128/3600)*T^2+deg2ra
d(0.0072/3600)*T^3; % Longman (1959);
%=======
```

```matlab
%== Mean longitude of lunar perigee
%pm_rad=dms2rad([334 19 40.87])+deg2rad((11*rev_sec+392515.94)/3600)*T-
deg2rad(37.24/3600)*T^2-deg2rad(0.045/3600)*T^3; % Longman (1959)
pm_rad=dms2rad([334 19 46.42])+deg2rad((11*rev_sec+392522.51)/3600)*T-
deg2rad(37.15/3600)*T^2-deg2rad(0.036/3600)*T^3; % Longman (1959):
%=======


%== Mean longitude of the Sun
%hs_rad=dms2rad([279 41
48.04])+deg2rad(129602768.13/3600)*T+deg2rad(1.089/3600)*T^2; % Longman (1959)
hs_rad=dms2rad([279 41
48.05])+deg2rad(129602768.11/3600)*T+deg2rad(1.080/3600)*T^2;  % Longman (1959)
[eq. 12']
%=======


%== Longitude of the Moon's ascending node in its orbit reckoned from the
reffered equinox
%N_rad=dms2rad([259 10 57.12])-
deg2rad((5*rev_sec+482912.63)/3600)*T+deg2rad(7.58/3600)*T^2+deg2rad(0.008/3600)
*T^3; % Longman (1959);
N_rad=dms2rad([259 10 59.81])-
deg2rad((5*rev_sec+482911.24)/3600)*T+deg2rad(7.48/3600)*T^2+deg2rad(0.007/3600)
*T^3; % Longman (1959);
%=======


%== Mean longitude of solar perigee
%ps_rad=dms2rad([281 13
15.00])+deg2rad(6189.03/3600)*T+deg2rad(1.63/3600)*T^2+deg2rad(0.012/3600)*T^3;
% Longman (1959) ;
ps_rad=dms2rad([281 13
14.99])+deg2rad(6188.47/3600)*T+deg2rad(1.62/3600)*T^2+deg2rad(0.011/3600)*T^3;
% Longman (1959);
%=======


%== Eccentricity of the Earth's orbit
es = 0.01675104 - 0.00004180*T - 0.000000126*T.^2;  % Longman (1959);
%=======


%== distance parameter
C_2=1/(1+e_crt_2*sin(lat_rad).^2);  % Longman (1959);
C=sqrt(C_2);
%== distance from observation point to the center of the Earth
r=C*a+H; % Longman (1959);
%=======


%== distance from Moon to the Earth's geocenter
a_moon=1/(c_m*(1-e_m^2)); % Longman (1959);
recip_d=1/c_m+a_moon*e_m*cos(sm_rad-pm_rad)+a_moon*e_m^2*cos(2*(sm_rad-
pm_rad))+(15/8)*a_moon*m_s2m*e_m*cos(sm_rad-
2*hs_rad+pm_rad)+a_moon*m_s2m^2*cos(2*(sm_rad-hs_rad)); % Longman (1959);
d=1/recip_d;
%=======


%== distance from Sun to the Earth's geocenter
a_sun=1/(c_s*(1-es^2)); % Longman (1959);
recip_D = ((1. / c_s) +a_sun * es * cos(hs_rad - ps_rad)); % Longman (1959);
D=1/recip_D;
%=======
```

```matlab
%== Inclination of the Moon's orbit to the equator
Im_rad=acos(cos(omega)*cos(i)-sin(omega)*sin(i)*cos(N_rad));% Longman (1959);
if (((Im_rad*180/pi)<18) || ((Im_rad*180/pi)>28)); warning('Inclination of the
Moon''s orbit to the equator is outside of the normal range.');end;
%======

%== Longitude in the celestial equator of its intersection A with the Moon's
orbit
nu=asin((sin(i)*sin(N_rad))/sin(Im_rad));% Longman (1959);
if (((nu*180/pi)<-15) || ((nu*180/pi)>15)); error('Longitude in the celestial
equator... is not correct.');end;
%======

%== Hour angle of mean Sun
t_rad = deg2rad(15. * (t0 - 12) - (-1*lon)); % Longman (1959);
%======

%== right ascension of meridian of place of observations reckoned from A
chi_m_rad=t_rad+hs_rad-nu; % Longman (1959);
%== right ascension of meridian of place of observations reckoned from the
vernal equinox
chi_s_rad=t_rad+hs_rad; % Longman (1959);
%======
%==
cos_alpha=cos(N_rad)*cos(nu)+sin(N_rad)*sin(nu)*cos(omega); % Longman (1959)
sin_alpha=(sin(omega)*sin(N_rad))/sin(Im_rad); % Longman (1959);
alpha=2*atan(sin_alpha/(1+cos_alpha)); % Longman (1959);
%======

%== Longitude in the Moon's orbit of its ascending intersection with the
celestial equator 'hi'
xi=N_rad-alpha; % Longman (1959),
%== Mean longitude of Moon in radians in its orbit reckoned from A
sigma=sm_rad-xi; % Longman (1959),
%======

%== Longitude of Moon in its orbit reckoned from its ascending intersection with
the equator

L_m_rad=sigma+2*e_m*sin(sm_rad-pm_rad)+5/4*e_m^2*sin(2*(sm_rad-
pm_rad))+15/4*m_s2m*e_m*sin(sm_rad-2*hs_rad+pm_rad)+11/8*m_s2m^2*sin(2*(sm_rad-
hs_rad)); % Longman (1959) [eq. 9]
%======

%== Longitude of Sun in the ecliptic reckoned from the vernal equinox
L_sun_rad=hs_rad+2*es*sin(hs_rad-ps_rad); % Longman (1959)
%======

%== Zenith angle of the Moon
cos_Zm=sin(lat_rad)*sin(Im_rad)*sin(L_m_rad)+cos(lat_rad)*((cos(Im_rad/2))^2*cos
(L_m_rad-chi_m_rad)+(sin(Im_rad/2))^2*cos(L_m_rad+chi_m_rad)); % Longman (1959)

%== Zenith angle of the Sun
cos_Zs=sin(lat_rad)*sin(omega)*sin(L_sun_rad)+cos(lat_rad)*((cos(omega/2))^2*cos
(L_sun_rad-chi_s_rad)+(sin(omega/2))^2*cos(L_sun_rad+chi_s_rad)); % Longman
(1959)
%======
```

```matlab
%==vertical component of tidal acceleration due to the Moon
gm=G*M_m*r*recip_d^3*(3*cos_Zm^2-1)+ 1.5*G*M_m*r^2*recip_d^4*(5*cos_Zm^3-
3*cos_Zm); % Longman (1959)
%== vertical component of tidal acceleration due to the Sun:
gs=G*M_s*r*recip_D^3*(3*(cos_Zs^2)-1);   % Longman (1959)

%== convert to gravity acceleration values using gravimetric factor
gm_gal=gm*beta;
gs_gal=gs*beta;
gm_mgal=gm_gal*1000;
gs_mgal=gs_gal*1000;
%======

%==
% Tidal acceleration due to the Moon and the Sun:
g0_gal=gm_gal+gs_gal; % Longman (1959) ;
g0_mgal=gm_mgal+gs_mgal;
%======
%%
output.input_variables=[lat,lon,H_msl,dd_ref,month_ref,yr_ref,sec,min,hour,utc_d
iff,day,month,year];
output.K5_a=a;
output.K6_m=m_s2m;
output.K7_e=e_m;

output.R1_T=T;
output.R2_t0=t0;
output.R20_s=sm_rad;
output.R21_p=pm_rad;
output.R22_h=hs_rad;
output.R23_N=N_rad;
output.R24_I=Im_rad;
output.R25_nu=nu;
output.R26_t=t_rad;
output.R27_chi_sun=chi_s_rad;
output.R28_chi_moon=chi_m_rad;
output.R29_alpha=alpha;
output.R30_xi=xi;
output.R30_sigma=sigma;
%output.el=el;

output.R31_p1=ps_rad;
output.R32_el=es;
output.R33_l_sun=L_sun_rad;
output.R34_l_moon=L_m_rad;
output.R70_C=C;
output.R71_r=r;
output.R72_am=a_moon;
output.R73_as=a_sun;

output.R75_cosZm=cos_Zm;
output.R76_cosZs=cos_Zs;

output.R80_d=d;
output.R81_D=D;

output.R85_gf=beta;
output.R90_gm=gm;
output.R91_gs=gs;
output.R98_gm_mGal=gm_mgal;
output.R99_gs_mGal=gs_mgal;
output.R100_g0_mGal=g0_mgal;
```

```matlab
function [angles_rad]=dms2rad(ddmmss)
    DD=ddmmss(:,1); % degrees
    MM=ddmmss(:,2); % minutes
    SS=ddmmss(:,3); % seconds
    angles_DEG=abs(DD)+abs(MM)./60+abs(SS)./3600; % decimal

    angles_rad=angles_DEG.*pi./180;
    indices=(DD<0 | MM<0 | SS<0); % if any element is negative, we suppose
the angle is negative
    angles_rad(indices)=-1.*angles_rad(indices);

end


function [angles_deg]=dms2deg(ddmmss)

    DD=ddmmss(:,1); % degrees
    MM=ddmmss(:,2); % minutes
    SS=ddmmss(:,3); % seconds
    angles_deg=abs(DD)+abs(MM)./60+abs(SS)./3600;
    indices=(DD<0 | MM<0 | SS<0);
    angles_deg(indices)=-angles_deg(indices);

end


function [angles_rad]=deg2rad(angles_deg)

    angles_rad=angles_deg.*pi./180;

end


function [ J_dn ] = Julian_day( yr,mnd,dan)

    % If the month is January or February, then 1 must be subtracted from
the year to get a new value of y, and 12 must be added to the month to get a new
value of m.

    % Thus, January and February of a given year are considered as being,
respectively, the 13th and the 14th month of the previous year.
    idx = (mnd <= 2);
    yr(idx)  = yr(idx)-1;
    mnd(idx) = mnd(idx)-12;
    %
    A=floor(yr/100.0);
    B=floor(A/4.0);
    C1=2-A+B;
    E=floor(365.25*(yr+4716.0));
    F=floor(30.6001*(mnd+1));

    J_dn=dan+C1+E+F-1524.5;

end
```

4.2   Magnetic Data Processing and Interpretation

As a gravimetric method also in magnetic method (as a total magnetic field survey is performed) there are some corrections related to the variation with the time of the magnetic field. The overall processing of magnetic data involves several steps. Data reduction can be separated into corrections and data enhancements.

There are three ways to remove the variations with the time: (i) the use of one magnetometer instrument as a base station. It allows recording the changes of the magnetic field as a function of the time. These data can be successively used to remove the change from the readings in the field magnetometer. As understandable, this is the most accurate way to use in total magnetic field measurements, but must also emphasise that it is more expensive, as two complete instruments are required; (ii) the use of a tie-point method which provides only one magnetometer instrumentation that returns to the base station at an established time interval. This assumes that the field is changing in a slowly way. This method is not as accurate as using a base station, but it is most cost-effective, as it only requires one magnetometer; (iii) is possible to perform a vertical gradient survey. Since one is measuring the rate of change between two sensors, any changes in the background field will apply to both sensors and one will not see any of these noise effects. This technique is quite effective for near-surface anomalies.

The base station computer clock is continuously synchronised with GPS time. The recorded data can be merged with the magnetometer data, and the diurnal correction is applied according to the equation

$$\mathbf{B}_{Tc} = \mathbf{B}_T + \left(\overline{B}_B - \mathbf{B}_B\right),$$

where
$\mathbf{B}_{Tc}$ = Corrected airborne total field readings

$\mathbf{B}_T$ = Airborne total field readings

$\overline{B}_B$ = Average datum base level

$\mathbf{B}_B$ = Base station readings

In a gradiometric acquisition, the above correction is automatically performed. In this way of data acquisition, other important data processing must be performed to improve magnetic data imaging: more magnetic map present spikes, stripes and zigzag effects (Fig. 4.17).

Spikes appear as localised black and/or white pixels if the data are visualised in greyscale and, to well interpreted data results, they have to be eliminated or attenuated. Spikes are related to the presence in the subsoil or soil surface of ferrous objects such as rods, wires, studs, etc. But could be related to the result of instrumental errors because of accidental failure of one of the two sensors.

Note that must pay close attention to the type of problems posed by investigators because if you are looking for buried metal drums, then the spikes are no longer a noise to be eliminated but a signal to be taken into consideration.

More commercial software allows eliminating the spike. This software uses the median filter or simple threshold methods with the threshold values chosen by the user.

Other method processes consider spikes as outliers that can be defined as observations (Barnett and Lewis 1984; Hawkins 1980).

In order to illustrate key concepts let z_1, z_2, ... z_i, ..., z_n be the Z-scores of the observations x_1, x_2, ..., x_i, ..., x_n.

The general expression for z_i is

Fig. 4.17 Raw data collected in the bi-directional mode

$$z_i = (x_i - \overline{x})/\mathrm{s}$$

where $\overline{x}$ is the mean value and s the standard deviation concerning the mean of the measurements acquired. This procedure works iteratively. After each iteration, the observation having the maximum Z-score is removed from the data set. The iterative procedure is labelled "the generalised extreme studentised deviate (GESD) procedure and more about it can be found on the site (https://www.itl.nist.gov/div898/handbook/eda/section3/eda35h3.htm).

Stripes are the effect that mainly occurs when data are collected in the bi-directional mode. The stripes appear as a series of alternating light and dark grey tones in neighbouring lines (Scollar et al. 1990; Ciminale and Loddo 2001). The setting at zero-mean value data collected along each line allows for correcting stripes (Ciminale and Loddo 2001).

The Zig Zag is related to the fact that data, collected along parallel profiles, do not always have the same length and this effect is amplified if data are acquired in the bi-directional mode (Becker 1995; Eder-Hinterleinter et al. 1996). One mode to improve data quality is to acquire them in a one-way mode acquisition, in which sensors are not rotated. Ciminale and Loddo (2001) suggest a useful method to remove the Zig-zag effect. It is a statistical procedure using a cross-correlation function. As write Ciminale and Loddo (2001), all parallel profiles are examined three at a time. Odd-numbered lines, for instance, lines 1 and 3 or lines 3 and 5, are kept fixed whereas the intermediate profile, surveyed in the opposite direction, is shifted by a lag yielding the best cross-correlation with the two neighbouring lines. It is interesting to note that to execute the cross-correlation, data along each profile are interpolated linearly, but the final shifting operation is applied to raw data.

The magnetogram shown in Fig. 4.18 illustrates the results obtained by processing the raw data showed in Fig. 4.17.

Interpretation of magnetic data: The magnetisation of most objects is complex, and the details of the magnetic anomalies are also complex. The inherent ambiguity makes the comprehensive interpretation of magnetic anomalies a complex *art*.

Major magnetic rock units commonly produce magnetic anomalies with characteristics that can be identified and used to infer the presence or absence of the rock. Volcanic rocks may produce high amplitude magnetic variations of very local extent. Negative magnetic anomalies produced by permanent magnetisation in a direction approximately opposite to the Earth's magnetic field may be associated with volcanic rocks. Large igneous intrusives produce anomalies with a wide range of amplitudes, but generally of greater extent than the anomalies associated with volcanic rocks. Mamor rocks may produce complex patterns and pronounced lineaments. Most sedimentary rocks are nonmagnetic, but magnetite, sands and gravels are a notable exception.

An experienced interpreter generally can identify rock type by inspection of the magnetic anomalies; however, such an interpretation is necessarily subjective. Figure 4.19 show typical magnetic susceptibility for some common mineral and rock types (modified Clark and Emerson 1991).

The magnetic anomalies are variable in shape and amplitude, are asymmetrical, and sometimes appear complex also if generated from simple sources. Since the term ambiguity is related to an infinite number of possible sources can produce a given anomaly. Magnetic results are usually displayed in the form of a contour map. The position and size of the anomaly depend on the position and size of the magnetic body. This allows interpreting the position of the body which has caused the anomalous reading. Often, however, the reading is complicated because of the position of the body concerning other rocks, its size, and what happens to the body at depth.

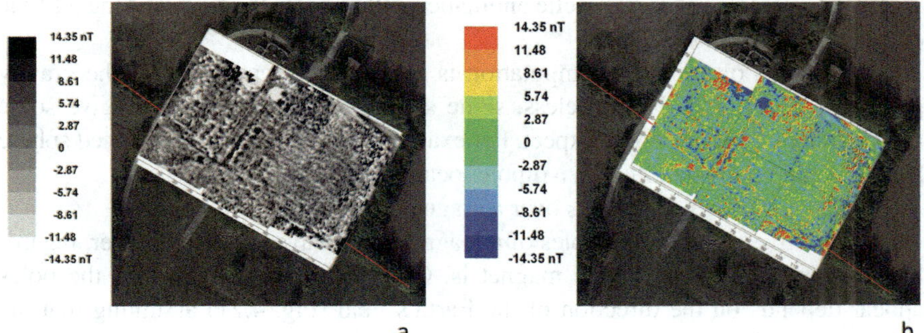

Fig. 4.18 Processed data collected in the bi-directional mode: **a** on a greyscale; **b** in a colour scale

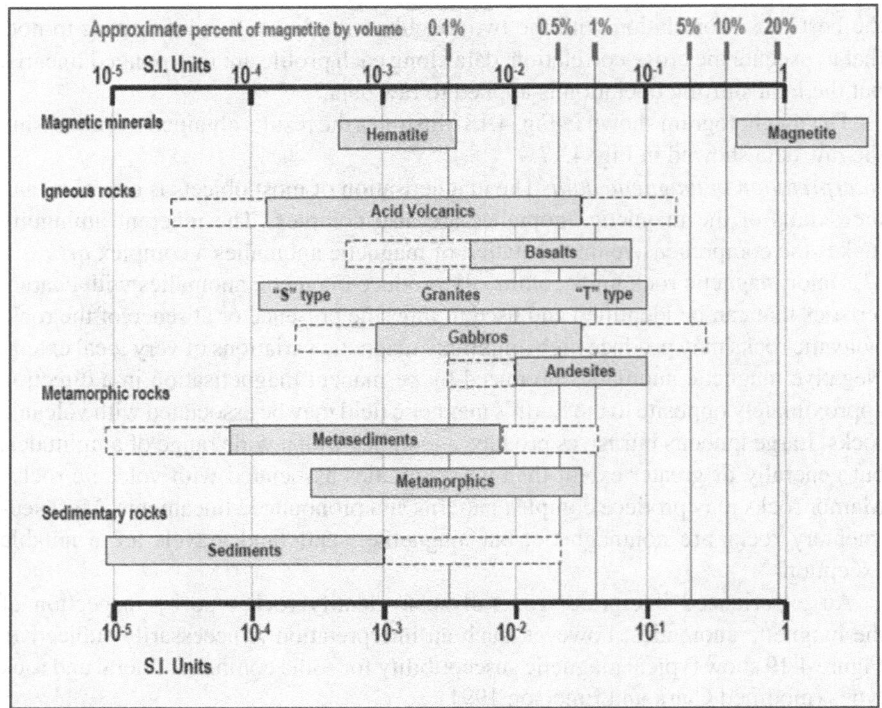

Fig. 4.19 Typical magnetic susceptibilities for some common mineral and rock types

As the gravity method, forward and inverse methods of magnetic interpretation are used to estimate the parameters of the source body, causing the anomaly.

For a forward model one can model the total magnetic field B and the horizontal magnetic field H. A magnetic body can be defined to be 2D when its strike length in the direction perpendicular to the profile is at least ten times its width (Telford et al. 1990).

The character of many magnetic anomalies will indicate the form and the attitude of the disturbing mass.

The process of anomaly computation is similar, as already seen in the gravity paragraph (Sect. 4.3). Nevertheless, some simple curves will show to give some impression of what one could expect. For example, the uniformly magnetised sphere has the same magnetic effect as a dipole located in its centre.

Figure 4.20 shows the curves over a magnetic dipole.

The positive and negative poles of a magnet are always present together, regardless of how large or small the magnet is. On which side of the body, the poles appear depends on the direction of the Earth's field (Fig. 4.21) assuming that the magnetisation is in the direction of the Earth's field.

The procedure is further complicated if a remanent magnetisation is present. Particularly, the thermoremanent magnetisation could be considerably high.

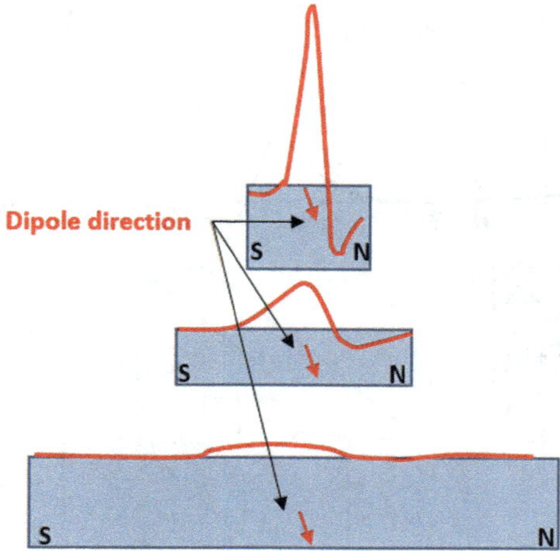

Fig. 4.20 Anomalies of dipoles at different depths. Note the changes in the amplitude and width of the anomaly

One can be modelled a probably magnetic anomaly due to several hidden objects using the simple geometrical bodies (Fig. 4.22).

In a detailed way, one can consider these field equation:

(1) *Dipping Dike (Prism)* (Fig. 4.23)

Figure 4.23 show the dipping dyke with all the considered geometrical parameters to write the analytical expression of the contribution as a magnetic source.

With referement to Fig. 4.23 the total magnetic field B and the horizontal component H can be written as:

$$B = 2kB_e \sin \xi \left\{ \left(\sin 2I \sin \xi \sin \beta - \cos \xi \left(\cos^2 I \sin^2 \beta - \sin^2 I \right) \right) \right.$$
$$\ln \left(\frac{r_2 r_3}{r_1 r_4} \right) + \left(\sin 2I \cos \xi \sin \beta + \sin \xi \left(\cos^2 I \sin^2 \beta - \sin^2 I \right) \right)$$
$$\left. (\phi_1 - \phi_2 - \phi_3 + \phi_4) \right\}$$

$$H = 2kB_e \sin \xi \left\{ (\cos I \sin \xi \sin \beta + \sin I \cos \xi) \ln \left(\frac{r_2 r_3}{r_1 r_4} \right) \right\}$$
$$+ (\cos I \cos \xi \sin \beta - \sin I \sin \xi)(\phi_1 - \phi_2 - \phi_3 + \phi_4) \right\}$$

where

'*B*' is the total magnetic field intensity;
'*H*' is the horizontal magnetic field intensity;
B_e is the EMF intensity;
k is the susceptibility contrast between the magnetic target and the host medium;
I is the earth's magnetic field inclination;

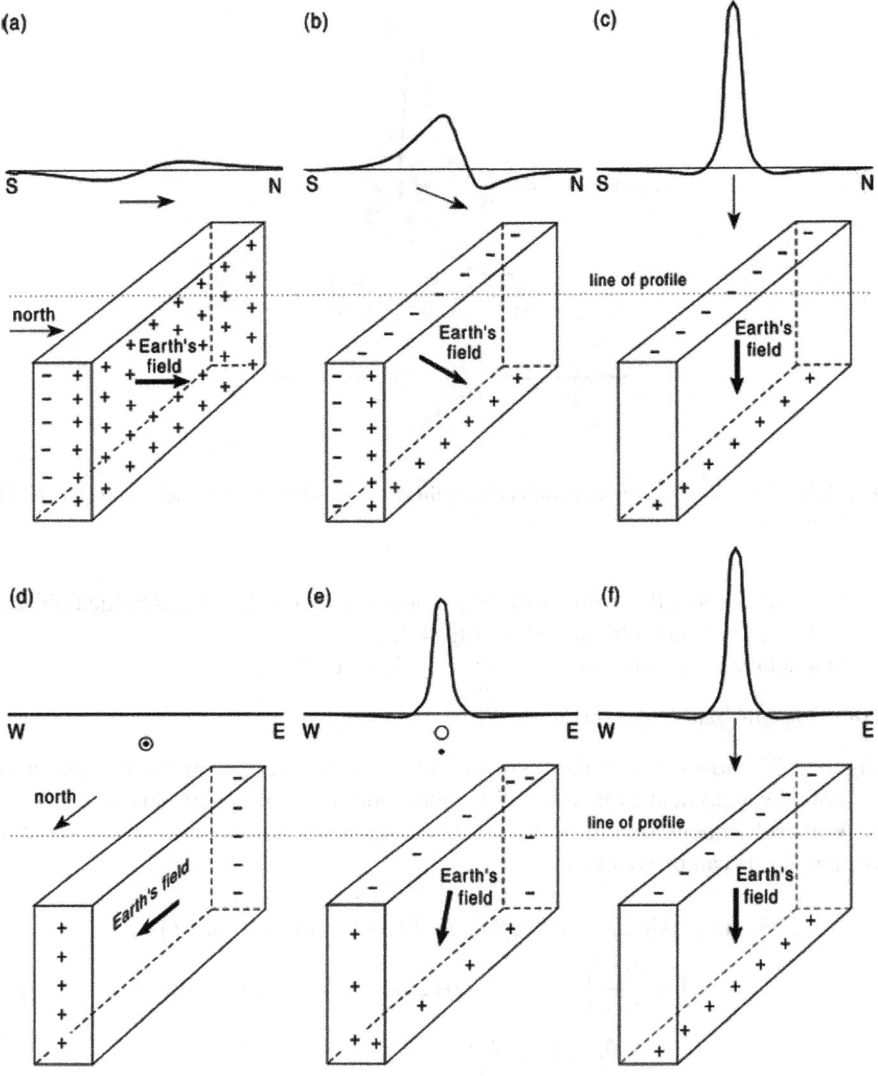

Fig. 4.21 Poles formed on a thin vertical sheet anomaly of a dipole at different latitudes (modified Musset and Khan 2000)

β is the strike angle of the dyke relative to the magnetic north;
r_i, d, D, h and φ_i are the geometric values of the distances and angles; and
x is the coordinate of the observation point.

Insert these equations in Matlab is possible to obtain the magnetic anomaly caused by dyke shaped bodies.

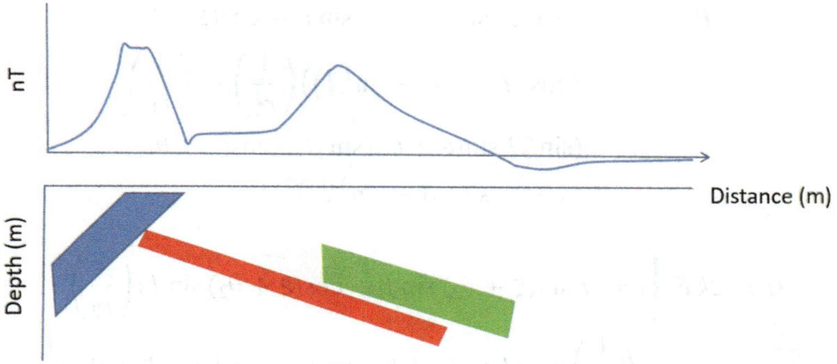

Fig. 4.22 An example of magnetic profile builds from the polygonal bodies with several susceptibilities

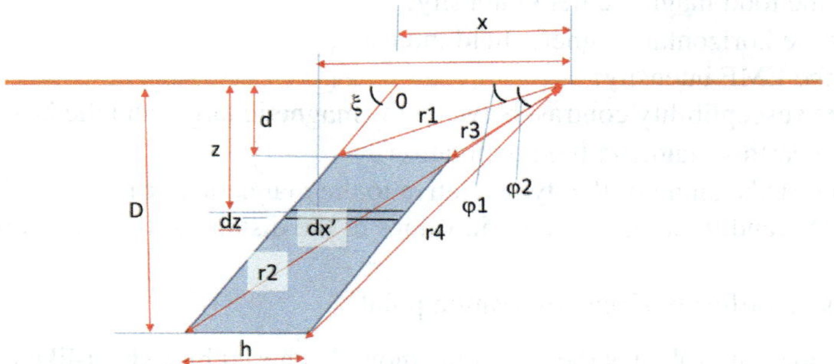

Fig. 4.23 An example of the dyke with dip ξ, strike β and infinite strike length

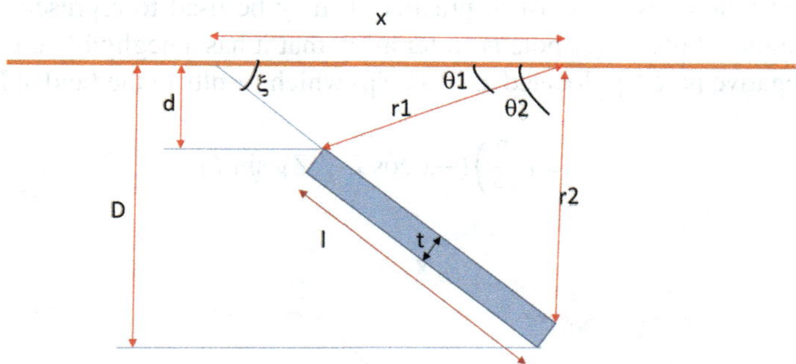

Fig. 4.24 The dipping sheet

(2) *Dipping Sheet* (Fig. 4.24):

The expressions for B and H profiles of the sheet are derived from dipping dyke by replacing the width h of dike model in terms of the thickness of the sheet. Profiles for both finite and infinite depth extent are determined.

$$B = 2kt\,B_e\{(\sin 2I\,\sin(\xi + \theta_2)\,\sin\beta - \cos(\xi + \theta_2)$$
$$\left(\cos^2 I\,\sin^2\beta - \sin^2 I\right))\left(\frac{1}{r_2}\right) - \left(\frac{1}{r_1}\right)$$
$$(\sin 2I\,\sin(\xi + \theta_1)\,\sin\beta - \cos(\xi + \theta_1)$$
$$\left(\cos^2 I\,\sin^2\beta - \sin^2 I\right))\}$$

$$H = 2k\,B_e\left\{(\cos I\,\sin(\xi + \theta_2)\,\sin\beta + \cos(\xi + \theta_2)\,\sin I)\left(\frac{1}{r_2}\right)\right.$$
$$\left. - \left(\frac{1}{r_1}\right)(\cos I\,\sin(\xi + \theta_1)\,\sin\beta + \sin I\,\cos(\xi + \theta_1))\right\}$$

where

'B' is the total magnetic field intensity;
'H' is the horizontal magnetic field intensity;
Be is the EMF intensity;
k is the susceptibility contrast between the magnetic target and the host medium;
I is the earth's magnetic field inclination;
β is the strike angle of the dyke relative to the magnetic north;
ri, d, D, t and θi are the geometric values of the distances, thickness and angles;
and
x is the coordinate of the observation point.

This equation evaluates the magnetic anomaly caused by a sheet-like body.

(3) *The isolated pole (Monopole)* (Fig. 4.25):

An isolated pole does not exist in practice. It may be used to represent a steeply dipping dipole whose lower pole is so far away that it has a negligible effect. If one takes, a negative pole '-p' located at (0, 0, Zp) which results in the field at P(x, y, 0):

$$B = \left(\frac{p}{r^3}\right)(-x\cos I + Z_P\sin I)$$

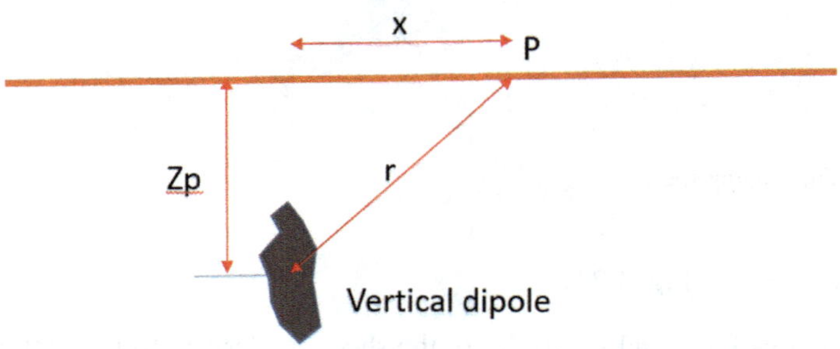

Fig. 4.25 The vertical dipole

$$H = \left(\frac{pZ_P}{r^3} \right)$$

where

'B' is the total magnetic field intensity;
'H' is the horizontal magnetic field intensity;
I is the earth's magnetic field inclination.

Free software for gravity and magnetic data can be found on http://bgi.omp.obs-mip.fr/links/free_gravity_software.

4.3 ERT Data Processing and Interpretation

After the acquisition of the ERT data, they must be inverted to make electrical-resistivity images. A variety of commercial codes are available for this purpose. Most codes follow the same inversion approaches, which are based on quasi-Newton algorithms. Based on the selection of a starting model and inversion parameters, these codes should produce similar results; default values differ greatly, however, and it is not always clear how parameters are used within the inversion. The selection of the inversion parameters is somewhat subjective and guided by the operator's intuition and/or the a priori information about the investigated medium or the nature of the targets. To obtain reproducible results, it is critical to (i) report all parameter selections and default values; (ii) document the algorithm used by the software; and (iii) archive a copy of the software code or executable. It is also important to underline that there is not enough information to find a unique solution for a data inversion, and calculating the true distribution of resistivity in the medium is a highly nonlinear problem because the current paths through the medium are dependent on the resistivity structure (Day-Lewis et al. 2005; Leucci 2015, 2019).

In geophysical tomographic, these problems are solved with an excess number of model parameters and use regularisation to create a mathematically stable solution (Constable et al. 1987) because this problem is solved using iterative inversion (Tripp et al. 1984). The solution to the inverse problem results in a 2D or 3D map that visualises the 2D or 3D distribution of the electrical resistivity values. To a great degree, this is based on the cases of nonlinear, least-squares minimisation of a two-part objective function. The misfit between the predicted and measured resistance values is the first part of this function. It minimises the discrepancy between experimental field data and the computed data based on Poisson's Equation ($\nabla^2 V = 0$). The term that minimises the roughness of the field measured electrical resistivity is the second part of this function, and tt allows for well-posedness of the inverse problem.

In the inversion problems, the regularisation term indicates the parameter that minimises the roughness of the electrical resistivity field and allows for well-posedness of the inverse problem. In the ERT data processing one search the objective function, O. It is given by

$$O = \left(d - g(\hat{m})\right)^T C_D^{-1}\left(d - g(\hat{m}) + \alpha \hat{m}^T D^T D \hat{m}\right)$$

where

d is the vector of electrical resistance measurements; $g(.)$ is the forward model for electric potential; $(\hat{m})$ is the vector of parameter estimates, (log electrical conductivity or resistivity); C_D is the covariance matrix of measurement errors; α is the regularisation parameter that determines the importance given to the smooth appearance of the electrical conductivity field relative to the misfit between calculated and observed resistances;

if one has small α the residual error between measured and modelled resistances will be minimised but may not converge to a unique solution;

large α will identify an overly smooth electrical conductivity field that may not fit the measured field data (resistances) well (Tikhonov and Arsenin 1977a, b);

D is the model-weighting regularisation matrix, which can be defined by either a discretised second-derivative filter or the covariance of the model parameters (Tarantola 1987a, b; Gouveia and Scales 1997; Kitanidis 1997; Vasco et al. 1997; Day-Lewis et al. 2007).

The model parameters are updated iteratively by the repeated solution of a linear system of equations for $\Delta \hat{m}$ at successive iterations. Such an approach results in the regularisation changing throughout the iterative process. The update appears like the following:

$$\left[J^T C_D^{-1} J \alpha D^T D\right]\Delta \hat{m} = J^T C_D^{-1}\left(d - g(\hat{m}_{k-1})\right) - \alpha D^T D \hat{m}_{k-1}$$

$$\hat{m}_k = \hat{m}_{k-1} + \Delta \hat{m}$$

where

J is the Jacobian matrix, with elements $J_{ij} = \frac{\partial \hat{d}_i}{\partial \hat{m}_j}$, $\hat{d}_i$ is the calculated value of measurement i; $\hat{m}_k$ is the vector of parameter estimates after updating in iteration k; and $\Delta \hat{m}$ is the vector of parameters updates for iteration k.

Data should be weighted by their measurement (reciprocal or stacking) errors, as stressed above, to prevent overfitting or underfitting of the data. In other words, the inversions will match the data to within the expected measurement errors, given errors seen in the field. Occam inversion refers to an approach that seeks to objectively set the tradeoff between the data match and the complexity (roughness) of the inverted resistivity tomograms (Constable et al. 1987). Some software packages leave it to the user to decide when the inversion has converged, placing the burden of balancing the tradeoff between the model and data misfit on the scientist's subjective judgment. All selections of inversion parameters should be recorded and reported.

In the equation that defines the objective function, O, the data misfit is based on an L2 norm, i.e., as the sum of weighted, squared differences between predicted and observed values. The L2 norm is highly sensitive to outlier data, hence the need to carefully edit datasets and remove data corresponding to electrodes with poor electrical contact or faulty channels. Many inversion packages also support the minimisation of the L1 data misfit, which is commonly referred to as robust inversion. The L1 norm is based on absolute differences instead of squared differences and is, therefore, less sensitive to outliers (Claerbout and Muir 1973). The L1 norm also can be used in defining the model misfit (i.e., roughness in the case where D is a second derivative filter), which is appropriate if sharp boundaries are expected. It is important not to confuse the use of L1 for data misfit with the L1 for model misfit.

The ERT inversion results are strongly affected by selected inversion parameters and regularisation criteria, especially in the presence of large measurement errors. In this case, it is important to perform multiple inversions to gain insight into the effects of different software settings.

The default setting, in more cases, results non-appropriate. The choice of inversion parameter is related to the apriori information that includes geological knowledge of the subsoil, drillers logs, characteristics of the expected target, past geophysical results, etc.

If during the inversion the symptoms are Minimum/maximum estimated resistivity too low/high compared to expected values, these could be done problems as the inversion may be overfitting the data or non-random outlier data may be present. In this case, the solution could be to stop inversion at an earlier iteration or increase the assumed measurement error, check the dataset for outliers and edit, try the L1 norm for data misfit.

If during the inversion the symptoms are that the tomogram is speckly or looks like a checkerboard, these could be done problems as the inversion may be overfitting the data; non-random outlier data may be present. In this case, the solution could be stop inversion at an earlier iteration or increase the assumed measurement error in Occam inversion, check the dataset for outliers and edit, try the L1 norm for data misfit.

If during the inversion the symptom is that the inversion can not match the data to within the reciprocal error, this could be done the problem that the optimisation algorithm may be caught in a local minimum, the finite-difference or finite-element grid may be too coarse, the inversion grid may be too coarse, non-random outlier data may be present. The solutions could change optimisation tolerances, increase the number of iterations, update the Jacobian more frequently, change the starting model, refine the grid or mesh, increase the number of inversion parameters, check the dataset for outliers and edit, try the L1 norm for data misfit.

If during the inversion the symptom is that the tomogram does not look like expected geology, this could be done the problem that the regularisation criteria may be smoothing/blunting the tomogram too much and resistivity may not correlate with lithology. The solutions could be to try robust model misfit instead of L2 model misfit, use anisotropic regularisation, try different regularisation criteria, another geophysical technique may be needed.

If during the inversion the symptom is that two (or more) tomograms that share a borehole appear inconsistent at the borehole, this could be done the problem that one has electrical anisotropy, outlier data are present in at least one dataset. The solutions could be to check the dataset for outliers and edit, try the L1 norm for data misfit.

If during the inversion the symptom is that the tomograms show vertical streaking, or high or low resistivity patches only at boreholes, this could be done the problem that resolution may vary greatly from the sides to the middle of the tomogram. The solutions could be to create synthetic or hypothetical forward models of the experiment to evaluate resolution and likely artefacts, examine plots of the inversion's sensitivity or resolution matrix.

An example of the possible results that should be obtained using Res2Dinv software (by Geotomo software http://www.geotomosoft.com/) is shown in Fig. 4.26. In this case, the inversion routine used by the program is based on the smoothness-constrained least-squares method (de Groot-Hedlin and Constable 1990; Sasaki 1992).

To prevent overfitting or underfitting, data should be weighted by their measurement errors. The inversions will match the data within the expected measurement errors starting from the errors measured in the field-data acquisition. Constable et al. (1987) described the Occam inversion, which is the approach used in the ERT data inversion that set the tradeoff between the data match and the complexity (roughness) of the inverted resistivity tomograms.

Due to the variability in resistivities of materials, interpretation of electrical-resistivity tomography data must be handled with caution. As seen in Chap. 2, there are overlaps of the resistivity values of earth materials that, in most cases, are given as ranges of values rather than absolute values. Other factors that affect resistivity changes are temperature, porosity, conductivity, salinity, clay content, saturation, and lithology.

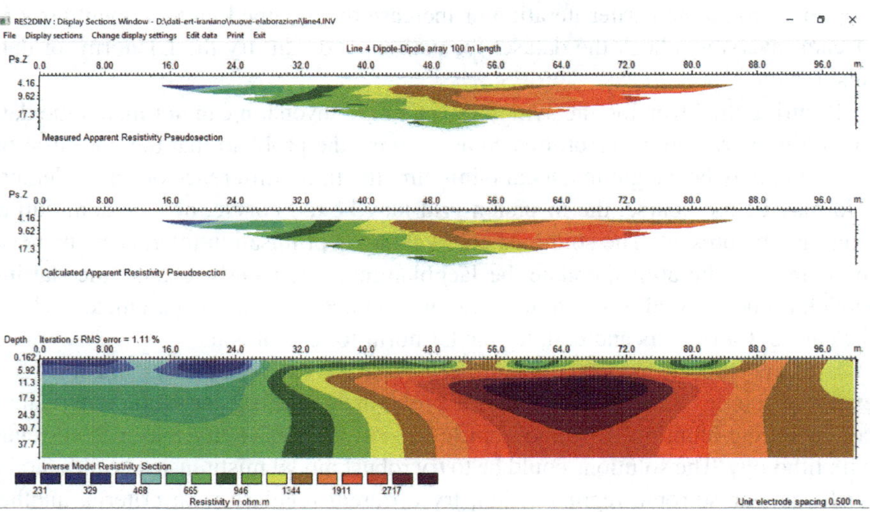

Fig. 4.26 The 2D ERT inversion results using the Res2Dinv software

Also, since, in the case of ERT data interpretation, the first step is related to the creation of "synthetic models" of 2D and 3D ERT profiles. Modelling can provide the interpreter with an idea of what to expect from resistivity data related to structures of archaeological interest and/or to the conservation state of monumental heritage, and therefore it enables more accurate interpretation after data has been processed.

Several software programs enable accurate ERT data modelling. Among them are: RES2Dmod, Res3Dmod (http://www.geotomosoft.com), and ErtLab (http://www.geostudiastier.it). In the model construction, it is important to define the shape and dimension of the bodies that simulate a target, their depth, and the associated resistivity values beyond also the geoelectrical properties of the materials and boundaries that are to be included in the simulation.

For example, synthetic ERT data were created to take into account the complexity of the subsoil in the case of a buried body. In the model, the uneven terrain of bodies is simulated using res3Dmod software. A grid of 23×11 m^2 was hypotized. A hidden body was hypotize at 1.5 m depth. The buried shape was rectangular 1×2 m^2. The filling material of the buried body consists of low resistivity (10 Ω m). Furthermore, the horizontal non-homogeneity caused by contact of natural soil and the artificial buried material was simulated by assigning the value of 400 and 2000 Ω m to the background. The thickness of the burial was 1.5 m (Fig. 4.27). Three types of arrays were array simulated, Wenner, Wenner-Schlumberger, and Dipole-dipole.

In Figs. 4.28, 4.29, and 4.30 the 3D resistivity inversion results using Res3Dinv software are shows.

The synthetic modelling approach showed that the dimensions of the burial were overestimated in the hypothesis of Wenner (Fig. 4.28) and Wenner-Schlumberger (Fig. 4.29) arrays.

Concerning the Dipole-dipole array (Fig. 4.30), as expected, gave a comparatively more resolvable inversion model concerning the other arrays. The horizontal dimensions and the location of the burial were reconstructed with great accuracy. On the other hand, another high resistivity anomaly (labelled A in Fig. 4.30) was created.

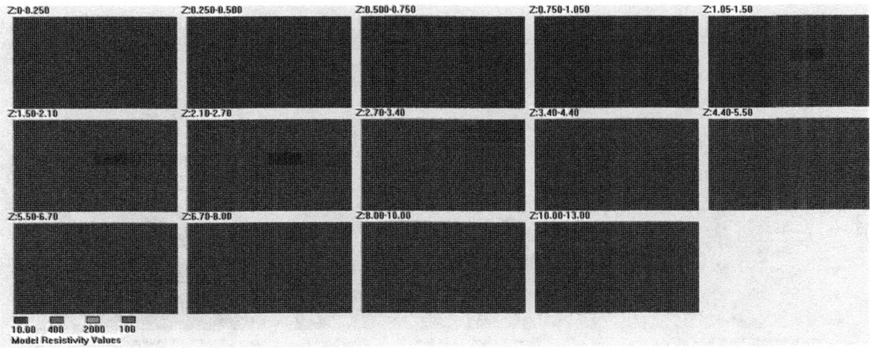

Fig. 4.27 3D resistivity model used to simulate the filling material and the buried body

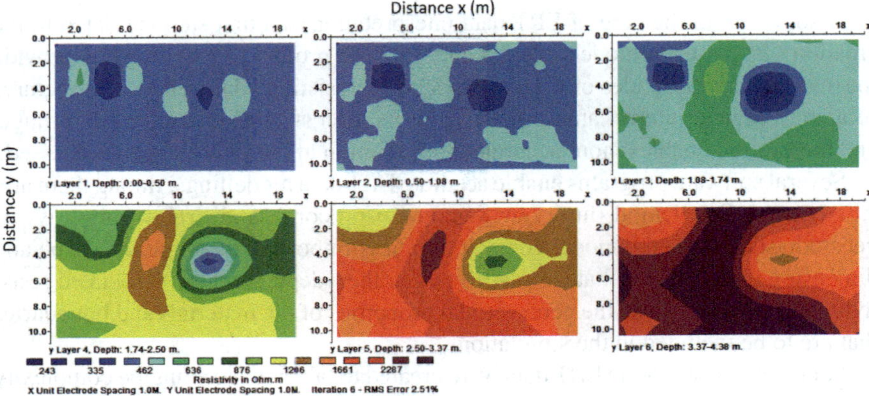

Fig. 4.28　3D resistivity distribution in the subsoil related to the Wenner array

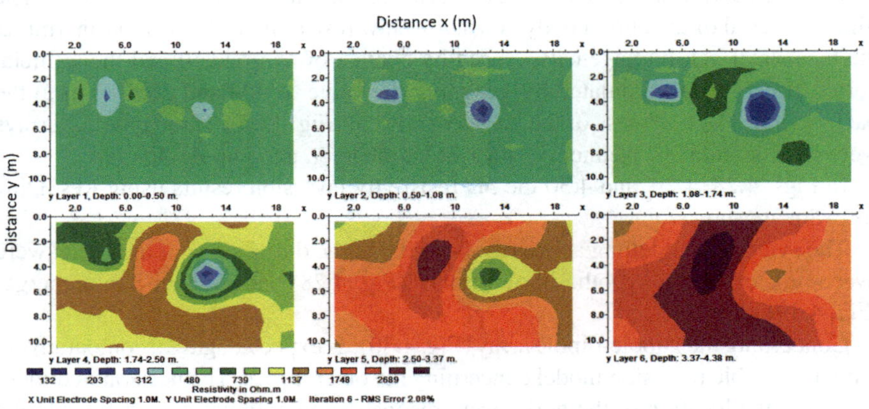

Fig. 4.29　3D resistivity distribution in the subsoil related to the Wenner-Schlumberger array

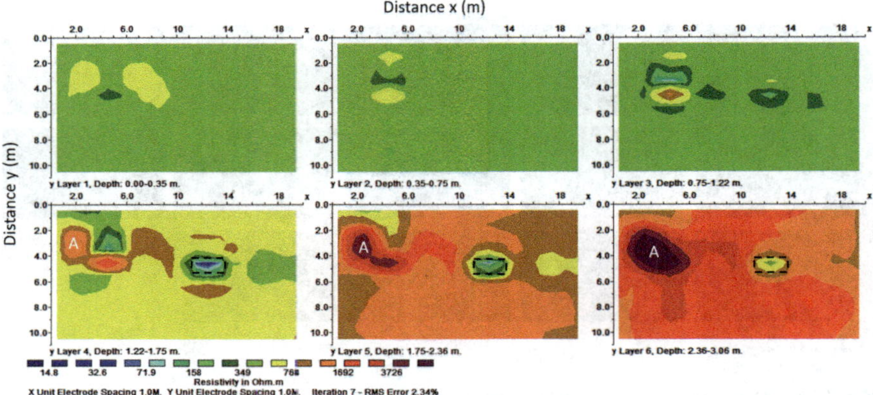

Fig. 4.30　3D resistivity distribution in the subsoil related to the Dipole-Dipole array

The anomaly A was related to the rectangular inserted in the second layer of the model (Fig. 4.27).

For Matlab routine in ERT depth slices see Leucci (2019).

4.4 SP Data Processing Interpretation

The classical SP inverse problem consists of determining the position and magnitude of SP sources. The inverse problem was the focus of many geophysical studies (Leucci et al. 2014; Fitterman and Corwin 1982; Patella 1997; Revil et al. 2001).

An elliptic partial differential equation governs this problem. It involves the measurement of a potential field (or its gradient) that can either be interpreted directly in dataspace or inverted to determine a source model that predicts the data using the underlying physics of the problem.

This paragraph attempts to contribute to the resolution of problems related to the inversion of SP data and their interpretation. Starting from the formulation of forwarding modelling solution for a target with known geometry (sphere, semi-infinite vertical cylinder, infinitely long horizontal cylinder) (Fig. 4.31).

The SP anomaly (V) due to the three simple geometrical bodies shown in Fig. 4.31 can be written as (Abdelrahman and Sharafeldin 1997; Abdelrahman et al. 2004; El-Araby 2004; Essa et al. 2008; Fedi and Abbas 2013; Monteiro Santos 2010)

$$V(x, z, q, \vartheta, k) = k \frac{x \cos \vartheta + z sen \vartheta}{\left(x^2 + z^2\right)^q}$$

where x is the coordinate of the measurement station (m), z is the depth of the body (m); the z-axis is chosen positive downward, q is the shape factor (dimensionless), θ is the polarization angle (degrees), and k (mV m^2 q^{-1}) is the electric dipole moment whose unit is shape factor-dependent to yield V in units of mV (Essa et al. 2008).

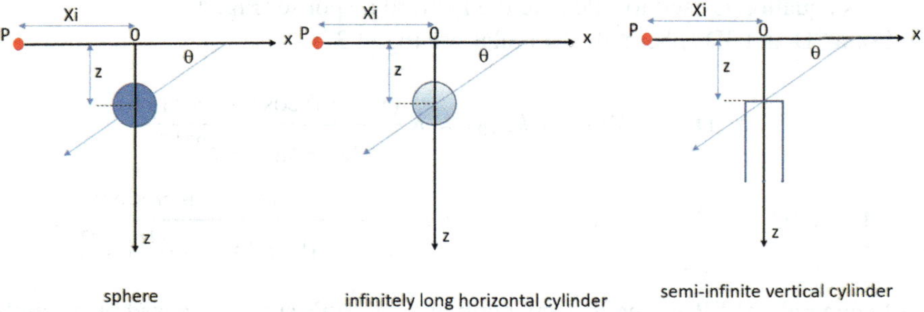

Fig. 4.31 Cross-sectional views (after El-Araby 2004), geometries and parameters of a sphere, a semi-infinite vertical cylinder, and an infinitely long horizontal cylinder (bottom panel)

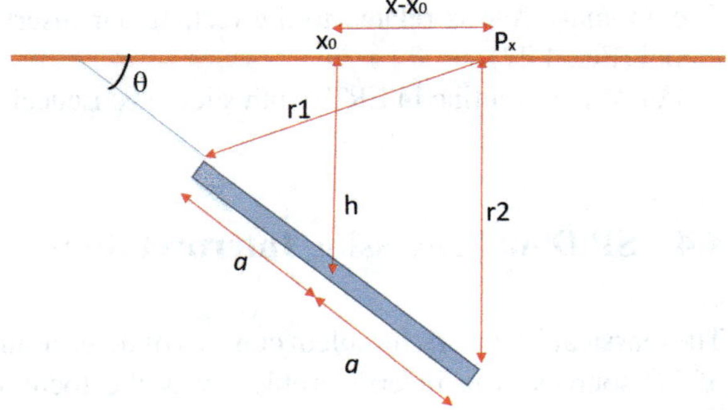

Fig. 4.32 Geometry of sheet which is clok-wisley inclined by θ

The shape factor of a sphere (3D), infinitely long horizontal cylinder (2D), and semi infinite vertical cylinder (3D) is 1.5, 1 and 0.5, respectively.

The SP field generated from a sheet which is clok-wisley inclined by θ (Fig. 4.32) can be modelled as follow:

SP anomaly V related to the structure showed in Fig. 4.32 is:

$$V(r_1, r_2, k) = K \log \frac{r_1^2}{r_2^2}$$

where r_1 and r_2 are the distances from top and bottom edges of the sheet to the measurement point; K is the electric moment dipole ($K = J\rho/2\pi$) where J is the current density and ρ is the resistivity of the surrounding medium; x_0 is the abscissa of the centre; h is the depth of the centre; a is the half-width of the sheet; θ is the inclination angle.

The general expression of the SP anomaly is (Asfahani and Tlas 2005):

$$V(x) = K \ln \left[\frac{\{(x - x_0) + a \cos\theta\}^2 + (h - a \sin\theta)^2\}}{\{(x - x_0) - a \cos\theta\}^2 + (h + a \sin\theta)^2\}} \right].$$

This equation is used to calculate the forward response (Fig. 4.33). For a 2D and 3D sphere, the formulas are Fig. 4.34:

$$\text{For a 2D} \qquad V(x, x_0, k, z_0) = K \frac{(x - x_0) \cos\theta + z_0 \sin\theta}{\left[(x - x_0)^2 + z_0^2 \right]^{1.5}}$$

$$\text{For a 3D} \qquad V(x, x_0, k, y, y_0, z_0) = K \frac{(x - x_0) \cos\theta + z_0 \sin\theta}{\left[(x - x_0)^2 + (y - y_0)^2 + z_0^2 \right]^{1.5}}$$

In general, the SP anomaly at any point of the Earth's surface, caused by a simple geometrical polarised body can be written as follow (Yungul 1950):

Fig. 4.33 SP anomaly

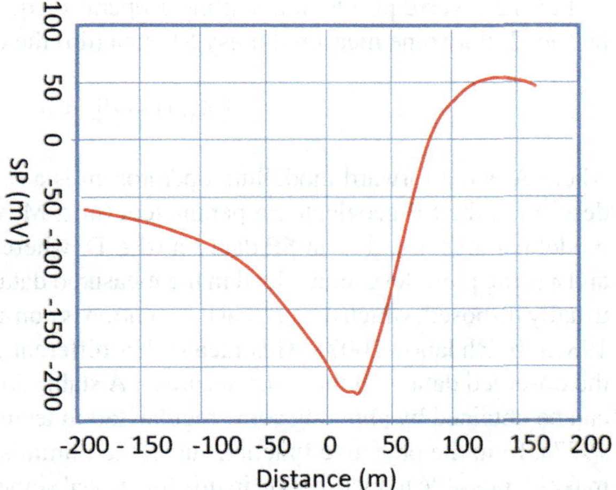

Fig. 4.34 SP anomaly related to a 3D single buried sphere model

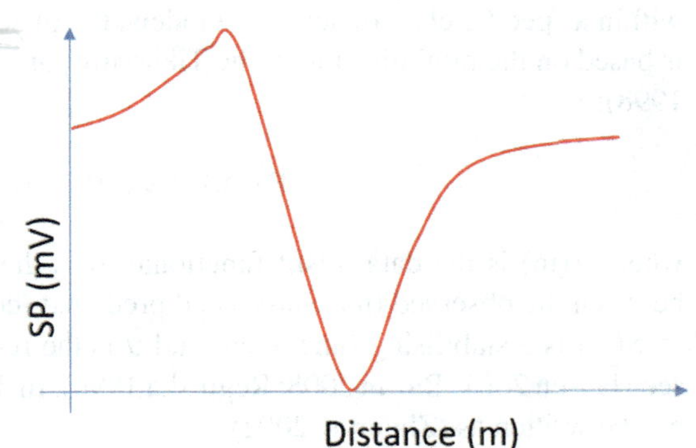

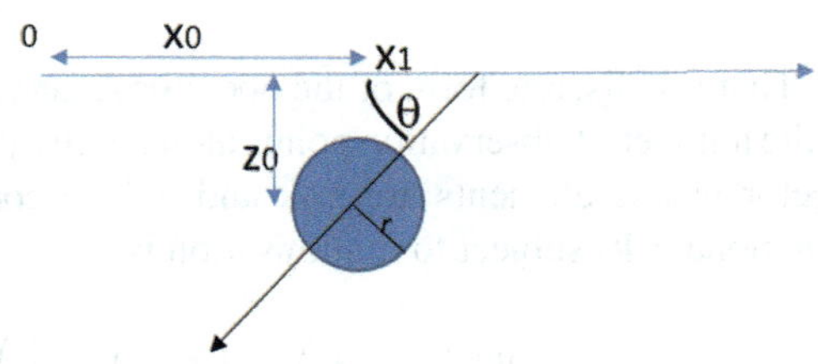

$$V(x, x_0, k, z_0, q) = K \frac{(x - x_0)\cos\theta + z_0 \sin\theta}{\left[(x - x_0)^2 + z_0^2\right]^q}$$

Since the burials (that hides body) structures are generally found in vertical and horizontal forms in the soil, the sources of such anomalies may be interpreted by assuming an inclined sheet model (Drahor 2004).

For the inverse problem, a nonlinear operator equation $A(m) = d$ solution must be found. For some measured noisy SP data (do) the equation is

$$\|A(m) - d\| \leq \delta$$

where A is the forward modelling operator, m is a scalar function (Zhdanov 2002) describing the SP geoelectrical parameters ($m \in M$, where M is a Hilbert space of model parameters), d is an SP data set ($d \in D$, where D is a Hilbert space of data), and δ is the noise level embedded in the measured data. Inverse problem $A(m) = d$ is usually ill-posed, which means that the solution is non-unique and unstable (Tarantola 1987a, b; Zhdanov 2002). This means that different geoelectrical models could fit the observed data with the same accuracy. A stable solution of an ill-posed problem can be obtained by applying some regularisation techniques (Tikhonov and Arsenin 1977a, b) in the objective function subject to minimisation. Also, the regularisation makes it possible to incorporate in this functional some a priori information about the model parameters (Zhdanov 2002). Since the regularisation is to search for a solution within a specific class of selected models, the of solving ill-posed inverse problems is based on the minimisation of the Tikhonov parametric functional (Tikhonov et al. 1998):

$$P\alpha(m) = \varphi(m) + \alpha S(m)$$

where $\varphi(m)$ is the data misfit functional. It is the squared norm of the difference between the observed (measured) and predicted (computed) data.

S(m) is a stabilising functional; and α is the regularisation parameter (for more see Hansen 2001; Bazán 2008; Reginska 1996). In discrete form, the inverse problem can be written as (Zhdanov 2002)

$$\hat{A}\left(\hat{m}\right) = \hat{d}$$

That is a discrete form of the operator A, and $\hat{d}$ is a finite set of data obtained at a finite number of observation points along with a profile, and $\hat{m}$ is the model parameter vector whose elements are z, k and θ. The corresponding parametric (objective) functional (Ψ) subject to minimisation is:

$$\Psi^\alpha\left(\hat{m}_\alpha, \hat{d}_0\right) = \left(\hat{A}\left(\hat{m}\right) - \hat{d}_0\right) + \hat{m}^T\hat{m} = \min$$

where T is the transpose operator.

The instability and non-convergence can be eliminated by transforming the parametric functional (above equation) into a new parametric functional using a mixed log space and linear space of the model parameters (log(z), log(k) and θ).

To examine and assess the practical applicability of the above-described inversion technique, real data were analysed. Data were acquired in a regular landfill of solid

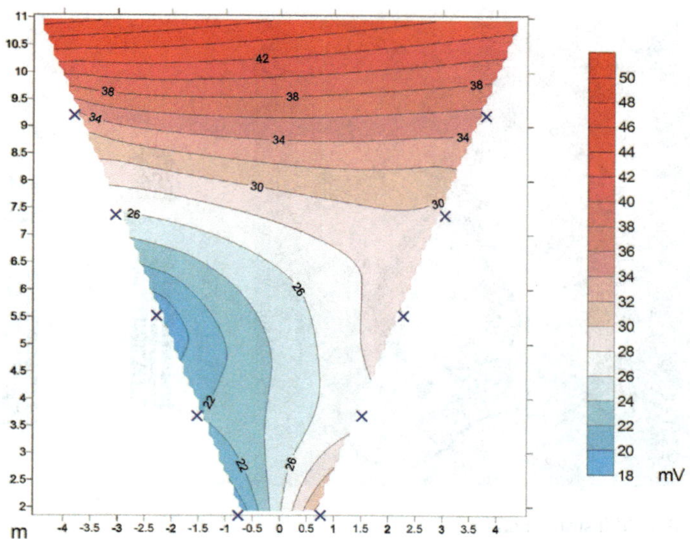

Fig. 4.35 SP map from a regular landfill of solid urban waste area. The x point indicates the base station positions

urban waste area. Data were acquired as described in Chap. 3. The results are shown in Fig. 4.35.

Results show in Fig. 4.35 allow to understand the distribution of the waste. SP values decrease in the areas in which are presents organic waste.

In the same area, spontaneous potential measurements were carried out using the "fixed base" technique through a base electrode and three measuring electrodes. The interelectrode distance used was 3 m, for a total of 64 measurements. The multimeter is a high-speed Fluke 77 (Fig. 4.36).

The measurements were repeated two times (with arid soil and after some rain) to highlight any differences and to verify the repeatability of the test (Fig. 4.37).

The measurements carried out, in addition to showing good results related to the inversion algorithm. The response of the method to the infiltration of meteoric waters in the subsoil has allowed highlighting the drainage action of the vegetation present in the area. In particular, the area is characterized by small trees planted with a regular 7 m grid. The draining action of the roots is responsible for bioelectrical potentials which, in the vicinity of the same, can reach several tens of mV due to the movement of water towards the surface. The study of this component, usually considered a disturbing factor for the quality of the measurements, can instead become interesting in the context of agricultural applications or in desertification studies.

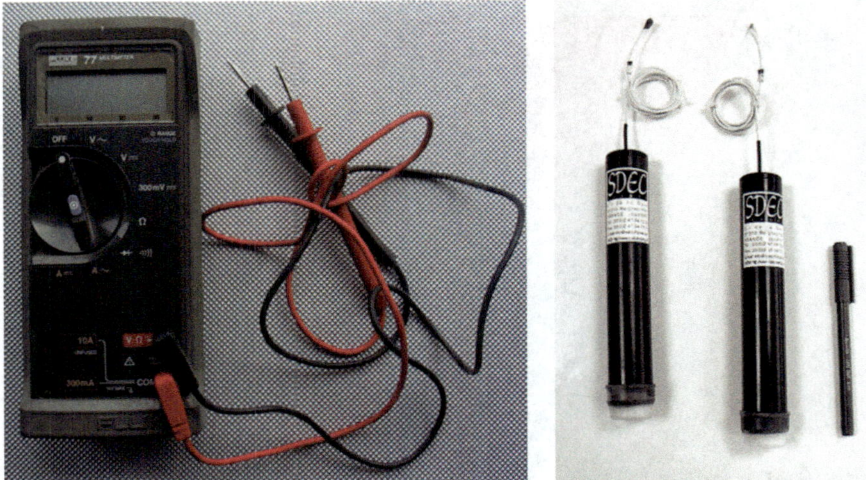

Fig. 4.36 SP field instrumentation

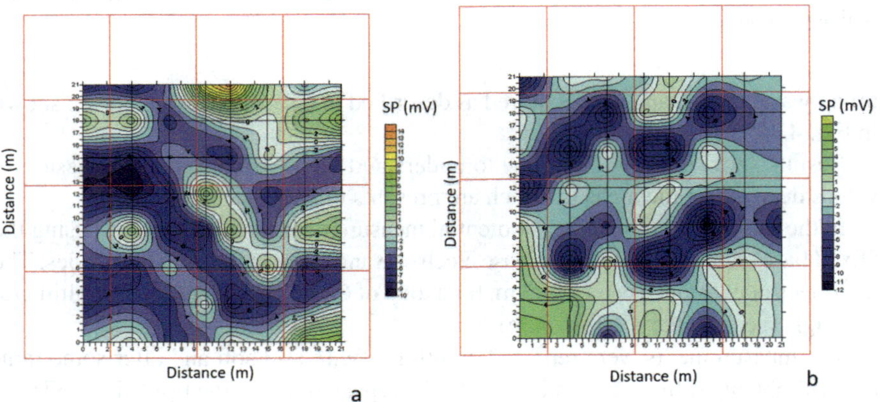

Fig. 4.37 SP map before (**a**) and after (**b**) the rain

4.5   Processing and Interpretation of Seismic and Ultrasonic Data

A general procedure for data inversion provides that one assume an initial model. The decision of the assumption of the initial mode is based on the expected anomalies. The initial model is then progressively refined in an iteration procedure until a proper fit is achieved between the predictions and observations. In every iteration step, the model parameters are modified to improve the fit between the observed and calculated data. The data misfit is characterised by an objective function which is minimised by a proper minimisation algorithm. The iteration procedure is continued until a stop criteria is met. The model obtained at the end of the iterative process can be considered the resolution of the inverse problem. Some considerations can be made. Always the exploration problems are of two types the over-determined and under-determined. The solution of even-determined case is a rare case.

The over-determined problem: it is solved using the least square method that involves the minimisation of the squared Euclidean length of the error end link to a perfect model resolution.

The under-determined problem: it is solved a minimisation of the model length and made a perfect data resolution.

In the determination of the starting model's parameters, one can have problems related to the over-knowledge of parameters (over-determined problem) or an under knowledge of parameters (under-knowledge). This lead to a non-uniqueness solution of the inverse problems.

The objective of the inverse problems is to find the best model m in which

$$d = G(m)$$

where G is an operator describing the explicit relationship between the observed data, d, and the model parameters.

G is the data kernel that was generated through the model (in this case can be the partial derivative of the seismic ray path concerning the ray travel time).

d (the data) and m (the best model) are vectors in the case in which a discrete linear inverse problem was describing a linear system. In this case, the problem can be written

$$d = Gm$$

where G is a matrix (Frechet matrix).

In order to obtain the best model one would solve the above equation as

$$m = G^{-1}d$$

Because of the over-determined or under-determined nature of the Frechet derivative matrix, not all matrices are invertible.

The matrix G may be rank deficient, have zero eigenvalues, and this lies to non-invertibility of the matrix. If an additional observation (i.e. more equations) to the matrix, G is no longer square. For these reasons, most inverse problems are undetermined and this means that do not exist an uniques solutions to the inverse problem. Only a full rank system allows an uniques resolution.

Over-determined system (more equations than unknowns) has the non-uniqueness issue and lead to an inconsistent set of equations.

The Frechet matrix must not directly inverted and for this a method that allows the optimisation the inverse problem must be used. For this purpose, an objective function for the inverse problem must be defined. It is a function that measures how close the predicted data from the recovered model fits the observed data. The objective function is written as

$$\varphi = d - Gm_2^2$$

Therefore the objective function φ has the scope to minimise the difference between the predicted and observed data.

To minimise a function a gradient of φ must be computed. From this, the above equations are simplified to

$$m = \left(G^T G\right)^{-1} G^T d$$

$$G^{-g} = \left(G^T G\right)^{-1} G^T$$

G^T is the matrix transposte of G

G^{-g} is the generalised inverse of the least square method.

For seismic tomography, the equation may be written as

$$\mathbf{Ax = b}$$

where A represents the matrix of the seismic ray length, x represents the inverse of the seismic wave velocity (the searched model) and b represent the seismic ray travel time.

To solve this equation, more commercial software uses the Simultaneous Iterative Reconstruction Technique (SIRT) algorithm.

As said above the system is typically solved by minimising some norm $\|Ax - b\|$. Each row of A describes the ray path in the subsoil or the studied material. SIRT alternates forward and back projections. Its update equation is

$$\mathbf{x}^{(t+1)} = \mathbf{x}^{(t)} + \mathbf{CA^T R}\left(\mathbf{b} - \mathbf{Ax}^{(t)}\right)$$

where C and R are diagonal matrices that contain the inverse of the sum of the columns and rows of the system matrix.

Iterations start with $x^{(0)} = 0$. The update equation then does the following things.

1. The current reconstruction $x^{(t)}$ is forward projected: $Ax^{(t)}$.
2. The result is subtracted from the original projections: $b - Ax^{(t)}$.
3. This difference is then back-projected. In essence, this is done by multiplying with A^T, but weighted with C and R.
4. This results in the correction factor $CATR\,(b - Ax^{(t)})$.
5. The correction factor is then added to the current reconstruction, and the whole process is repeated from step 1.

Below is a MATLAB code (matrix A is known).

```
% Input: sparse system matrix A, data b.
% Output: SIRT reconstruction x.
x = zeros(d * d, 1);
[rows cols] = size(A);
C = sparse(1 : cols, 1 : cols, 1 ./ sum(A));
R = sparse(1 : rows, 1 : rows, 1 ./ sum(A'));
CATR = C * A' * R;
for i = 1 : 100
  x = x + CATR * (b - A * x);
end
```

In order to calculate the misfit function below a matlab code where N is the input Number of equations in matrix-vector Ls = t; slope is the input Slope of line; L is the variable $N \times 2$ matrix; s is the variable 2×1 vector of unknowns; t is the variable $N \times 1$ vector; eps is the output misfit function $\|Lx\text{-}t\|$;

```
t=0;L=0;s=0;
for iter=1:7;
N=5;slope=.1*iter
xstart=-30;xend=30;
x=xstart:1:xend;y=x;
x0=0;y0=1;subplot(111)
% Compute Nx2 L matrix and Nx1 t vector
for m=1:N
    mm=(m-3)*slope;
    if m==1;L=[-mm 1];t=L*[x0 y0]';else;
    L=[L;-mm 1];t0=[-mm 1]*[x0 y0]';t=[t;t0];end
end;
s = inv(L'*L)*L'*t;e = (L*s-t)'*(L*s-t)
%
% Compute Misfit Function for different values s(1) & s(2)
%
for i=xstart:xend
ii=i-xstart+1;
for j=xstart:xend
jj=j-xstart+1;xx=[i j];eps(ii,jj)=(L*xx'-t)'*(L*xx'-t);
end;end
imagesc([xstart:xend],[xstart:xend],eps');colorbar;hold on
title('Misfit Function: ||Ls-t||');xlabel('s(1)');ylabel('s(2)');hold on;
axis([xstart xend xstart xend])
%% Plot Lines
%
for m=1:N;plot(x,-L(m,1)*x+t(m),'-w');end;hold off;c1=cond(L);
text(xstart+2,xstart+2,['condition # =',num2str(c1)]);
pause(1);end
```

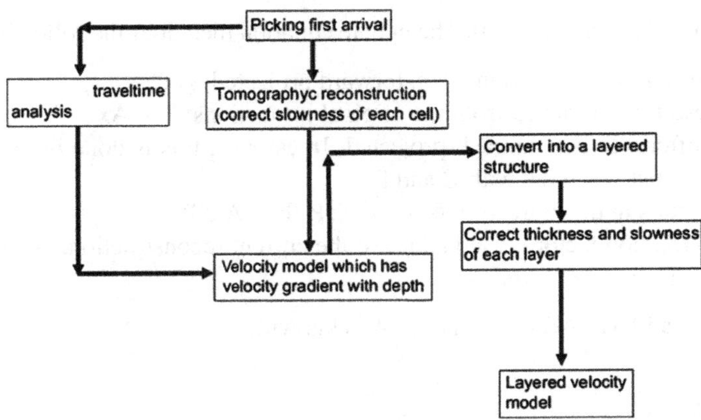

Fig. 4.38 The seismic-tomography, data-inversion scheme

Clearly the suggestions that data processing must be performed by" an expert in sonic-test processing using dedicated software products. This entails evaluating the pulse transit time by measuring the elapsed time of the recorded waveforms, that is to say, by picking, on the amplitude-versus-time diagram the starting point of the hammering pulse and the first arrival recorded by the accelerometer. In the case that the time selection has been performed with automated procedures, by using a threshold value to find the wave onset, the reliability of the procedure and appropriateness of the threshold value should be checked. After evaluating the transit time, the pulse velocity is calculated as follow: $v = l/t$ where v is the pulse velocity, [m/s]; l is the distance between the transducer, [m]; and t is the transit time, [s]. This simple procedure should be followed for S- and US-pulse velocity tests with direct, semi-direct, or indirect transmission methods.

Data processing for tomography requires specific software codes. The tomographic inversion is a challenging problem, and the performances and the flexibility of tomographic programs can vary significantly. The standard procedure in a tomographic reconstruction consists of first picking the travel times of the direct wave travelling from TX to RX positions through the masonry section. The transit times are then inverted by the tomographic software to obtain a velocity map. The problem is typically nonlinear, and the final solution is the result of several iterations where ray-paths are iteratively updated by using a ray-tracing module. The critical issues in this procedure are the design of the inversion grid and the control of the stability and reliability of the inversion. The grid size should be defined with special care, considering the expected resolution of the experiment and the spacing between acquisition points. Pillars with complex geometries may create data where the direct wave partially or totally travels in the air. These data should be removed before the inversion stage to prevent artefacts in the final velocity map. The processing flow used in the method is shown in Fig. 4.38.

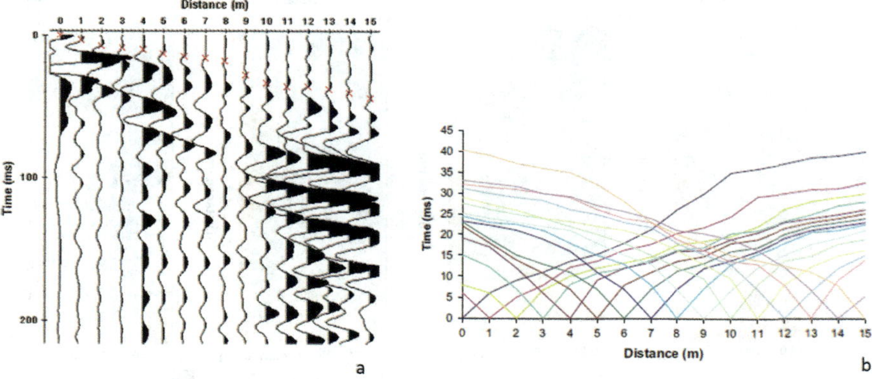

Fig. 4.39 **a** Picked seismogram; **b** first-arrival travel-time curves (Leucci et al. 2007)

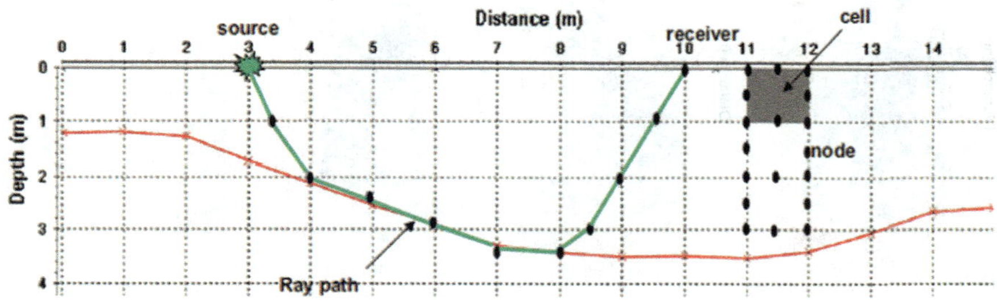

Fig. 4.40 Principle of ray tracing (Leucci et al. 2007)

In the tomography survey, the travel times of first arrivals of sonic waves related to source-receiver distances located along the profile is measured (Fig. 4.39). The travel times must be picked manually on a PC. The travel times of each source and each receiver must be successively combined to achieve an initial velocity model. In this case, one can assume a homogeneous medium. As shown in Fig. 4.40 the model is represented by N × M cells. Rays are traced through this model to give calculated travel times. The squared differences between the observed and computed travel times representing the misfit function will be calculated. The model is adjusted until the misfit function is minimised and the iterations are stopped when the root-mean-square (RMS) travel time residual (the difference between the calculated travel times for the initial model and the observed ones) is less than the average pick error of the travel times.

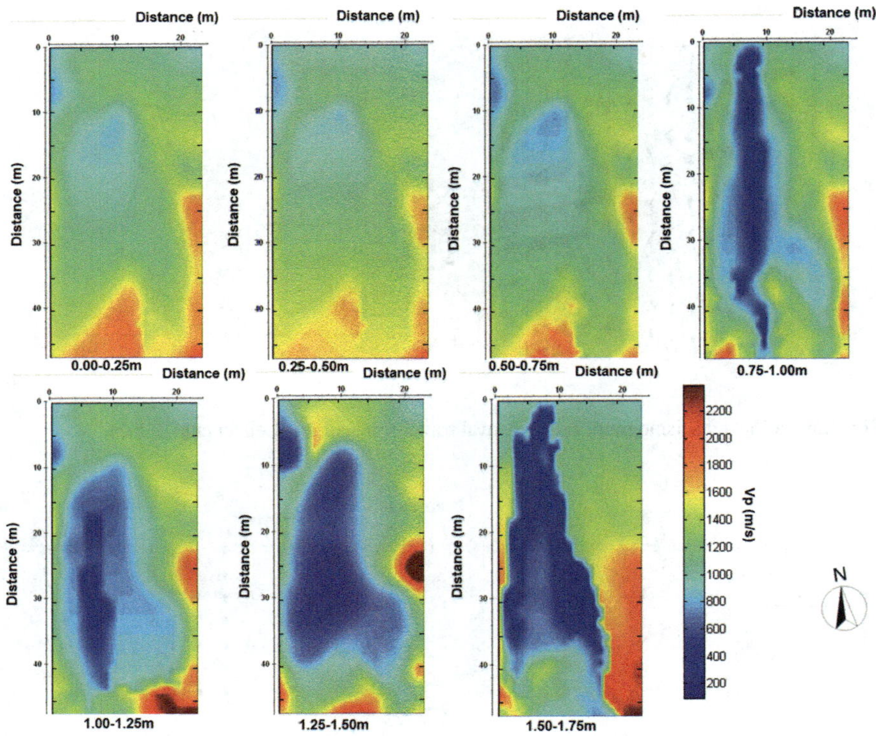

Fig. 4.41 The velocity distribution visible as 2D depth slices (Leucci and De Giorgi 2015)

The results of data processing may be visualised in the form of 2D or 3D contour maps in grey or colour scales, or other (Figs. 4.41 and 4.42).

Tomographic results can be presented as velocity maps or colour images where the colour is associated with a given velocity. This visualisation enables direct evaluation of the internal composition of a surveyed material structure, such visualisations being, in general, areas with higher/lower velocities representing changes in terms of material, the presence of flaws/voids, or diffuse crack patterns. Whatever the representation is, it is essential to include a drawing of the structural element (wall, pillar, etc.) with an accurate indication of the location of the experiment.

For easier visualisation and the correct interpretation of the results of the test, it is also necessary to use the same colour scale throughout the entire testing report, even if the values of S-pulse velocity are very different (e.g., if the quality of different structural elements vary a lot, the same test is carried out before and after injection, etc.). This may need a recalibration of the colour scale after all the tests are finished and the data processed.

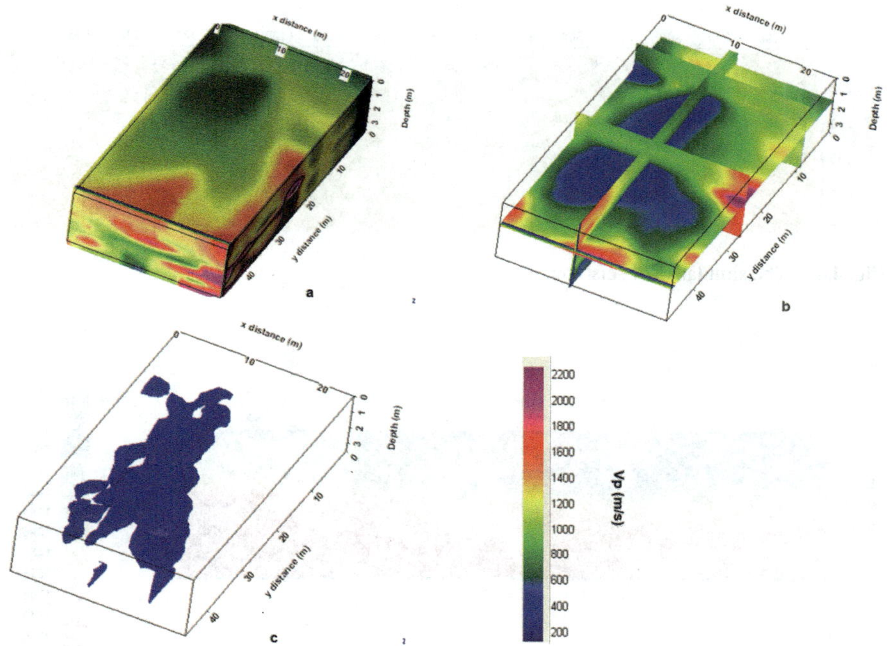

Fig. 4.42 The velocity distribution is shown as 3D volumes (Leucci and De Giorgi 2015)

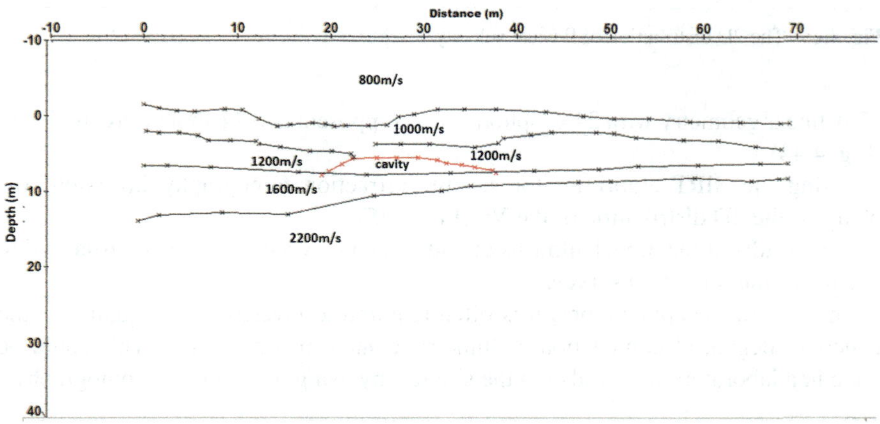

Fig. 4.43 The stratified subsoil model

Interpretation of the results is a fundamental and critical step. The models can useful in help interpretation. As an example, a stratified subsoil was modelled using the reflex software (Fig. 4.43).

The model contains a lithology with five layers with Vp that increase with depth. A cavity was inserted below the third layer.

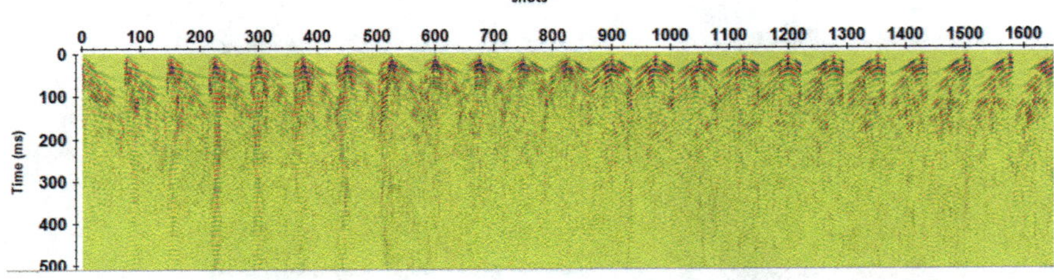

Fig. 4.44 The simulated 24 seismograms

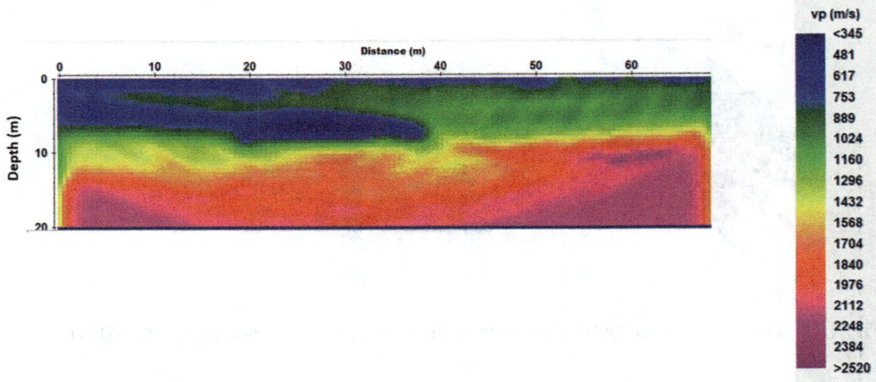

Fig. 4.45 The 2D distribution of P wave velocity

A linear geometry with 24 geophones was supposed, and 24 shots were modelled (Fig. 4.44).

Using the SIRT algorithm for seismic refraction tomography inversion was obtained the 2D distribution of the Vp (Fig. 4.45).

The results of the model allow us to understand how a cavity can be visualised in a real seismic refraction survey.

In the forensic applications, it is often required to investigate the quality of the concrete (degree of compaction, volumetric content in water, etc.). This could be done in a laboratory test or also in the situ test by using the S and US tomography.

Table 4.2 Correlation of seismic wave velocity with the quality of concrete

Apparent transmission velocity (m/s)	Concrete condition
4575 and above	Excellent
3660–4575	Good
3050–3660	Questionable
2135–3050	Poor
Below 2135	Very poor

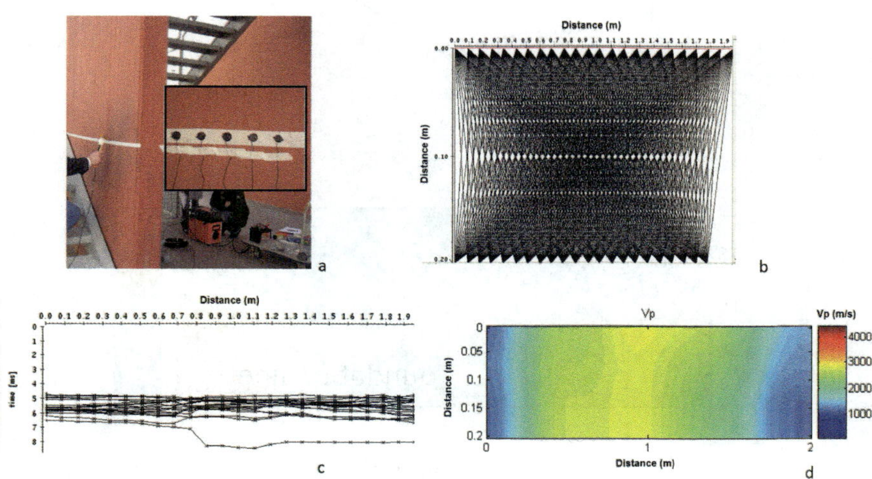

Fig. 4.46 **a** Sonic data acquisition on a pillar; **b** seismic ray tracing; **c** dromocrones; **d** P-wave velocity distribution

In this way it is possible to determine concrete uniformity, to control its quality, to follow up the deterioration, to check the presence of internal flaws and voids, and to make comparisons with reference specimens, it may estimate potential compressive strength. Finally, when regularly used, it may provide data on the development of problems. Table 4.2 shows the correlation of seismic wave velocity with the quality of concrete (Source from Testconsult CEBTP Ltd, UK).

An example of the application of this method in situ is shown in Fig. 4.46.

Results (Fig. 4.46) show the central part with $2500 < Vp < 3000$ m/s that indicates poor quality of the concrete (Table 4.2).

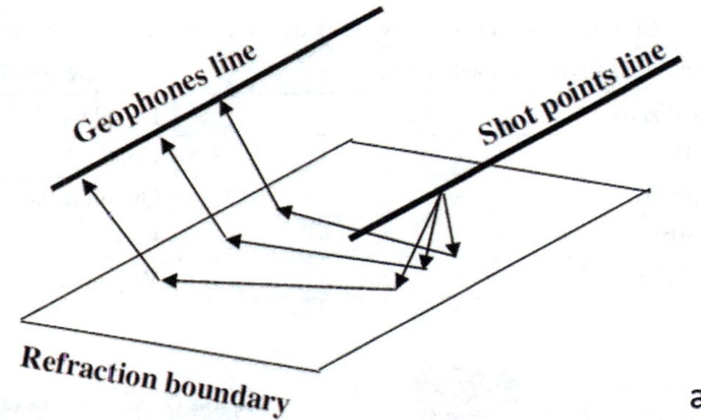

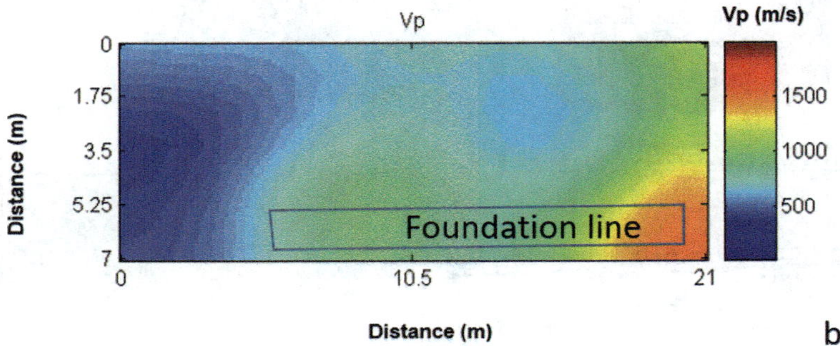

Fig. 4.47 **a** Sonic data acquisition geometry; **b** P-wave velocity distribution

Another problem is the foundation. Foundation concrete containing foreign materials (i.e. materials other than concrete such as soil inclusions, cobbles, bentonite cake or honeycombing, etc.) has a much lower propagation velocity. In this case, it could be needed to use a non-standard acquisition of S data. The geometry of the survey could be for example, that shown in Fig. 4.47a, geophones and shot points located respectively along with two, spaced parallel lines.

Results (Fig. 4.47b) show the low quality of the concrete foundation.

Below a matlab code for traveltime tomography with a straight ray. It solves the equation Ls = t by s = (R * L$'$ * L + alpha * I) * R * L$'$ * t where R is regularizer matrix [i.e., inverse of diag(L$'$ * L)] and alpha is damping parameter.

```
(nx,nz)     - input- # of gridpoints along (x,z) axes
dx          - input- spacing of pixels in meters
firstsx     - input- first source position along top boundary
firstgx     - input- first geophone position along bottom boundary
 (ds,dg)    - input- (source,geophone) increments in meters
(ns,ng)     - input- # of (sources, geophones)
slow1(nxm,nxm) - input- slowness model where nxm=nx-1
alpha       - input- damping parameter
slow1       -output- slowness tomogram
%
clear all;hold off;
nx=20;nz=nx;dx=10;dxx=dx;firstsx=0;ds=10;ns=20;firstgx=0;dg=10;ng=20;
[SLL,LL]=raypath(nx,nz,dx,firstsx,ds,ns,firstgx,dg,ng);alpha=.01;
%
% Create a Vertical Layer Slowness Model
slow=ones(nx,nz)/2000;
slow(round(nx/2):nx,:)=slow(round(nx/2):nx,:)*.5;
nxm=nx-1;slow1(1:nxm,1:nxm)=slow(1:nxm,1:nxm);
%slow1=slow1';%                                    Model becomes 2 layer
horizontal model
%nst=round(nx/2);nen=round(nst*1.3);%              Add a slowness anomaly
%for i=nst:nen;slow1(i,nst:nen)=1/2000;end%        Add a slowness anomaly
subplot(211);imagesc([1:nx-1]*dxx,[1:nx-1]*dxx,1./slow1');
colorbar;ylabel('Z (m)');title('Velocity Model');
text(dxx*nx,-1.5*dxx,'Velocity (km/s)')
%
% Solve for least square solution of Ls=t;
%
TIME=SLL*slow1(:);LTL=SLL'*SLL;LTt=SLL'*TIME;
d=diag(LTL);D=diag(1./d,0);ID=eye(length(d))*alpha;
appslow=inv(D*LTL+ID)*D*LTt;
%
% Plot Tomogram
%
appslow=reshape(appslow,nx-1,nx-1);subplot(212);
imagesc([1:nx-1]*dxx,[1:nx-1]*dxx,1./appslow');
colorbar;xlabel('X (m)');ylabel('Z (m)');title('Slowness Tomogram');
text(dxx*nx,-dxx,'Velocity (m/s)')

function [cc]=coord(xs,xg,zg,nz,dx,icase)
  xx=xg-xs;
  k1=xx/zg;
  iii=1;
  if xs>xg
    iii=-1;
  end
  if icase==1
    cc=zeros(nz-2,2);
    for i=1:1:nz-2
      cc(i,2)=i*dx;
      cc(i,1)=k1*cc(i,2)+xs;
```

```matlab
%         [i,cc(i,1),cc(i,2)]
      end
   else
      if abs(xx)>0.00001*dx
        xx1=floor(xs/dx)*dx;
        if abs(xx1-xs)<0.00001*dx
           xx1=xx1+iii*dx;
        end
        xx2=floor(xg/dx)*dx;
        if abs(xx2-xg)<0.00001*dx
           xx2=xx2-iii*dx;
        end
        n1=floor(xx1/dx);
        n2=floor(xx2/dx);
        cc=zeros(sign(iii)*n2+sign(-iii)*n1+1,2);
        nn=0;
        for i=n1:iii:n2
          xx=i*dx;
          nn=nn+1;
          cc(nn,1)=xx;
          cc(nn,2)=(cc(nn,1)-xs)/k1;
        end
      end
   end


   function [cc3]=merge(cc1,cc2,dx)
if isempty(cc2)
  cc3=cc1;
else
  nn=0;
  [n1,x]=size(cc1);
  [n2,x]=size(cc2);
  m=zeros(n2);
  for i=1:1:n2
    x=cc2(i,1);
    z=cc2(i,2);
    if z-floor(z/dx)*dx > 0.00001*dx
      nn=nn+1;
      m(nn)=i;
    end
  end
  cc3=zeros(n1+nn,2);
  nnn=1;
  mm=0;
  for i=1:1:n1
    mm=mm+1;
    if nnn <= nn
      if cc1(i,2) < cc2(m(nnn),2)
        cc3(mm,1)=cc1(i,1);
        cc3(mm,2)=cc1(i,2);
      else
        cc3(mm,1)=cc2(m(nnn),1);
        cc3(mm,2)=cc2(m(nnn),2);
        mm=mm+1;
        nnn=nnn+1;
        cc3(mm,1)=cc1(i,1);
        cc3(mm,2)=cc1(i,2);
      end
    else
        cc3(mm,1)=cc1(i,1);
        cc3(mm,2)=cc1(i,2);
    end
```

```
   end
   if nnn < nn
     for i=nnn:1:nn
       mm=mm+1;
       cc3(mm,1)=cc2(m(i),1);
       cc3(mm,2)=cc2(m(i),2);
     end
   end
end


%nx---the number of grid points in x-direction; cell number is nx-1
%nz---the number of grid points in z-direction; cell number is nz-1
%
%grid:    (1,1)  (2,1)  (3,1)  (4,1)  (5,1)    (nx,1)
%cell:       1      2      3      4      5
%         (1,2)  (2,2)  (3,2)  (4,2)  (5,2)    (nx,2)
%            6      7      8      9     10
%         (1,3)  (2,3)  (3,3)  (4,3)  (5,3)    (nx,3)


function [LL]=raymat(nx,nz,dx,firstsx,ds,ns,firstgx,dg,ng)
  icase=1;
  num_cell=(nx-1)*(nz-1);
  num_ray=ns*ng;
  LL=zeros(num_ray,num_cell);
  for i=1:1:ns
    xs=(i-1)*ds+firstsx;
    for j=1:1:ng
      [i,j]
      xg=(j-1)*dg+firstgx;
      [lc,indc,nc,cc3]=ray(xs,xg,icase,nx,nz,dx,ns,ng);
%      [size(lc),size(indc),nc]
%      indc
      for k=1:1:nc
        LL((i-1)*ng+j,indc(k))=lc(k);
      end
    end
  end
function [SLL,LL]=raypath(nx,nz,dx,firstsx,ds,ns,firstgx,dg,ng);

[LL]=raymat(nx,nz,dx,firstsx,ds,ns,firstgx,dg,ng);
for i=1:nx*nz;
x=(LL(i,:));
y=reshape(x,nx-1,nz-1);imagesc([1:nx-1]*dx,[1:nx]*dx,y');pause(.5);
title('A Raypath');xlabel('X (m)');ylabel('Z (m)');
end;
SLL=sparse(LL);
```

4.6 GPR Data Processing and Interpretation

In GPR data acquisition, raw data are easily viewed in real-time on a computer screen. This is one of the advantages of the GPR method. Generally, the GPR data processing step does not require standard but depends on the data quality, and it is important to underline that no amount of processing will save poor quality data (Cassidy 2009).

The first step of GPR data processing is standard, and it is the most time-consuming part of a processing sequence (data editing). It involving data reorganisation (spatial position of the profiles), data file merging (some profile can be interrupted for the presence of an obstacle and successively continued after the obstacle, and therefore the two profiles must be merged to prepare they for 3D analysis). The occurrence of inevitable errors during data acquisition means some traces may need to be reversed, merged or omitted. This first step is important when trying to obtain good quality interpretations.

An important phase in GPR data processing is the determined of the electromagnetic wave velocity to convert two-way travel-time into depth. This value can be obtained in several modes. One is to measure it directly in the field excavating to determine the depth to a known reflector (Fig. 4.48a), then using the depth and measured reflection time to calculate the velocity (Leucci 2019). Others mode are the Common mid-point (CMP) sounding (Fig. 4.48b) and wide-angle reflection and refraction (WARR) (Fig. 4.48c) surveys estimate signal velocity by increasing the separation between the transmitter and receiver in steps at a fixed location and measuring the change of the two-way travel-time to reflections (Annan 2005; Leucci 2019). CMP surveys are generally considered to be more precise than WARR surveys and should be the first survey completed on arrival at a site. This, however, is only appropriate when performed above a horizontal reflector in multiple directions.

Another way to measure electromagnetic velocity is related to the measure, on the acquired radar profile, the angle of hyperbolic reflections as a proxy for the moveout

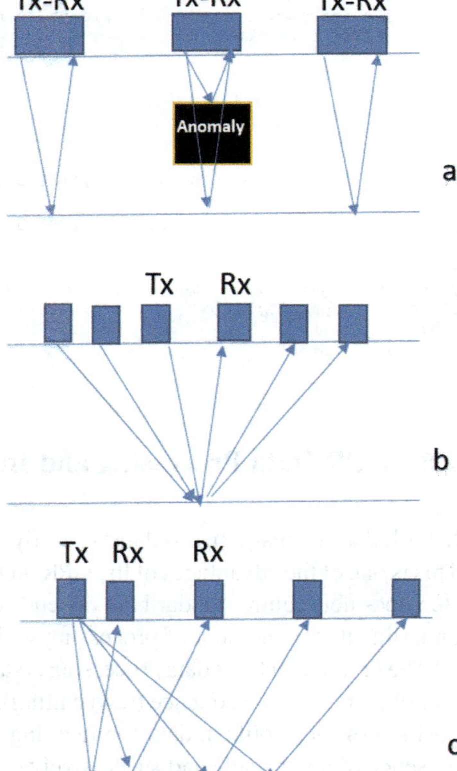

Fig. 4.48 **a** Measure of the electromagnetic wave velocity directly in the field excavating to determine the depth to a known reflector; **b** common mid-point (CMP); **c** wide-angle reflection and refraction (WARR)

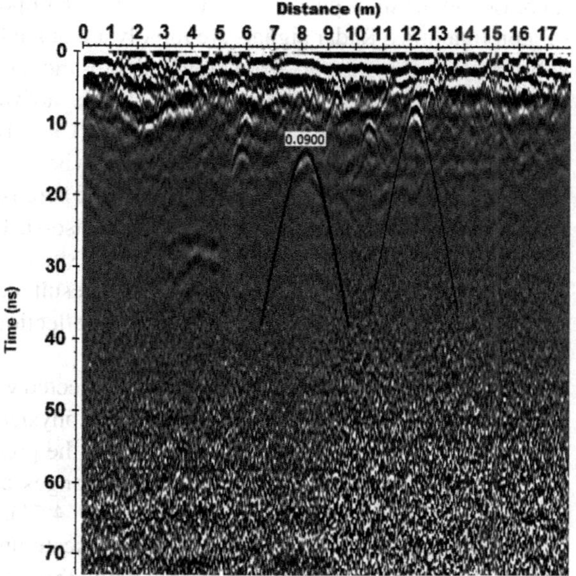

Fig. 4.49 The measure of the electromagnetic wave velocity from diffraction hyperbola

encountered in CMP, calculating velocity by fitting curves to observed hyperbolas as illustrated in Fig. 4.49.

Due to the airwave first breakpoint or the first negative peak of the trace, traces need to be adjusted to a common time-zero position. Successful realignment causes all reflections beneath to become correctly aligned. Successively temporal filtering to remove very-low-frequency components from the data (Annan 2005). It reduces the data to a mean zero level. This filter is known as de wow filter.

Several types of filters can be used to improve the visual quality of the GPR data. They can be the band-pass filters or sophisticated domain and transform filters. Temporal filters are used to remove noise at frequencies that are higher or lower than the main GPR signal, ultimately acting as clean-up filters which make the GPR section visually better. Spatial filtering is applied to either suppress or emphasise specific features. They are often used to remove the strong airwave response and ringing from GPR data. For more see Leucci (2019). Another filter that removes the low-frequency component is the background removal filter. It often takes the form of an average trace removal (a form of spatial filtering). This step allows subtle weaker signals to become visible in the processed section (Annan 2005).

Topographic correction is required to place the GPR data within its correct spatial context. When the surface and subsurface stratigraphy are horizontal, elevation static corrections can be used to topographically correct data, repositioning the time zero in the vertical axis and adjusting reflections accordingly.

Another processing step is to select a gain function for the data. Gains improve the visual form of the GPR sections, with most techniques altering the data structure in some form. Therefore the gain is only a way to enhance the visualisation of the radar section, and it does not allow to increase the electromagnetic signal to increase the depth of penetration. It is, therefore important to understand the effects of gain functions. Radar signals are rapidly attenuated as they propagate through the subsurface, making events from greater depths more difficult to discriminate (Annan 2005). Gains enhance the appearance of later arrivals due to the effect of signal attenuation and geometrical spreading losses (Cassidy 2009).

After processing GPR, results need to be visualised in a 3D way to understand the spatial distribution of the anomalies well. There is two way to visualise the GPR data in a 3D way: the amplitude slice and the iso-surfaces.

The amplitude slice creates images of the spatial distribution of reflected wave amplitude differences within a grid. The result can be a series of slices that illustrate the three-dimensional location of reflection anomalies derived from the two-dimensional profiles.

An analysis of the spatial distribution of reflected wave amplitudes helps in understand subsurface changes in lithology or other physical properties of materials in the ground. Amplitude changes can be related to the presence of important buried features and therefore, the location of those changes can be used to reconstruct the subsurface in three dimensions (Figs. 4.50 and 4.51).

Areas of low-amplitude waves usually indicate uniform matrix material or soils (the blue colour in Figs. 4.50 and 4.51) while those of high amplitude denote areas of high subsurface contrast such as buried bodies, voids, or important stratigraphic

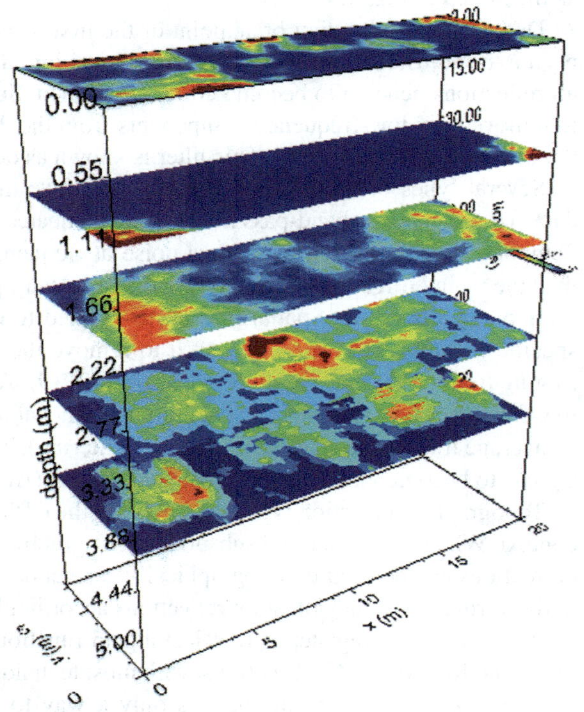

Fig. 4.50 Amplitude slice maps distributed in a 3D volume visualisation

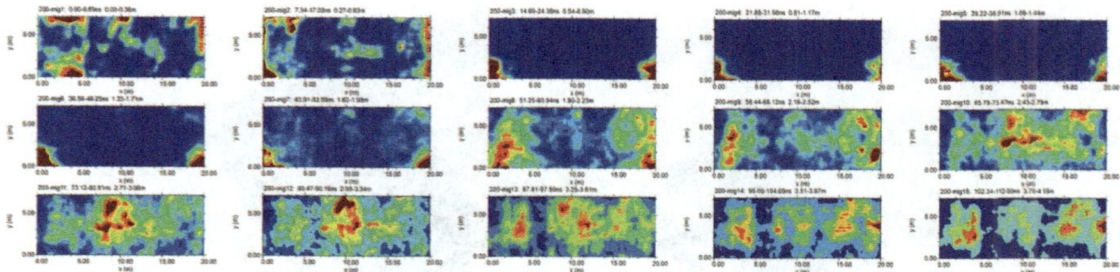

Fig. 4.51 Amplitude slice maps

changes (from green to red in the Figs. 4.50 and 4.51). Amplitude changes or differences can be interpreted analysing in the form of *depth slices* that examine only changes within specified depths in the ground. Each depth slice represents the spatial distribution of all reflected EM wave amplitudes. This distribution can be successively related to changes in sediments, soils, and buried materials. Amplitude slices can vary in thickness and orientation, depending on the questions being asked.

There are several software that allows producing horizontal amplitude slice-maps. They compare amplitude variations within traces that were recorded within a defined time window.

For small-size zones, a complete understanding of the subsurface can be achieved utilising various 3D data presentations, including 3D cubes. Nuzzo et al. (2002) describe two 3D visualisation approaches. In the first approach, the processed GPR data can be combined in a 3D image of the sought diffracting or reflecting objects. This image could be obtained by extraction of a particular complex signal attribute (trace envelope); thresholding; 3D contouring using the iso-amplitude surface.

The second approach consisting of extraction, from the pre-processed GPR data, of the trace energy and trace envelope; three stacks separately performed along each coordinate axis, providing separate 2D results: stacking of the energy along the depth or Z-axis, to obtain a plan view of the high-energy suspected zones; stacking of the trace envelope along X; stacking of the trace envelope along Y; thresholding; 3D rendering of the presumed target by projection in 3D space of the automatically selected thresholded data. In both cases, the threshold calibration is a very delicate task.

As example Figs. 4.52 and 4.54 show the three modes of 3D visualisation.

However, since the 3D data volume is sliced at fixed time samples, without any averaging, the images appear slightly noisier.

To understand the importance of the threshold values the same dataset of Fig. 4.52 is displayed with iso-amplitude surfaces using two threshold values: 50% (a) and 40% (b) of the maximum complex-trace amplitude in Fig. 4.53. Lowering the threshold value increases the visibility of the main anomaly and smaller objects (the linear anomaly denoted by an arrow in Fig. 4.53b), but also heterogeneity noise.

Following the third approach, visualised as the three coordinate planes of a 3D volume (Fig. 4.54a) and after a careful selection of the threshold value (Fig. 4.54b), data were projected in 3D space to produce the final display (Fig. 4.54c).

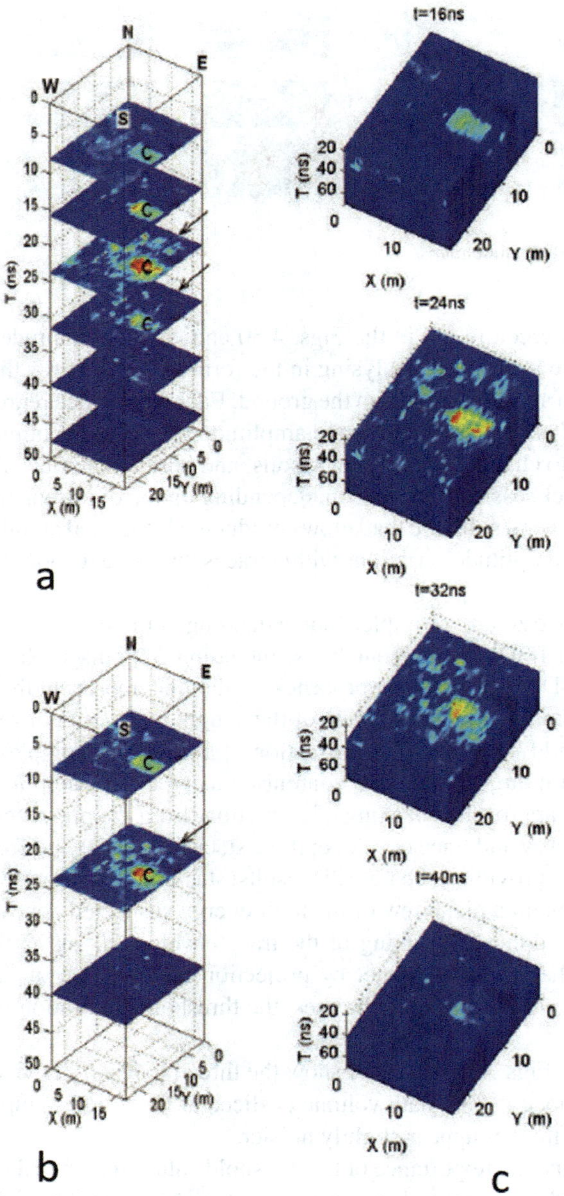

Fig. 4.52 Amplitude slices obtained averaging the absolute amplitude of the data within different time windows: 8 ns (**a**) and 16 ns (**b**). Arrows denote a weak and thin linear anomaly probably due to a ditch in the bedrock. The same lineament, although slightly noisier due to the lack of averaging, are also visible on the 3D volume presentations (modulus of the Hilbert transform) sliced at fixed times (**c**)

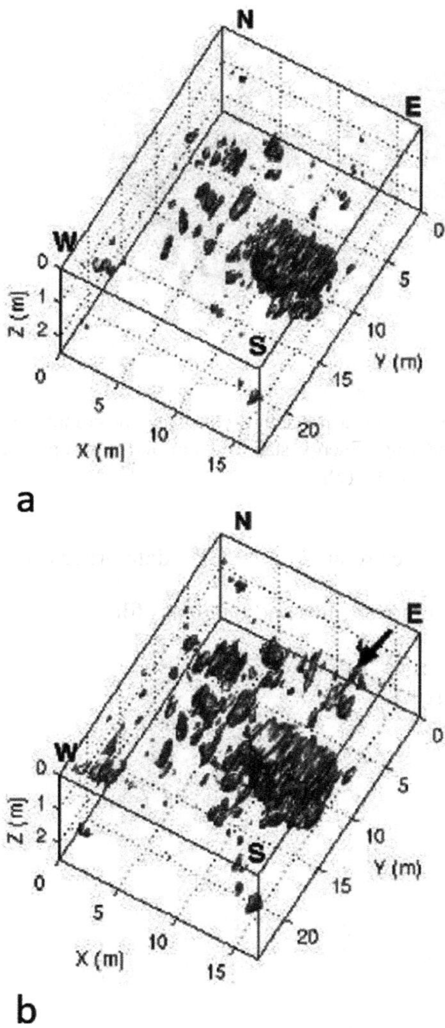

Fig. 4.53 3D visualisation using iso-amplitude surfaces of the complex-trace amplitude. Perspective views (from the south) using different thresholds: 50% (**a**) and 40% of the maximum (**b**)

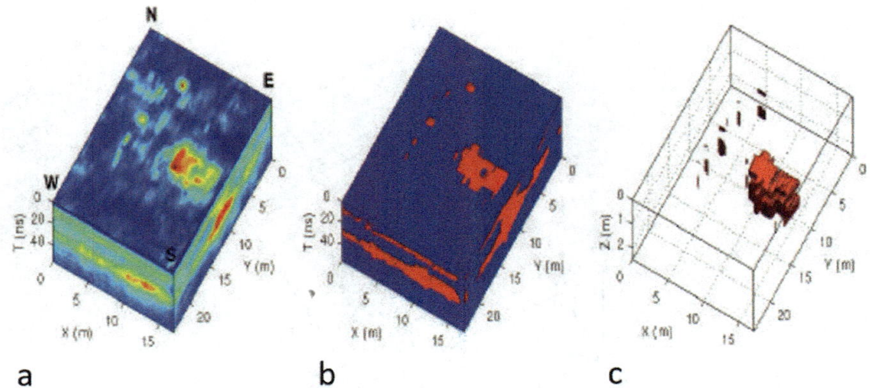

Fig. 4.54 3D visualisation using a procedure consisting of complex-trace attribute extraction, thresholding and 3D rendering. Energy stack on top and envelope stack on lateral sides (**a**), thresholding (**b**) and 3D rendering (**c**)

Following Matlab codes related to the GPR data processing.

(1) For background removal filter and bandpass filter

```
clear all
close all
t=linspace(0,8.15e-9,164);
t1=t';
x=linspace(0,2.4,241);
fmin=1000e6;
fmax=3200e6;
NF=33;
f1=linspace(fmin,fmax,NF);
load ET_16.ASC
dati=ET_16;
pp=size(dati);
NTR=pp(2);
for k=1:NTR
    v=dati(:,k);
    v1=fouriert(v,t1,f1);
    Esf1(:,k)=v1.';
%    k
end
%filter
for k=1:NTR
    v1=Esf1(:,k);
    v2=ifouriert(v1,f1.',t1);
    dati_filtrati(:,k)=2*real(v2.');
end
figure
subplot(2,1,1)
imagesc(x,t1,dati)
subplot(2,1,2)
imagesc(x,t1,dati_filtrati)

%bkg removal
N=input('numero tracce mediate');
%average trace
v=dati(:,1);
for k=2:NTR
```

```
      v=v+dati(:,k);
end
v=(1/NTR)*v;
trm=v;
prima(:,1)=trm;
for k=2:N
    prima(:,k)=trm;
end
dopo=prima;
datiforbkg=[prima, dati, dopo];
datiforbkg1=0*datiforbkg;
for k=N+1:N+NTR
    trml=datiforbkg(:,k-N);
    for h=k-N+1:k+N
        trml=trml+datiforbkg(:,h);
    end
    trml=(1/(2*N+1))*trml;
    datiforbkg1(:,k)=datiforbkg1(:,k)-trml;
end
datibkg=datiforbkg1(:,N+1:N+NTR);
figure

subplot(2,1,1)
imagesc(x,t1,dati)
subplot(2,1,2)
imagesc(x,t1,datibkg)
```

(2) Fourier function

```
function f=ifouriert(funzione,var_spettrale,var_spaziale)
M=length(var_spaziale);
dx=var_spettrale(2)-var_spettrale(1);
for k=1:M
    integrando=exp(i*2*pi*var_spaziale(k)*var_spettrale).*funzione;
    f(k)=dx*sum(integrando);
end
```

(3) Read data matrix

```
function matr = readmatrix(filename,m,n)

% named A=readmatrix(filename,m,n) is possible to read A(m,n)

%m number of row

%n number of columns

% read files write x row

    matr=zeros(m,n);

    fiid = fopen(filename,'r');

    fprintf('\n');fprintf('   Reading %s\n', filename);

    matr = (fscanf(fiid,'%g',[n,m]))';

    fclose(fiid);
```

(4) In order to visualize the matrix, it is possible to use the following .m file:

```
File: plot.m
filename = input(' nome file dati:  ');
 m=input('number of row ?: ');
 n=input('number of column ?');
 ngraf=input('number of graphics?');
 y=(readmatr(filename,n,m))';
 x=1:1:n;
 for j=1:ngraf;
  figure;
 m1=input('num. 1¡ trace to be plot ?: ');
 m2=input('num. last trace to be plot ?: ');
 ASCHIFT=input('SHIFT ALONG Y ?');
   for i=m1:m2;
   n1=(i-1)*n+1;n2=n*(i-1)+n;
   plot(x,y(n1:n2)+ASCHIFT*(i-1),'b');hold on;
   end;
end;
```

(5) Now, it is possible to write a .m file that can help in the matrix visualisation:

```
File: visualizzation.m
% read the GPR files and visualise they
ns=252; %(number of samples in GPR section)
nt1=558; %(number of traces in GPR section)

 z=readmatrix(radar1.ASC',nt1,ns); %(radar section that must be read)

x=[1:nt1];
a1=min(min(z));
a2=max(max(z));
a=[a1 a2]; %(allow to set max and min of electromagnetic wave reflection events amplitude)
figure(1);subplot(211);imagesc(z');caxis(a);colorbar;title('inserttitle);%set(gca,'YTicklabelMode','manual','YTick',[1 60 130],'YTicklabel',[0 3 6]);set(gca,'XTicklabelMode','manual','XTick',[1 60 130],'XTicklabel',[0 3 6]);
```

(6) For circular radargram

```
dati=load('R1.asc'); % file ascii
[r,c]=size(dati)
L=19.667;    % length of the circular line
rag=L/(2*pi); % radius
R=rag*ones(1,r);                 % cilinder
vel=0.098;   % electromagnetic wave velocity (m/ns)
s0=0.1;      % initial curvilinear abscissa (m)
sf=20.56;    % final curvilinear abscissa (m)
ds=0.01;
S=[s0:ds:sf];
size(S)
t0=0;
tf=18;
%dt=0.05859375;
dtt=(tf-t0)/(r-1);
T=[t0:dtt:tf];
size(T)
Zeta=0.5*vel*T; %Zeta=vel*T; (depth)
size(Zeta)
Zmax=max(Zeta)

N=ceil(L/ds+1)
N1=s0/ds
C=zeros(r,N);
C(:,N1+1:N)=dati(:,1:N-N1);
C(:,1:N1)=dati(:,N-N1+1:N);
cmax=max(max(C))
clim=[-cmax/4 cmax/4];
%figure;
%subplot(211);imagesc(dati(:,1:N),clim);
%subplot(212);imagesc(C,clim);

[X,Y,Z] = cylinder(R,N);
Z=Zmax*Z;
size(X)
%size(Y)
%size(Z)
sw=1;
```

```
if sw==0
figure;
%subplot(121);surf(X,Y,C)%subplot(122);
surf(X,Y,Z,C); shading flat;axis equal;caxis([-1000 1000]);view(38,40);
colormap(pink);
%colormap(flipud(gray))
%colormap(jet)


figure;
%CC=abs(C);
CC=abs(hilbert(C));          % envelop
surf(X,Y,Z,CC); shading flat;axis equal;caxis([0 1000]);view(38,40);
colormap(flipud(gray))
%colormap(jet)
%colormap(gray);%colormap(bone);
%colormap(pink);%colormap(hot);colormap(copper);
%colormap(spring);colormap(summer);colormap(autumn);colormap(winter);


elseif sw==1
figure;
surf(X,Y,Z,C); shading flat;caxis([-1000 1000]);axis tight;%view(38,40);
colormap(pink);
set(gca,'Xdir','reverse','Ydir','normal','Zdir','reverse','FontSize',[10],'XTick',[-3:0.5:3],'XTickLabel',[-3:0.5:3],'YTick',[-3:0.5:3],'YTickLabel',[-3:0.5:3],'ZTick',[0:0.25:1.5],'ZTickLabel',[0:0.25:1.5]);
%xlabel('X (m)','Position',[0.4 4.4 1.56],'FontSize',[10]);ylabel('Y (m)','Position',[4.4 1.1 1.14],'FontSize',[10]);zlabel('Z (m)','FontSize',[10]);
xlabel('X (m)','Position',[0.4 4.4 0.78],'FontSize',[10]);ylabel('Y (m)','Position',[4.4 1.1 0.57],'FontSize',[10]);zlabel('Z (m)','FontSize',[10]);
text(0.39,0.92,0,'bottom','FontWeight','bold','FontSize',[12],'FontName','Arial');
text(0.47,0.07,0,'top','FontWeight','bold','FontSize',[12],'FontName','Arial');
daspect([1 1 0.4]); view(-157,32);grid on;box on;%daspect([1 1 0.75]);
colorbar


else
```

```
p = patch(isosurface(eX,eY,eZ,Cubor,isovalue));

isonormals(eX,eY,eZ,Cubor, p)

set(p, 'FaceColor','red', 'EdgeColor', 'none');axis([x1 x2 y1 y2 0 0.5*vel*et2]); %axis([x1 x2 y1 y2 0.5*vel*et1
0.5*vel*et2]);%axis tight;  %% RED

%%set(p, 'FaceColor',[.9 0 0], 'EdgeColor', 'none');axis([x1 x2 y1 y2 0 0.5*vel*et2]); %axis([x1 x2 y1 y2 0.5*vel*et1
0.5*vel*et2]);%axis tight;  %% RED (light)

%set(p, 'FaceColor', [.5 .5 .5], 'EdgeColor', 'none');axis([x1 x2 y1 y2 0 0.5*vel*et2]);     %% GRAY

xlabel('X (m)','Position',[2.5 -3 2],'FontSize',[8]);ylabel('Y (m)','Position',[-2.5 4 2],'FontSize',[8]);zlabel('Z
(m)','FontSize',[8]);


set(gca,'Xdir','normal','Ydir','reverse','Zdir','reverse','FontSize',[8],'XTick',[0:2:x2],'XTickLabel',[0:2:x2],'YTick',[0:2:y2],'Y
TickLabel',[0:2:y2]);

text(8.2,-3,4.5,'c)','FontWeight','bold','FontSize',[12],'FontName','Times');

daspect([1 1 dverz]); view(-150,60);grid on;box on;%view(30,60);

camlight

lighting phong;%lighting gouraud;%lighting flat;

%title('3D rendering','FontSize',[10]);

end
```

(7) For iso-surfaces using Tiff files

```
clear all
%process a bunch of TIFF files into a volume

filebase = 'e:\Audax';
startFrame = 1;
endFrame = 74;
%read frames, reduce size, show frames, and build volume
for i=startFrame:endFrame
    filename=[filebase, num2str(i,'%2d'),'.tif']
    temp=double(imresize(imread(filename), 0.5));
    slice(:,:,i) = (temp(:,:,1)==255) + 2*(temp(:,:,2)==255) +
4*(temp(:,:,3)==255);
    imagesc(slice(:,:,i));
    colormap('gray')
    drawnow
end
The display program computes an isosurface for each of the hand-traced
structures, smooths the resulting isosurfaces, colors them, then displays them
as a 3D object.
clf
clear all

%load the file with the preprocessed images
load 'c:\my documents\matlab\nichole\slice.mat'
%the original images were had-coloured by region to one of 7 colours
```

```
%patch smoothing factor
rfactor = 0.125;
%isosurface size adjustment
level = .8;
%useful string constants
c2 = 'facecolor';
c1 = 'edgecolor';

%build one isosurface for each of 7 different levels
%The "slice" matrix takes on one of 7 integer values,
%so each of the following converts the data to a binary
%volume variable then computes the isosurface between the
%1 and 0 regions

p=patch(isosurface(smooth3(slice==1),level));
reducepatch(p,rfactor)
set(p,c2,[1,0,0],c1,'none');

p=patch(isosurface(smooth3(slice==2),level));
reducepatch(p,rfactor)
set(p,c2,[0,1,0],c1,'none');

p=patch(isosurface(smooth3(slice==3),level));
reducepatch(p,rfactor)
set(p,c2,[1,1,0],c1,'none');

p=patch(isosurface(smooth3(slice==4),level));
reducepatch(p,rfactor)
set(p,c2,[0,0,1],c1,'none');

p=patch(isosurface(smooth3(slice==5),level));
reducepatch(p,rfactor)
set(p,c2,[1,0,1],c1,'none');

p=patch(isosurface(smooth3(slice==6),level));
reducepatch(p,rfactor)
set(p,c2,[0,1,1],c1,'none');

p=patch(isosurface(smooth3(slice==7),level));
reducepatch(p,rfactor)
set(p,c2,0.8*[1,1,1],c1,'none');

%lighting/image control
set(gca,'projection','perspective')
box on
lighting phong
light('position',[1,1,1])
light('position',[-1,-1,-1])
light('position',[-1, 1,-1])
light('position',[ 1,-1,-1])

%set relative proportions
daspect([1,1,1])
axis on
set(gca,'color',[1,1,1]*.6)

view(-30,30)

rotate3d on
```

For more on Matlab codes, see Leucci (2019).

GPR Traveltime Tomography: As seismic, the GPR data can also be acquired in a tomography mode using two antennas, one as transmitter and one as a receiver. In this case, an important factor is a resolution in the EM wave velocity distribution into the investigated materials. The resolution depends on the dimensions of the cells in which the investigated area is divided. This parameter is linked to the fundamental frequency of the EM wave that travels in the investigated area. The Fresnel ray theory suggests the minimum dimension of the object that we can detect by

$$RF = \frac{1}{2}\sqrt{\frac{d\,v}{f}}$$

where d is the distance between the transmitter and receiver antennas, v is the mean EM velocity of the investigated area, f is the central frequency of the transmitter antenna, and RF is the Fresnel's ray. For this reason, RF represents the minimum dimension of the cell.

GPR travel time tomography data can be acquired by positioning the transmitting antenna on one side of the medium and collecting data at several receiver depths in the opposite side of the medium. The transmitting antenna must then moved to a new position, and the data collected in the same manner. Repeating this process creates a dense array of intersecting ray paths, where each ray path represents the shortest path from the source to the receiver and is perpendicular to the EM wavefront (Fig. 4.55).

The main objective of the tomography is to know the internal structure of an object, starting from the data observed on the surface. Physical parameters are continuous functions of spatial coordinates and are generally linked to data by non-linear equations. The data, instead, are acquired on some surface, according to specific geometries, in a finite number of discrete points. The problem thus formulated is a classical inverse, non-linear, subdetermined problem (in which, that is, the number of equations is less than the number of unknowns) and which admits infinite solutions. The acquired data allow obtaining only exemplified models of the internal structure of the object, described by a finite number of parameters. To realise a possible model, which satisfies the experimental data, the inverse methods are mainly used. The reversal process is a problem that involves several aspects:

Mathematical, which concerns the choice of algorithm and mathematical parameters;
Physical, which includes the initial model and the a priori information. Among the different inversion methods, we will analyse first those of direct inversion and then the iterative ones.

For the numerical resolution of the problem, it is necessary, first of all, to discretise the space of the parameters, which are generally continuous functions. This is done by subdividing the area of investigation into square cells, of side $\Delta\lambda$ and assigning the values of the parameters in each node of the grid; if the latter consists, for example, of L rows and K columns, the values that the parameter assumes are $M = L \times K$. Once the discretisation has been performed, vectors of the type represent the data and parameters:

$$\mathbf{d} = [d_1, d_2, \ldots, d_N]^T, \ \mathbf{m} = [m_1, m_2, \ldots, m_M]^T$$

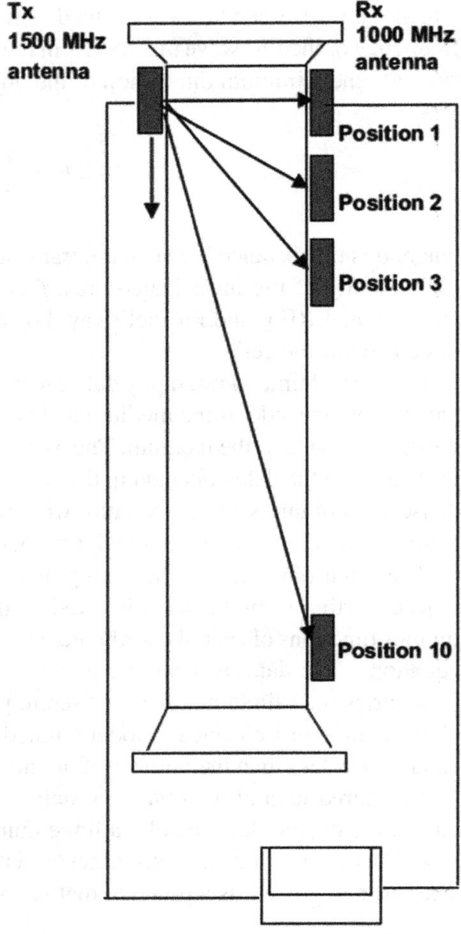

Fig. 4.55 Schemes of data acquisition for ground-penetrating radar traveltime tomography

where N is the data number and M the number of the parameters. A non-linear operator, g generally represent the link between the data and the parameters, called forward modelling operator, which depends on the physics of the problem and is such that:

$$\mathbf{d} = g(\mathbf{m}).$$

To solve the problem, it is necessary to formulate hypotheses on the operator g.

If we assume that the data are a linear function of the model parameters, the nonlinear operator, g, can be replaced by a matrix G, such that:

$$\mathbf{d} = \mathbf{Gm}.$$

Another hypothesis is represented by the request that the perturbations of the data be a linear function of those of the parameters of the model. Suppose that the observed data can be obtained from the action of the operator g on a model realised by applying a small perturbation Δm to an initial model, m_0:

$$\mathbf{d}_{oss} = g(\mathbf{m}_0 + \Delta \mathbf{m});$$

the synthetic data are, instead, those obtained by the application of g on m_0:

$$\mathbf{d}_{sint} = g(\mathbf{m}_0).$$

Developing the equation $\mathbf{d}_{oss} = g(\mathbf{m}_0 + \Delta \mathbf{m})$ in Taylor series around the value m_0 and stopping at the first order, is obtain:

$$g(\mathbf{m}_0 + \Delta \mathbf{m}) = g(\mathbf{m}_0) + \left. \frac{\partial g(\mathbf{m})}{\partial \mathbf{m}} \right|_{\mathbf{m}=\mathbf{m}_0} \Delta \mathbf{m};$$

using the equations $\mathbf{d}_{oss} = g(\mathbf{m}_0 + \Delta \mathbf{m})$ and $\mathbf{d}_{sint} = g(\mathbf{m}_0)$, we have:

$$\mathbf{d}_{oss} = \mathbf{d}_{sint} + \left. \frac{\partial g(\mathbf{m})}{\partial \mathbf{m}} \right|_{\mathbf{m}=\mathbf{m}_0} \Delta \mathbf{m},$$

or

$$\Delta \mathbf{d} = \mathbf{G}_0 \Delta \mathbf{m}$$

where $\Delta \mathbf{d} = \mathbf{d}_{oss} - \mathbf{d}_{sint}$ e $\mathbf{G}_0$ is the matrix of the partial derivatives of the synthetic data concerning the parameters of the model. Now we have a linear relationship between the perturbations of the data and those of the parameters of the model. The last above equation is formally identical and, therefore, the existence of one or more solutions of the linear or linearized problem depends on whether the problem is well-posed or not.

The physical properties of materials are continuous functions of the coordinate space; in inversion, we try to derive these functions from a finite number of measures, which leads to not having a single result.

The non-uniqueness of the solution may also be due to the limited capacity of the data to identify the model. In tomography problems, regions that do not have good coverage of the rays cannot be adequately resolved and therefore, the parameters estimated there might not be unique. Therefore, resolution and uniqueness are closely linked.

The solution depends continuously on the data (stability); a system is said to be unstable when small variations in the errors correspond to large variations in the solution, which may even reach values that are not physically acceptable. Consequently, a stable solution is not affected by the presence of small data errors. A system (Menke

1989), not very sensitive to the presence of a small number of strong errors (outliers), implies a solution that remains almost identical (robustness).

To resolve the inverse problem, for example, the software reflex uses the Simultaneous Iterative Reconstruction Techniques (SIRT) (Nolet 1987).

As an example, the results obtained on a column will be shown. Here the TT data were collected at source and receiver intervals of 0.05 m along the sides of the column. To track any fluctuations in the GPR transmitter output during data collection, a time zero calibration was conducted every eight transmitter depths. This involves placing the transmitting and receiving antenna a known distance apart in the air, which allows the user to determine the arrival time and to correct for system error. First-arrival time was used as data to invert for EM wave velocity propagation. The inversion domain consists of a discrete grid of 2D cells that are 0.05 m on both sides. These cell dimensions are a good choice because the source and receiver interval is 0.05 m higher than RF. The Reflexw software (Sandmeier 2013) was used to produce images of velocity. Reflexw employs simultaneous iterative reconstruction techniques (SIRT) and ray tracing methods to generate tomographic images. The reference model has the dielectric properties of the ($\varepsilon_r = 7$, $\mu = 1$, $\sigma = 0.1$ mS/m). The Dijkstra algorithm (Dijkstra 1995) allows performing ray tracing through this model.

The Dijkstra algorithm allows calculating ray paths by minimising the travel time between each transmitter-receiver pair. The resulting ray path satisfies Snell's law. An iterative process updated EM wave velocities by adjustments that move the calculated times towards the observed times. The iterations are stopped when the traveltime residuals were no longer decreased. Since EM waves tend to travel through high-velocity regions rather than low-velocity regions, the ray paths in an inhomogeneous media tend to curve around the low-velocity regions, and thus they are no longer straight lines. Therefore the curved ray inversion was used. The inversion results for the data set (Fig. 4.56) show a major high dielectric constant (ε_r) (low velocity) zone (labelled H and F3). In the zones labelled F1 and F2 the ε_r decreases (EM wave velocity increase).

The interpretation of a GPR profile dependent on the user's knowledge, skill and experience of Annan (2005).

The GPR profiles contain a series of "anomalies" that could have a precise meaning but could be related to a change in wave velocity along with buried interfaces.

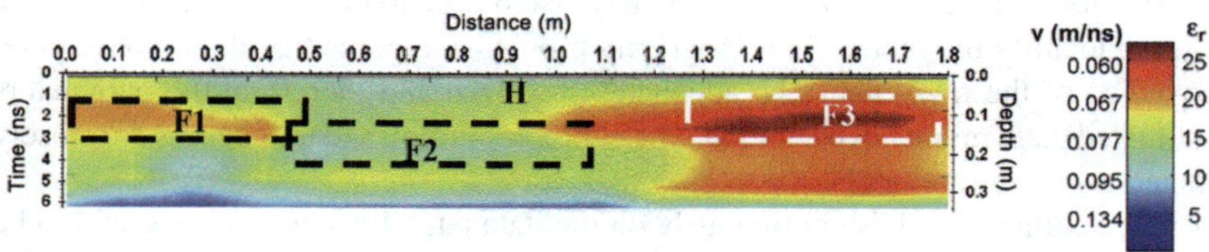

Fig. 4.56 GPR traveltime tomography: the electromagnetic velocity (v) and dielectric constant (ε_r) model distribution in the column

Any change in the electromagnetic properties within the surveyed material will produce reflections that are not necessarily anomalies. Therefore, one must be able to understand which of these reflected events can be linked to structures of interest for the preordained purposes of the GPR survey.

The first step in GPR data interpretation could be the creation of "synthetic models" of 2D-GPR profiles. Modelling can provide the interpreter with an idea of how real-world GPR reflection data should be done and therefore produce a more accurate interpretation of GPR reflection profiles once they have been processed.

Several software programs produce accurate data-reflection modelling. Among them are: GPRSim (http://www.gpr-survey.com/gprsim.html), GPRmax (http://www.gprmax.com/), and Reflexw (http://www.sandmeier-geo.de/reflexw.html). In the model construction, it is critical to define: (i) the electrical properties of the materials and boundaries that are to be included in the simulation, and (ii) the shape and dimension of the bodies that simulate a target and their depth. A real case can have infinitely variable dielectrics and conductivities located at all locations in the surveyed material (heterogeneity), but the model should contain an idea of the expected reflection events.

In the forensic application of GPR method, an important case is related to the presence of hidden bodies: that of a 2D-GPR profile acquired to cross the buried body in a hypothetical cavity. The 2D synthetic model was produced using GPRSim (Goodman 2013).

A model of stratified subsoil with the presence of a void contained a body was assumed (Fig. 4.57). Three homogeneous layers with dielectric constants of 14, 9, and 6 were modelled for surface soil, calcarenite and hard bedrock, respectively. In the calcarenite, a void space, with a dielectric constant of 1, with an inside buried body with a dielectric constant of 20 were modelled. The synthetic reflections demonstrated that the hidden body was defined. Other reflection events were generated at the contact between soils 1 and 2. Finally, lower amplitude reflections produced from the soil 3.

It is important to underline that, when radar energy is reflected from a buried interface where the wave velocity decreases, the polarity of the reflected wave will be the same as the direct-wave generated from the transmitting antenna (Conyers 2015a, b). However, if a drastic increase in velocity occurs at a boundary, such as a void space where propagating radar waves increase again to the velocity of light, a reflection will be generated that is visible in traces as a reversed-polarity sine wave (Conyers 2015a, b).

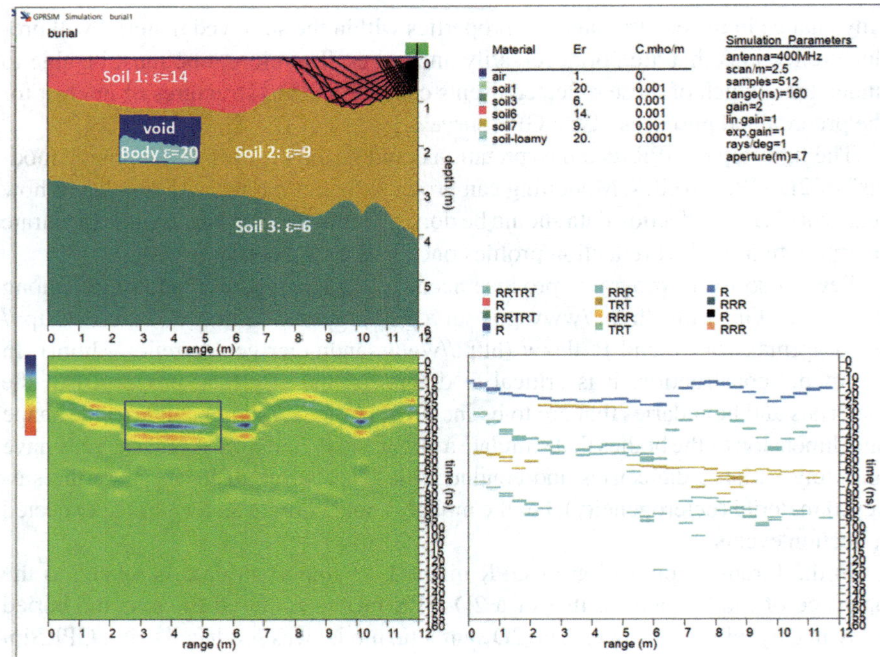

Fig. 4.57 Synthetic model: three homogeneous layers with a dielectric constant of 14, 9, and 6 with a modelled hidden body

Below a matlab code in order to calculate the volumetric water content distribution using the Topp formula:

```
close all
 clear all
% filename = input(' nome file dati:  ');
% n=input('numero righe ?');
% m=input('numero colonne ?: ');
% y=readmatr(filename,n,m);

n=3381; number of traces
m=481; number of samples
z_00=readmatr('secco.ASC',n,m); the name of the EM wave matrix

x=[1:m];
a1=min(min(z_00));
a2=max(max(z_00));
a=[a1 a2];

c=0.3; (EM wave velocity in the vacum space)
K_00=((c)*(c))./(z_00.*z_00); %dielectric constant matrix
x=[1:m];
m1=min(min(K_00));
m2=max(max(K_00));
m=[m1 m2];
```

```
w00=((-0.053)+(0.029*K_00)-((0.00055*(K_00.^2)))+((0.0000043*(K_00.^3))));  %topp
formula
x=[1:m];
c1=min(min(w00));
c2=max(max(w00));
c=[c1 c2];

figure(1);subplot(321);imagesc(z_00');caxis(a);colorbar;title('');set(gca,'YTick
labelMode','manual','YTick',[1 74 148 222 296 370 444 512],'YTicklabel',[0 10 20
30 40 50 60 70]);set(gca,'XTicklabelMode','manual','XTick',[1 100 200 300 400
500],'XTicklabel',[0 2 4 6 8 10]);
```

Also for GPR traveltime tomography is possible to create a model to allow an easy interpretation of real data.

The software REFLEXW allows choosing different procedures for solving the direct problem, which, given the complete model of electromagnetic constants and positions of sources and receivers, calculates the arrival times to the latter, using appropriate algorithms, which find the foundations in theory discussed in Chap. 1. The methods made available in the program are:

- Network ray tracing, which is based on the Dijkstra algorithm;
- FDM-Vidale, which is based on the solution of the iconal equation;
- FDM (Finite difference method), which derives directly from the discretisation of Maxwell's equations.

Since the first two methods allow the calculation of only the first arrivals, for what concerns the forward problem, will be used the FDM, which allows the calculation of the total wavefield and the times after the first arrivals. Therefore follows a detailed practical description of the finite difference method proposed by REFLEXW for the treatment of the forward problem and that concerning the data inversion.

Generation of synthetic radar sections: Before starting any simulation, a directory of the project to be analysed must be created. In this directory, some file folders will be automatically inserted, which will be filled up appropriately, during data processing.

The first step is to create a template. Once in the modeling option, Reflexw allows to set appropriate parameters, depending on the geometry of the investigated system and the type of survey carried out; for this purpose, select the desired wave-type (in this case, electromagnetic) and enter the dimensions of the explored area, which are plotted in a reference system (x, z). In the particular case of travel time tomography, the x coordinate is referred to the length of the investigated object, while z determines its thickness. Therefore, having established the outer edges of a two-dimensional model, it is possible to insert a certain number of discontinuities into it, consisting of stratifications or isolated objects of various shapes and sizes, which differ from the surrounding medium for the values assumed by some electromagnetic parameters, which the software allows to introduce. It is possible to assign to the individual parts of the model a dielectric constant ε_r, a relative magnetic permeability μ_r and an electrical conductivity σ, which, determine the propagation of an electromagnetic wave in the medium.

It is also possible to associate vertical gradients to the aforementioned physical quantities. Completed this procedure and assigned a filename to the model, with the option save the model, it will be automatically saved in "Model", one of the folders of the main directory, and will have a *.mod file extension. At this point, the simulation of the propagation of electromagnetic waves can be started. Therefore, start the calculation of theoretical travel times, via the FDM, the FDM Vidale or the Network ray tracing.

The simulation, based, in the electromagnetic case, on the direct discretisation of the Maxwell equations (see Chap. 2), will result in the total wavefield, as a function of the x, z and t coordinates. The program requires to specify the central frequency of the antenna, as well as the components of the electromagnetic field to be simulated and rasterises the model, according to the spatial increments ($\Delta x = \Delta z$) and temporal Δt, set by the user. The program automatically determines the critical values of these increments; the maximum value of Δt depends on the maximum velocity, v, of the wave in the middle, according to the relation:

$$\Delta t_{crit} \leq \Delta x / \left(\sqrt{2} v \right)$$

in which, "crit" stands for the critic.

The maximum value of the spatial increase is given by:

$$\Delta x = \frac{\lambda}{12} = \frac{1}{12} \frac{v}{f_c},$$

where f_c is the antenna frequency.

The model is thus discretised into several points given by:

$$N_x = \frac{x_{max}}{\Delta x} + 1 \quad e \quad N_z = \frac{z_{max}}{\Delta z} + 1.$$

Ultimately, the model will be represented by three matrices (one for ε_r, one for μ_r and one for σ), each of size Nx × Nz. Another parameter to be supplied to the software is the maximum time, tmax (expressed in nanoseconds), up to which to extend the simulation. In the case of transmitted waves, since the object of interest is the first arrivals, this time corresponds to that which the electromagnetic wave takes to travel the space between that particular source-receiver pair that is more distanced. If, for example, we consider a homogeneous model, with the simple geometry of equispaced sources and receivers, represented in Fig. 4.58, the maximum distance travelled is that indicated by the arrow and goes from S1 to R6; therefore, the time taken to cover this distance is given by:

$$t_{max} = \frac{1}{v} \cdot \sqrt{\overline{S_1 R_1}^2 + \overline{R_1 R_6}^2}.$$

In practice, it is preferable to enter a slightly longer time than the one calculated in the above equation, to be sure not to cut the interest signal. Furthermore, it is possible to insert absorption modes at the edges of the *.material to prevent the formation of significant reflections, which can disturb the signal of interest. To this end, a fictitious

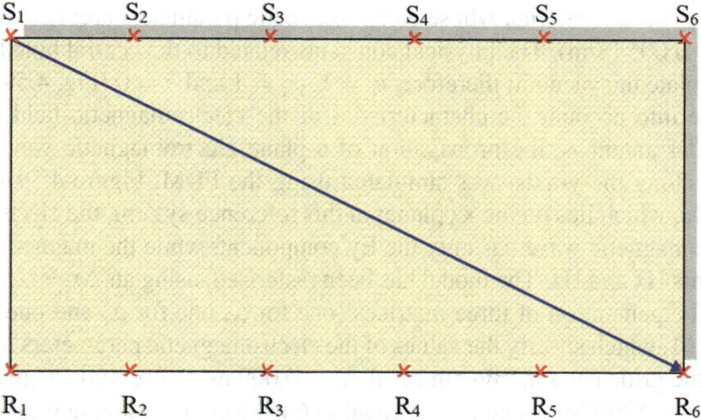

Fig. 4.58 Maximum distance between sources and receivers

zone is defined around the model, in which the conductivity is increased, up to a maximum value, starting from the one set in the modelling, linearly or exponentially. Once the type (point source) and the coordinates of the source to be simulated have been set, select the output parameters, according to the result of the simulation, which you want to store:

the total wavefield in all points (xi, zi) and for all times ti from 0 to tmax (total wavefield);

Etica, the synthetic radar section for all times ti 0 to tmax, corresponding to the position of the receivers (single line).

Several synthetic radar sections, corresponding to different geometries of sources and receivers (several single lines); in this case, a file of shots is required in input, in which the following information is reported in the order: coordinates (x, z) of the source, abscissa of the first receiver and the last receiver, ordered of the first receiver and the last receiver, the name of the input file. The receivers are considered equispaced in x with the same sampling step set in the rasterisation of the model. The generated field can be stored with a coarser sampling step, equal to:

$$\Delta t' = t_S \Delta t,$$

where t_S is an integer number called "time scaling". This greatly reduces the disk space required for storing files. This operation involves cutting all the frequencies above the new Nyquist frequency:

$$\Delta t' = 1/(2t_S \Delta t).$$

The simulation was performed considering a block of Lecce stone, with a square section of 30 cm side, with an empty square central hole of 10 cm side. In the synthetic model, this situation has been reproduced by assimilating the Lecce stone to a homogeneous material, whose electromagnetic parameters are: $\varepsilon_r = 8.85$, $\mu_r = 1$ and $\sigma = 0.005$ (S/m). The physical constants related to the central hole have been set to simulate the vacuum; therefore: $\varepsilon_r = 1$, $\mu_r = 1$ and $\sigma = 0$ (Fig. 4.59).

To take into account the characteristics of the electromagnetic field generated by the radar antennas, the propagation of a plane electromagnetic wave, linearly polarised along the y-axis, was simulated using the FDM. Figure 4.59 shows the survey area, which lies on the xz plane; in this reference system, the electric field of the electromagnetic wave has only the Ey component, while the magnetic field has components Hx and Hz. The model has been rasterised, using an $\Delta x = 0.50$ cm; this involves the generation of three matrices (one for ε_r, one for μ_r and one for σ), of size 61×61, which specify the values of the electromagnetic parameters in the 3721 nodes of the grid. The sampling interval $\Delta t = 0.009$ ns, the central frequency of the antenna, $fc = 1200$ MHz and a t_{max}, equal to 6 ns, were also supplied as inputs. An antenna with $fc = 1200$ MHz generates an electromagnetic wave of wavelength $\lambda = v/fc$, where v is the velocity of the wave in the medium; in this case, $v = 10.08$ cm/ns, and therefore $\lambda \cong 8$ cm.

Concerning Fig. 4.60, a name was given to each output file, which was indicative of the acquisition geometry. For example, if the source is located on the B side and the receivers on the C side, the file name will be BC, followed by a number representing the position of the source on the B side. The shots files are, in the case

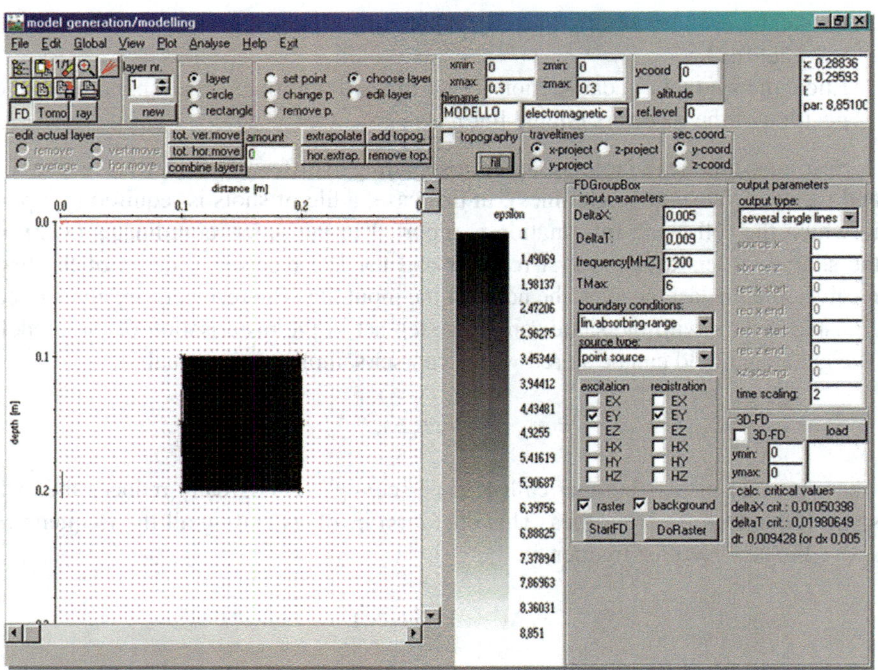

Fig. 4.59 Reflex window to calculate the FDM, relative to a homogeneous rasterised block model with an empty hole in the centre

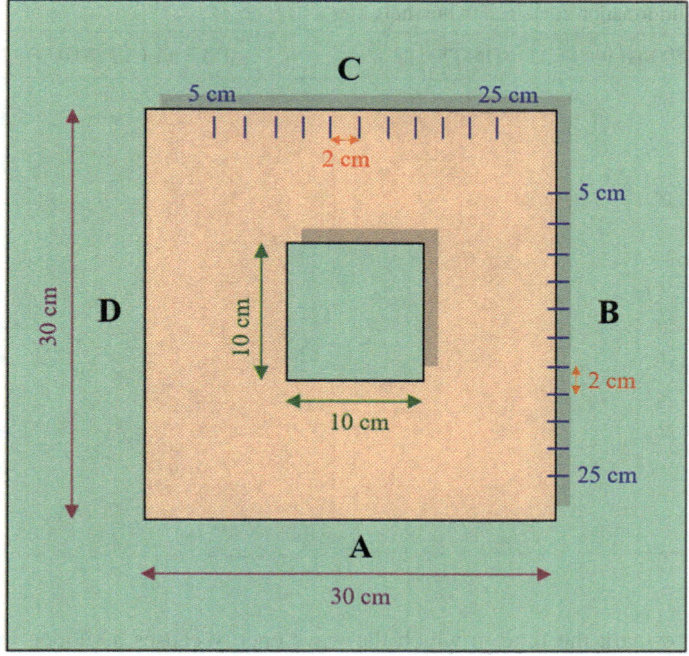

Fig. 4.60 Schematization of the homogeneous block with a hole in the centre, on whose sides sources and receivers are considered to be located

under examination, two: one is related to the configuration of sources and receivers orthogonal to each other (BC and BA), while the other contains the positions of the sources and receivers that are parallel to each other (CA and BD) (Table 4.3). At this point, the simulation can start, using the FDM, at the end of which, is possible to view and treat the results.

The *.dat files created using the FDM (Finite Differences Method)—according to the procedure described in the previous section—can be opened, one by one, in the 2D data-analysis, where they are displayed in a Cartesian reference system with the distance expressed in meters (in abscissa) and the times expressed in nanoseconds. In the simulation carried out, 44 output files were generated, each of which contains 61 tracks. The display can be made in one of the following two ways:

wiggle mode (in which, the individual tracks are positioned next to each other); point mode (in which, small bands alternate in different shades of colours, depending on the intensity of the signal).

In the case of tomography, you are interested in the first arrivals, which can be hit by activating the pick option; at this point, you can choose between manual pick, which allows you to select the first arrival times manually; phase follower, which automatically pings all impulses with the same phase. The correct to zero option is

Table 4.3 Information contained in the shots files

ORTHOGONAL SHOTS							PARALLEL SHOTS						
0.30	0.05	0.00	0.30	0.00	0.00	BC05	0.05	0.00	0.00	0.30	0.30	0.30	CA05
0.30	0.07	0.00	0.30	0.00	0.00	BC07	0.07	0.00	0.00	0.30	0.30	0.30	CA07
0.30	0.09	0.00	0.30	0.00	0.00	BC09	0.09	0.00	0.00	0.30	0.30	0.30	CA09
0.30	0.11	0.00	0.30	0.00	0.00	BC11	0.11	0.00	0.00	0.30	0.30	0.30	CA11
0.30	0.13	0.00	0.30	0.00	0.00	BC13	0.13	0.00	0.00	0.30	0.30	0.30	CA13
0.30	0.15	0.00	0.30	0.00	0.00	BC15	0.15	0.00	0.00	0.30	0.30	0.30	CA15
0.30	0.17	0.00	0.30	0.00	0.00	BC17	0.17	0.00	0.00	0.30	0.30	0.30	CA17
0.30	0.19	0.00	0.30	0.00	0.00	BC19	0.19	0.00	0.00	0.30	0.30	0.30	CA19
0.30	0.21	0.00	0.30	0.00	0.00	BC21	0.21	0.00	0.00	0.30	0.30	0.30	CA21
0.30	0.23	0.00	0.30	0.00	0.00	BC23	0.23	0.00	0.00	0.30	0.30	0.30	CA23
0.30	0.25	0.00	0.30	0.30	0.00	BC25	0.25	0.00	0.00	0.30	0.30	0.30	CA25
0.30	0.05	0.00	0.30	0.30	0.30	BA05	0.30	0.05	0.00	0.00	0.00	0.30	BD05
0.30	0.07	0.00	0.30	0.30	0.30	BA07	0.30	0.07	0.00	0.00	0.00	0.30	BD07
0.30	0.09	0.00	0.30	0.30	0.30	BA09	0.30	0.09	0.00	0.00	0.00	0.30	BD09
0.30	0.11	0.00	0.30	0.30	0.30	BA11	0.30	0.11	0.00	0.00	0.00	0.30	BD11
0.30	0.13	0.00	0.30	0.30	0.30	BA13	0.30	0.13	0.00	0.00	0.00	0.30	BD13
0.30	0.15	0.00	0.30	0.30	0.30	BA15	0.30	0.15	0.00	0.00	0.00	0.30	BD15
0.30	0.17	0.00	0.30	0.30	0.30	BA17	0.30	0.17	0.00	0.00	0.00	0.30	BD17
0.30	0.19	0.00	0.30	0.30	0.30	BA19	0.30	0.19	0.00	0.00	0.00	0.30	BD19
0.30	0.21	0.00	0.30	0.30	0.30	BA21	0.30	0.21	0.00	0.00	0.00	0.30	BD21
0.30	0.23	0.00	0.30	0.30	0.30	BA23	0.30	0.23	0.00	0.00	0.00	0.30	BD23
0.30	0.25	0.00	0.30	0.30	0.30	BA25	0.30	0.25	0.00	0.00	0.00	0.30	BD25

activated to mark the time in which the wave energy causes a sudden increase in amplitude.

Since the picking operation is carried out also on real data, in which the recognition of the first arrivals can sometimes be problematic, the program allows for correcting the results obtained by the automatic procedure, according to the judgment of the operator and his experience. Figure 4.61 shows the REFLEXW window relating to the picking phase of the first arrivals of file CA05; the display is set in wiggle mode to allow visualisation of the pulse shape. The data relating to picking must, therefore, be saved in a file with a .tom extension; the format is the one requested in input by the tomographic inversion program in two dimensions. The saved file can be displayed in a table, whose columns contain, in order, the following information: travel times (ns), coordinates (x, z) of the transmitting antenna and finally coordinates (x, z) of the receiving antenna.

The Tomography option allows for solving the inverse problem using the iterative SIRT method. Since the SIRT envisages the introduction of an initial model, a starting model is created, generally homogeneous, since, in the inversion of experimental data, presumably, one is not aware of the inhomogeneities within the investigated area. In radar tomography, the aim is to obtain, from the time of the first arrival, all possible information on the inhomogeneities presents within the investigated medium. The algorithm reverses a system of the type:

$$d = G\,m,$$

where, d is the vector of times of the first arrival, m of slowness and G the tomographic matrix, in which each element Gij represents the path of the i-th radius in the j-th

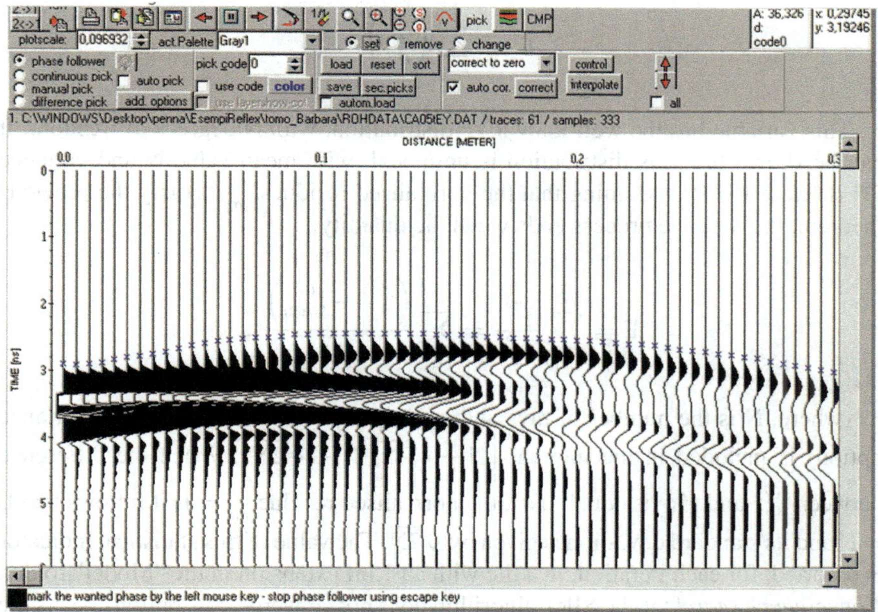

Fig. 4.61 Example of first arrival picking

cell of the grid. REFLEXW reverses the system $d = G\ m$ not as a function of slowness (defined as the reciprocal of velocities), but as a function of the dielectric constants. The parameters required by the reversal program are shown below, in which the criteria determining the end of the inversion process are indicated by ê Max. Iterations: denotes the maximum number of iterations, which the algorithm must execute. Max. Time: specifies the maximum time, in seconds, of the CPU (Central Processing Unit). Threshold indicates the numerical value of the threshold such that, if the residuals, satisfy the condition:

$$\frac{r_i^{(q)} - r_i^{(q-1)}}{r_i^{(q-1)} - r_i^{(q-2)}} < treshold,$$

the iteration process stops.

Default data-variance: it represents a measure of data uncertainty; in reality, it is necessary to insert the standard deviation, σd, of the data and not the variance, since the program automatically calculates the square.

Statistical criterium: by setting this option, a stop criterion is activated, which is based on certain statistical assumptions. Consider a set of N random, x_i, Gaussian, independent variables with the same variance σ_d^2; suppose we build a new random variable:

$$\chi^2 = \sum_{i=1}^{N} \left(\frac{x_i}{\sigma_d} \right)^2 .$$

This variable has the well-known χ^2 distribution, with N degrees of freedom. It can be shown that this distribution is unimodal, with mean value N and variance 2N (Menke 1989). Assuming that the normalised residues, $\frac{r_i}{\sigma_d}$, satisfy the previous theorem, the iterative process ends when the quantity:

$$E = \sum_{i=1}^{N} \frac{r_i^{(q)2}}{\sigma_d^2} = \sum_{i=1}^{N} \frac{\left(t_{oss_i}^{(q)} - t_{calc_i}^{(q)} \right)^2}{\sigma_d^2},$$

(where, N is the number of rays and σ_d is specified in the default data-variance option) is included in the interval $\left[N - \sqrt{2N}, \ N + \sqrt{2N} \right]$. When the difference between $t_{oss_i}^{(q)}$ and $t_{calc_i}^{(q)}$ is equal to σ_d, the normalised residue $\frac{r_i}{\sigma_d} = 1$; it follows that E = N and its standard deviation is given by $\sqrt{2N}$. The value of this statistical indicator is reported, for each iteration, in a file with an *.inf extension in the "Model" folder.
Convergence search: if the SIRT algorithm does not converge, this option checks the maximum number of changes in the model before stopping the procedure.
Model change A: this is a multiplicative factor of the slowness model when the SIRT algorithm converges. If the value is smaller than one, a greater number of iterations is necessary before obtaining the final model, but the search for convergence, being done in smaller steps, could lead to a better result; otherwise, the necessary steps are considerably reduced. However, the model change A must be a positive quantity.
Modelchange B: has the same function as the model change A, but is used in the case of non-convergence; in particular, if the SIRT algorithm does not converge after a certain number of iterations, the last of these is eliminated, while the previous step is repeated using the multiplicative factor in question.

The modelchange A and B are equivalent to the parameter ω of the SIRT family, for which convergence is theoretically demonstrated.
Max. Def. change: allows you to enter the maximum variation, expressed as a percentage, of the values of the current model compared to those of the initial model. This option must be set in such a way that the parameters of the model, at each iteration, fall within the real physical limits (for example, the dielectric constant, εr, must be between 1 and 80 because it has a physical meaning).
Ved Curved ray: if activated, the algorithm for calculating travel times uses curvilinear radii, rather than straight ones. The option can be deactivated; if you are dealing with a roughly homogeneous model; therefore, it is sufficient to use only straight rays. By rectilinear radii, it is meant that the path of the radius consists of the line segment connecting the source to the receiver. By selecting this option, the path of the rays is calculated at each iteration.
Start curved ray: the number of the iteration is inserted, to which you want to start the calculation of travel times by using curved rays. To minimise the calculation time,

the program provides a "mixed" approach, in which, in the first iterations, straight radii are used, while, starting from a certain point, curvilinear ones are used.

Average x: allows you to specify the number of cells, in the x-direction, which will automatically average the result of each iteration.

Average z: it is similar to the previous one, in the z-direction.

Min. Velocity: indicates the minimum value of the speed used in the inversion.

Max. Velocity: a maximum speed value is entered.

Check no ray area: by activating this option, the region of the model that is not covered by the rays is considered separately; in this area, model changes are calculated by averaging the data contained in the surrounding cells.

Force 1. Iter: if activated, the model obtained after the first iteration is taken as the starting one for the next one; in this way, the algorithm is subjected to a forcing, to change, in any case, the initial model.

Using the load data option, the input set consisting of data relating to travel times previously loaded is loaded; the initial model is discretised, and the start tomography is performed. For the duration of the inversion, two windows are displayed simultaneously:

That of Processing status updates the number of iterations and, at the end of the process, provides information on the outcome of the inversion (i.e., whether convergence has been achieved or not) and the number of iterations carried out;

That is about the visualisation of the model obtained after each iteration. The final image is relative to the last model, in the case of convergence; when the algorithm does not converge, the program displays the best model obtained and the number of the corresponding iteration.

It is possible to have further and more detailed information through the .inf file, which has already been discussed, and which also indicates the best iteration relating to the final model, the result of the inversion.

The outcome of the tomographic inversion depends on a very high number of parameters which, if not correctly set, can compromise the validity of the final model; this required the execution of multiple tests, to verify the influence of each of these.

The picking of the synthetic radar sections constitutes the experimental data to be inverted. Figure 4.62 shows the realisation of a homogeneous geometric model, flanked by the Tomography Group Box, containing all the parameters previously described.

In tomographic problems, the path of the rays can be considered rectilinear if and only if particular conditions are valid, dictated by Dines and Lytle (1979) in the course of their studies concerning the subject in question. In particular, there are two conditions, under which the hypothesis of straight rays is to be considered valid:

- the refractive index of the investigated medium varies slowly; that is when, in the investigated medium, there is a small speed gradient. Also, Dines and Lytle (1979) have shown that, for the radii to be considered straight, the speed variations should not exceed 15%.
- the distance between source and receiver is greater than $\lambda/2\pi$, where λ is the length of the electromagnetic wave in the middle.

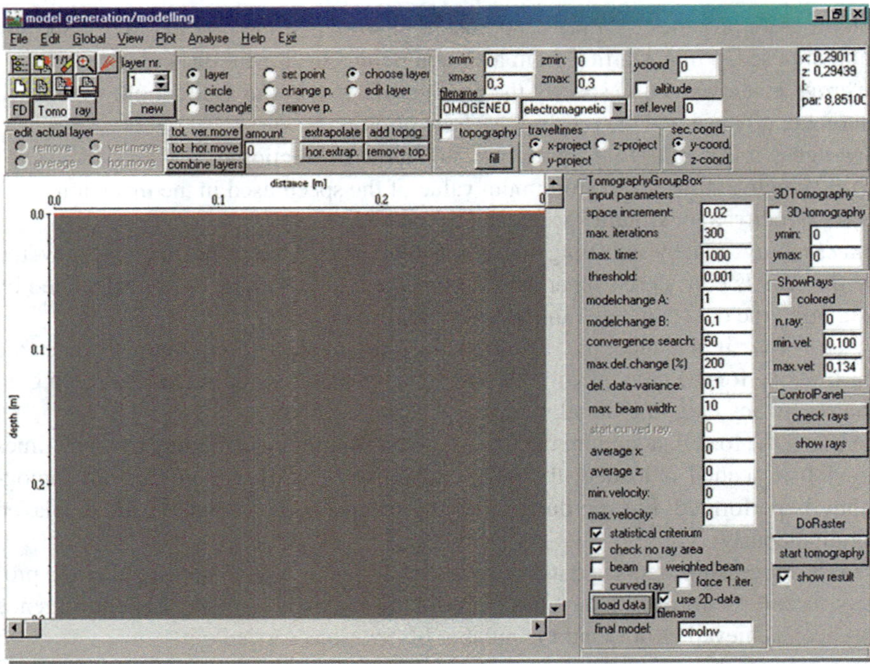

Fig. 4.62 REFLEXW window related to a homogeneous rasterised block ($\varepsilon_r = 8.85$), used as an initial model for the inversion

In the case in question the distance between source and receiver is indeed greater than $\lambda/2\pi$ ($\cong 1$ cm), but the refractive index of the medium undergoes abrupt variations; however, from the statistical point of view, the region of inhomogeneity constitutes approximately 10% of the survey area and, therefore, propagation takes place mainly in a homogeneous medium.

The remaining parameters are the default ones. The result of the inversion (obtained after four iterations) is shown in Fig. 4.63. In the visualisation phase, to the result of the inversion, the contours of the original model were superimposed, consisting of the homogeneous block with the empty central hole (dotted line in Fig. 4.63).

Note how, qualitatively, the tomographic inversion returns a central zone with a lower dielectric constant (4–5), surrounded by a region with an ε_r that varies between 5 and 10, although, on the left side, an artefact appears with low values of that above constant. Quantitatively, on the other hand, satisfactory restitution of the central anomaly is not obtained, either in terms of form or terms of ε_r values.

Acting on the average x and z option, which allows mediating the outcome of each iteration, (for example, on two cells), an improvement of the inversion result is obtained (Fig. 4.64). In this new view, compared to Fig. 4.63, the range of the ε_r has been restricted to an interval ranging from 3.72 to 10. The minimum and maximum speeds, automatically calculated by the program in the instant in which the input set

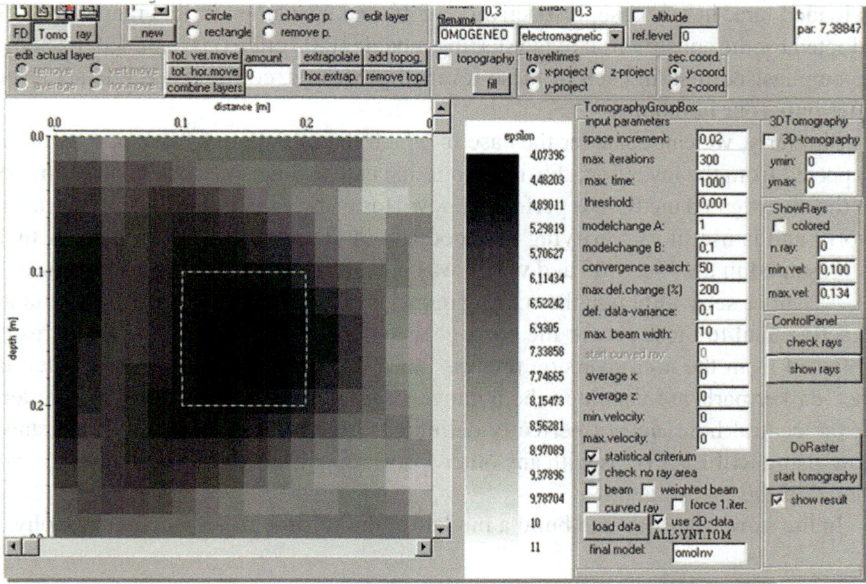

Fig. 4.63 Result of the inversion, starting from a homogeneous block ($\varepsilon_r = 8.85$), with the cells of the tomographic grid of size equal to 2 cm

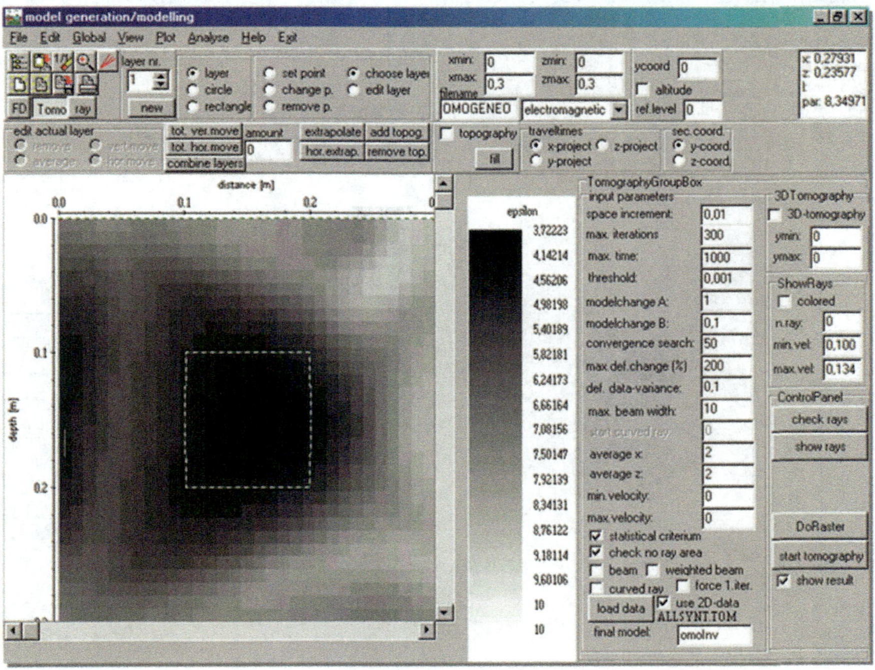

Fig. 4.64 Result of the inversion, starting from a homogeneous block ($\varepsilon_r = 8.85$), with the cells of the tomographic grid equal to 1 cm and average parameters equal to 2

is loaded, are, in any case, respectively, of 10 and 13 cm/ns; in this range, the one adopted for the construction of the starting model (10.08 cm/ns) is included. It should be emphasised that, in each of these reversal processes, convergence was achieved after only four iterations.

Similarly, we can consider the case in which the initial model is exactly what is expected from the inversion; this is a proof that can be done only when the geometry of the investigated medium is perfectly known (as in the situation treated in this thesis work) and is useful for verifying the goodness of the result of the inversion, which is presented in the Fig. 4.65 and which was obtained, again, after four iterations.

As can be seen, although the starting model is characterised by physical constants exactly equal to those set for the homogeneous block with a central hole the result obtained from the non-inversion process faithfully reproduces the initial model; in fact, darker parts are visible in the homogeneous area of the block, and the borders of the central hole are not perfectly delimited. All this can be used to understand how, using REFLEXW, as with any other software, you can only get an approximate model of the real situation.

In the same way, one can build a model for the seismic traveltime tomography.

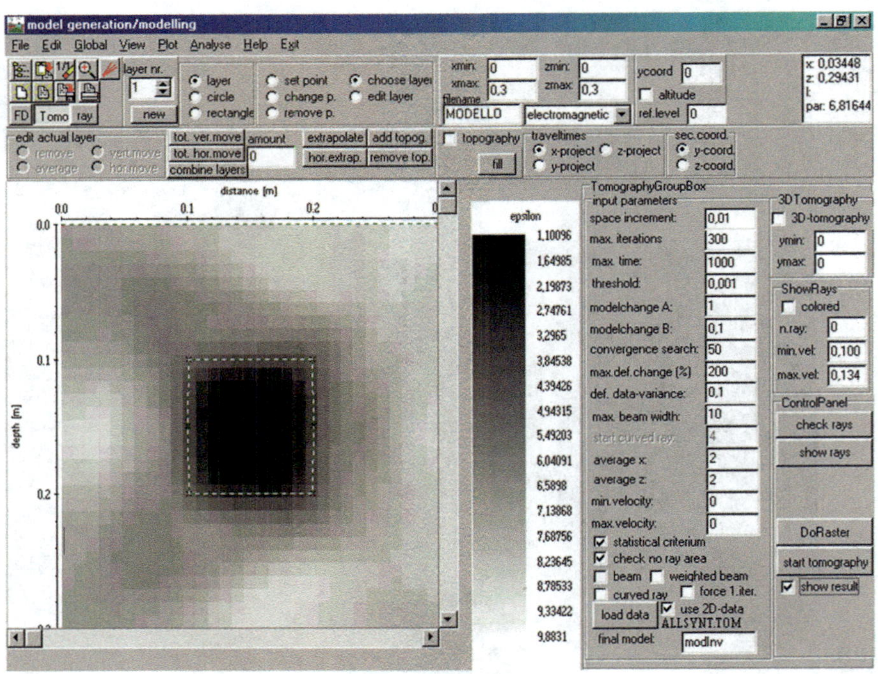

Fig. 4.65 Result of the inversion, starting from a homogeneous block ($\varepsilon_r = 8.85$) with a central hole ($\varepsilon_r = 1$), with the cells of the tomographic grid equal to 1 cm and average parameters equal to 2

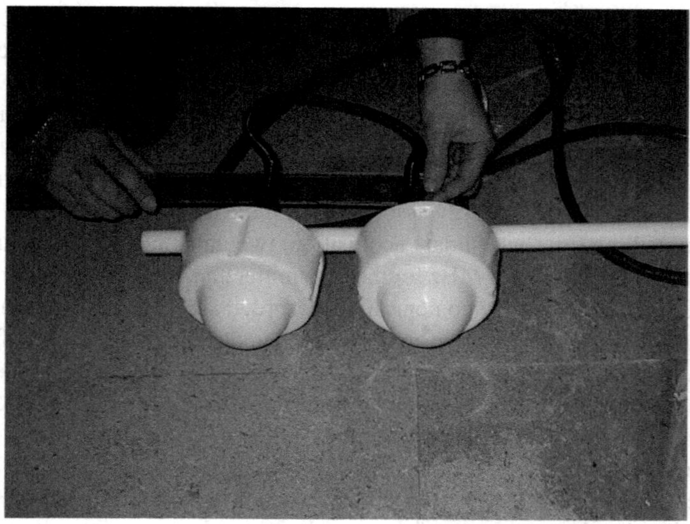

Fig. 4.66 The 1800 MHz antenna

At this point, the real data were considered. Data were acquired using two antennas, the 1000 MHz antenna used as transmitter and 1800 MHz (Fig. 4.66) antenna used as a receiver.

The chose of the 1000 MHz antenna as a transmitter, was due to the fact to be sure that the signal crossed the investigated object since the depth of penetration of an electromagnetic wave is inversely proportional to its frequency.

The use of antennas of different frequencies was determined by the fact that the 1800 MHz antenna consists of two elements, as shown in Fig. 4.66, joined together by a cable of fixed length (about 60 cm), which physically prevented the use of the antenna pair, for tomographic investigation. The radar antennas are essentially dipole antennas, designed to generate a unicycle pulse of duration $T_c = 1/f_c$, where f_c is the central frequency of the antenna, which has a band length $\Delta f = f_c$. For example, an antenna with a centre band frequency of 1800 MHz, should emit pulses with a frequency between 900 and 2700 MHz, with a distribution like the one shown in Fig. 4.67.

Fig. 4.67 The ideal frequency distribution for an 1800 MHz antenna

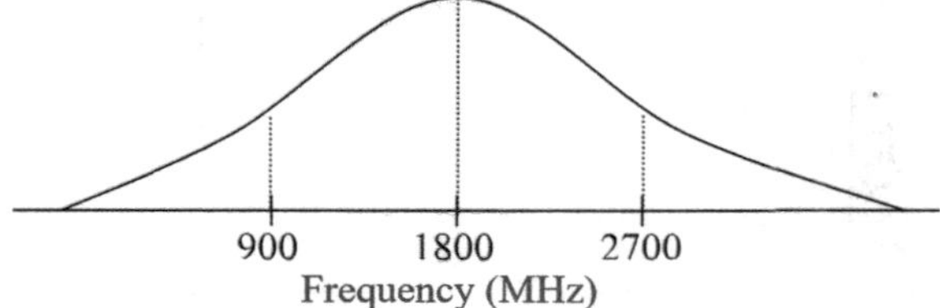

In reality, the frequency spectrum of the acquired traces is presented as an asymmetric distribution, with many peaks, around a dominant frequency, which, sometimes, does not coincide with the nominal frequency of the antenna. The variations of the dominant frequency are caused by multiple factors, which depend on the electromagnetic characteristics of the medium crossed. When the radar pulse propagates in the ground, the centre band frequency can be moved below the dominant one, due to a dissipation of the energy associated with the high-frequency tail. To this, it must be added the fact that, in data acquisition, two antennas with different central frequencies were used (1000 MHz for the transmitter and 1800 MHz for the receiver) and, therefore, the frequency content of the recorded signal depends on the characteristics of both. Since the resolution of the method depends on the centre frequency of the recorded signal, it is necessary to determine it. By selecting, therefore, the ASCII files related to the actual spotted times, the average frequency spectrum of all the tracks were calculated, taking into consideration the entire range of times. Figure 4.68 shows an example.

The result in Fig. 4.68 shows a peak corresponding to around 1000 MHz, which confirms that the characteristics of the signal sent by the transmitting antenna have not undergone substantial changes in crossing the material.

Four blocks of Lecce stone represent the object of investigation—originating from a processing company of the material mentioned above, appropriately assembled, to

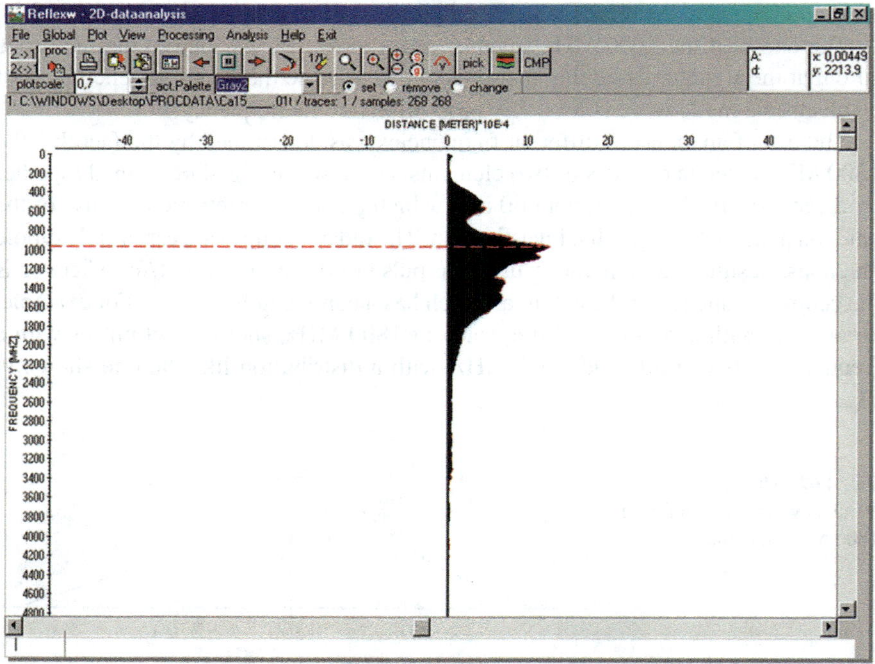

Fig. 4.68 The average spectrum corresponding to 61 tracks, in the time interval between 0 and 20 ns

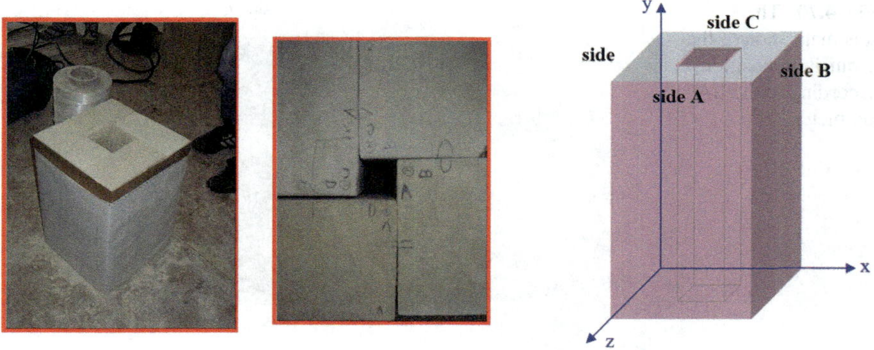

Fig. 4.69 The result of the assembly of four blocks of Lecce stone, which is the object of tomographic investigation

produce a simple geometry, that is, a single square section block side 30 cm, with a central empty hole with a square section of 10 cm side, as shown in Fig. 4.69.

Before starting the phase of the measurements, the instrumentation was set to guarantee a correct data acquisition (the time-window set at 20 ns, the gain function (to amplify the signal) and to the number of samples per track—512). This last parameter, therefore, takes into account the time taken by the electromagnetic wave to travel, inside the cables, from the control unit to the transmitter and, from the surface of the investigated medium to the receiver. It should also be pointed out that the antenna must be dragged as slowly as possible (at an almost constant speed) and above all away from the cables, to obtain a good sampling of the signal and good coverage of the investigated medium. Also, the manual marker (which is used to record the position of the antenna along the survey surface during data acquisition) must be emitted whenever the exact half of the antenna crosses the marked point (which, in the case of question, it is fixed every 5 cm). To get an idea of how the measurements on the block were performed, consider Fig. 4.69.

Since the focus was on a tomographic analysis in two dimensions, a median plane was investigated at the height of about 50 cm from the ground, whose perimeter is delimited by the red line indicated in Fig. 4.70.

To understand the acquisition mode, keep in mind Fig. 3.7: once the transmitting antenna (1000 MHz) is positioned on the face B of the block, a 5 cm from the edge and at the red line, the receiving one (1800 MHz) is dragged—at the same height—for the entire length of side A (30 cm), in the x-direction. Subsequently, the transmitter is moved 2 cm—always on the face B, while the receiver travels the entire A-side once more; iterate this procedure until the transmitting antenna reaches the position at 25 cm from the edge, thus covering, on the B side, a total distance of 20 cm. This configuration is called BA (BA05, …, BA25) and is, therefore, characterised by 11 positions of the springs spaced out progressively by 2 cm at a time. Similarly, repeating the spacing of the transmitting antenna described above, again on the face B, the receiver is dragged along the entire side C and, subsequently, along the side

Fig. 4.70 The schematisation of the acquisition position, according to the BA geometry

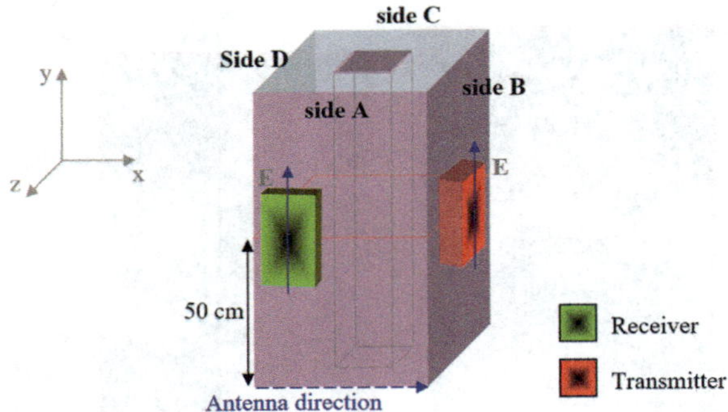

D, thus giving rise to the configurations BC and BD, which, like that previous, they are again distinguished by 11 positions of the sources, always spaced from each other by 2 cm. As for the CA configuration, the transmitting antenna is, as before, progressively shifted by 2 cm at a time, but on the C side, while reception takes place over the entire length of the A-side; also, in this case, the positions assumed by the transmitter are 11. The acquisition phase is shown in Fig. 4.71.

Given the limited dimensions of the object investigated, it was decided to use acquisition in continuous rather than by points, to obtain a good resolution and avoid progressively positioning the receiver, since it would have been complicated to obtain, in a precise manner, extremely small movements. Moreover, the small size of the block has caused the transmission of the signal, between transmitter and receiver, into the air, which has compromised, as will be seen later, the interpretation of the radar profiles. In this context, the simulation of synthetic data has been of great help to identify the wave of real interest. In Fig. 4.72, some of the experimental data are shown, acquired in the four configurations.

Figure 4.73 shows, on the left, the picking of an experimental data file and, on the right, a synthetic file, both related to one of the most unfavourable configurations (BC 07).

Fig. 4.71 The acquisition phase, in which the transmitting antenna is held fixed on one side, while the receiving one is dragged along a face of the block

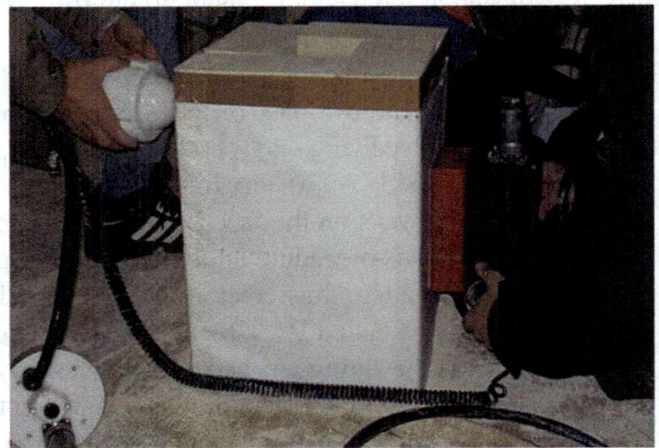

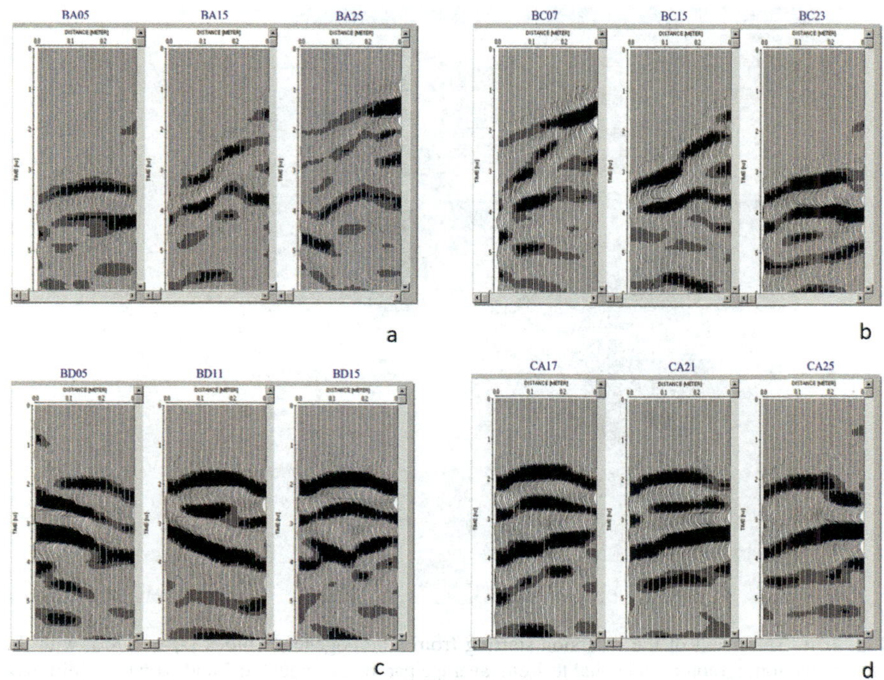

Fig. 4.72 The experimental data acquired according to the geometry **a** BA05, BA15 and BA25; **b** BC07, BC15 and BC23; **c** BD05, BD11 and BD15; **d** CA17, CA21 and CA25

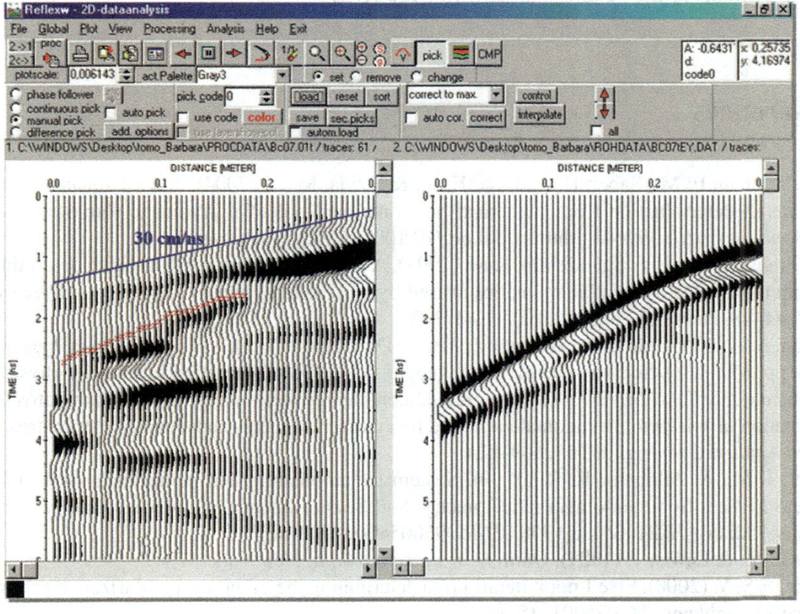

Fig. 4.73 The example of BC 07 file picking related to experimental and synthetic data

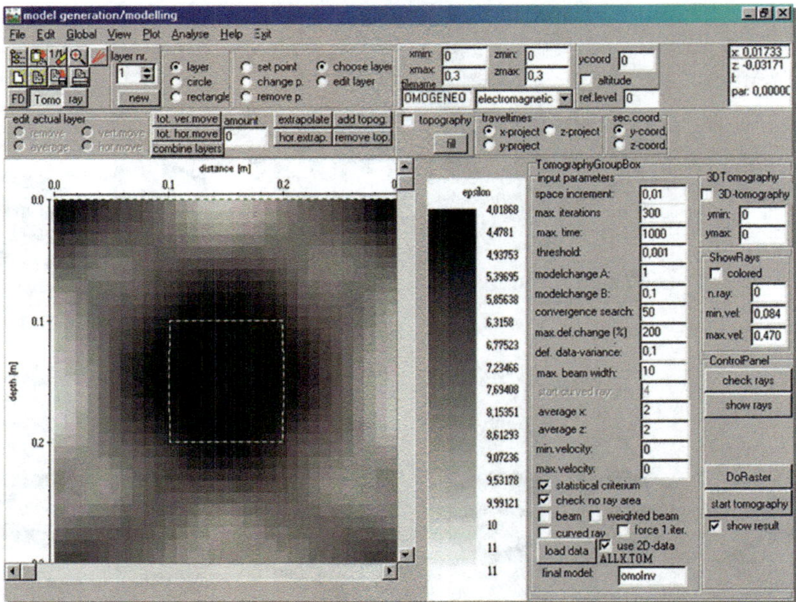

Fig. 4.74 The result of the inversion starting from a homogeneous block ($\varepsilon_r = 8.85$), with the cells of the tomographic grid equal to 1 cm, average parameters equal to 2 and with the addition of symmetrisation

As expected, the result is shown in Fig. 4.74.

References

Abdelrahman, E. M., Saber, H. S., Essa, K. S., & Fouda, M. A. (2004). A least-squares approach to depth determination from numerical horizontal self-potential gradients. *Pure and Applied Geophysics, 161,* 399–411. https://doi.org/10.1007/s00024-003-2446-5.

Abdelrahman, E. M., & Sharafeldin, S. M. (1997). A least-squares approach to depth determination from residual self-potential anomalies caused by horizontal cylinders and spheres. *Geophysics, 62,* 44–48. https://doi.org/10.1190/1.1444143.

Annan, A. P. (2005). Ground-penetrating radar. In D. K. Butler (Ed.), *Near surface geophysics* (Vol. 13, pp. 357–438). Society of exploration geophysicists: Tulsa, Investigations in Geophysics.

Asfahani, J., & Tlas, M. (2005). A constrained nonlinear inversion approach to quantitative interpretation of self-potential anomalies caused by cylinders, spheres and sheet-like structures. *Pure and Applied Geophysics, 162,* 609–624.

Baker, J. M., & Allmaras, R. R. (1990). System for automating and multiplexing soil moisture measurement by time-domain reflectometry. *Soil Science Society of America Journal, 54,* 1–6. https://doi.org/10.2136/sssaj1990.03615995005400010001x.

Barnett, V., & Lewis, T. (1984). *Outliers in statistical data.* New York: Wiley.

Bazán, F. S. V. (2008). Fixed-point iterations in determining the Tikhonov regularization parameter. *Inverse Problems, 24,* 035001. 15 pp.

Becker, H. (1995). From nanotesla to picotesla—A new window for magnetic prospecting in archaeology. *Archaeological Prospection, 2*, 217–228.

Cassidy, N. J. (2009). Ground penetrating radar data processing, modelling and analysis. In H. M. Jol (Ed.), *Ground penetrating radar: Theory and applications* (pp. 141–176). Amsterdam: Elsevier.

Ciminale, M., & Loddo, M. (2001). Aspects of magnetic data processing archaeological prospection. *Archaeological Prospection, 8*, 239–246.

Claerbout, J. F., & Muir, F. (1973). Robust modeling with erratic data. *Geophysics, 38*, 826–844.

Clark, D. A., & Emerson, D. W. (1991). Notes on rock magnetization characteristics in applied geophysical studies. *Exploration Geophysics, 22*(3), 547–555.

Constable, S. C., Parker, R. L., & Constable, C. G. (1987). Occam's inversion: A practical algorithm for generating smooth models from electromagnetic sounding data. *Geophysics, 52*, 289–300.

Conyers, L. B. (2015a). Analysis and interpretation of GPR datasets for integrated archaeological mapping. *Near Surface Geophysics, 13*, 645–651.

Conyers, L. B. (2015b). Multiple GPR datasets for integrated archaeological mapping. *Journal of Near Surface Geophysics, 13*(3).

Dalton, F. N., & van Genuchten, M Th. (1986). The time-domain reflectometry method for measuring soil water content and salinity. *Geoderma, 38*, 237–250.

Day-Lewis, F. D., Chen, Y., & Singha, K. (2007). Moment inference from tomograms. *Geophysical Research Letters, 34*, L22404. https://doi.org/10.1029/2007gl031621, 6 p.

Day-Lewis, F. D., Singha, K., & Binley, A. M. (2005). Applying petrophysical models to radar traveltime and electrical resistivity tomograms—Resolution-dependent limitations. *Journal of Geophysical Research, 110*, B08206. https://doi.org/10.1029/2004jb005369, 17 p. 37.

de Groot-Hedlin, C., & Constable, S. (1990). Occam's inversion to generate smooth, two dimensional models from magnetotelluric data. *Geophysics, 55*(12), 1613–1624.

Dijkstra, E. W. (1995). A note on two problems in connection with graphs. *Numeriskche Mathematik, 1*, 269–271.

Dines, K., & Lytle, J. R. (1979). Computerized geophysical tomography. In *Proceedings of the IEEE* (Vol. 67, No. 7), Luglio.

Drahor, M. G. (2004). Application of the self-potential method to archaeological prospection: Some case histories. *Archaeological Prospection, 11*(2), 77–105.

Eder-Hinterleiter, A., Neubauer, W., & Melichar, P. (1996). Restoring magnetic anomalies. *Archaeological Prospection, 3*, 185–197.

El-Araby, H. M. (2004). A new method for complete quantitative interpretation of self-potential anomalies. *Journal of Applied Geophysics, 55*, 211–224. https://doi.org/10.1016/j.jappgeo.2003. 11.002.

Essa, K., Mehanee, S., & Smith, P. D. (2008). A new inversion algorithm for estimating the best fitting parameters of some geometrically simple body to measured self-potential anomalies. *Exploration Geophysics, 39*, 155–163.

Evett, S. R. (2000). The TACQ computer program for automatic time domain reflectometry measurements: II. Waveform interpretation methods. *Transactions of the ASAE, 43*, 1947–1956. https://doi.org/10.13031/2013.3100.

Fedi, M., & Abbas, M. (2013). A fast interpretation of self-potential data using the depth from extreme points method. *Geophysics, 78*(2), E107–E116. https://doi.org/10.1190/geo2012-0074.1.

Fitterman, D. V., & Corwin, R. F. (1982). Inversion of self-potential data from the Cerro Prieto geothermal field, Mexico. *Geophysics, 47*, 938–945.

Goodman, D. (2013). *GPR sim manual.* http://www.gprsurvey.com/. Accessed June 2013.

Gouveia, W. P., & Scales, J. A. (1997). Resolution of seismic waveform inversion–Bayes versus Occam. *Inverse Problems, 13*, 322–349.

Grant, F. S., & West, G. F. (1965). *Interpretation theory in applied geophysics.* New York: McGraw-Hill.

Hammer, S. (1965). Terrain corrections for gravimeter stations. *Geophysics, 4*, 184–194.

Hansen, P. C. (2001). The L-curve and its use in the numerical treatment of inverse problems. In P. Johnston (Ed.), *Computational inverse problems in electrocardiology* (pp. 119–142). Southampton: WIT Press. (invited chapter).

Hawkins, D. M. (1980). *Identification of outliers*. New York: Chapman and Hall.

Kitanidis, P. K. (1997). The minimum structure solution to the inverse problem. *Water Resources Research, 33*(10), 2263–2272.

Leucci, G. (2015). *Geofisica Applicata all'Archeologia e ai Beni Monumentali* (p. 368). Palermo: Dario Flaccovio Editore. ISBN: 9788857905068.

Leucci, G. (2019). *Nondestructive testing for archaeology and cultural heritage: A practical guide and new perspectives*. Springer International Publishing.

Leucci, G., & De Giorgi, L. (2015). 2D AND 3D seismic measurements to evaluate the collapse risk of cave in soft carbonate rock. *Central European Journal of Geosciences, 7*(1), 84–94. https://doi.org/10.1515/geo-2015-0006.

Leucci, G., De Giorgi, L., & Scardozzi, G. (2014). Geophysical prospecting and remote sensing for the study of the San Rossore area in Pisa (Tuscany, Italy). *Journal of Archaeological Science, 52*, 256–276. https://doi.org/10.1016/j.jas.2014.08.028.

Leucci, G., Greco, F., De Giorgi, L., & Mauceri, R. (2007). Three-dimensional image of seismic refraction tomography and electrical resistivity tomography survey in the castle of Occhiola (Sicily, Italy). *Journal of Archaeological Science, 34*, 233–242. https://doi.org/10.1016/j.jas.2006.04.010.

Longman, I. M. (1959). Formulas for computing the tidal accelerations due to the Moon and the Sun. *Journal Geophysical Research, 64*(12), 2351–2355. https://doi.org/10.1029/JZ064i012p02351.

Menke, W. (1989). *Geophysical data analysis: Discrete inverse theory*. International Geophysics Series (Vol. 45, 1989). Academic Press.

Monteiro Santos, F. A. (2010). Inversion of self-potential of idealized bodies anomalies using particle swarm optimization. *Computers & Geosciences, 36*, 1185–1190.

Musset, A. E., & Khan, M. A. (2000). *Looking into the Earth*. New York, USA: Cambridge University Press.

Nettleton, L. L. (1976). *Gravity and magnetics in oil prospecting* (2nd ed). New York: McGraw-Hill, Publishing Co.

Nolet, G. (1987). *Seismic tomography with applications in global seismology and exploration geophysics*. D. Reidel Publishing Company.

Nuzzo, L., Leucci, G., Negri, S., Carrozzo, M. T., & Quarta, T. (2002). Application of 3d visualization techniques in the analysis of GPR data for archaeology. *Annals of Geophysics, 45*(2), 321–337.

Or, D., Jones, S. B., Van Shaar, J. R., Humphries, S., & Koberstein, L. (2004). User's guide WinTDR. Version 6.1. Utah State University, Logan. http://www.usu.edu/soilphysics/wintdr/documentation.cfm.

Patella, D. (1997). Self-potential global tomography including topographic effects. *Geophysical Prospecting*, 843–863.

Reginska, T. (1996). A regularization parameter in discrete ill-posed problems. *SIAM Journal on Scientific Computing, 3*, 740–749.

Revil, A., Ehouarne, L., & Thyreault, E. (2001). Tomography of self-potential anomalies of electrochemical nature. *Geophysical Research Letters, 28*, 4363–4366.

Sandmeier, K. J. (2013). *Reflexw 7.0 manual*. Karlsruhe: Sandmeier Software.

Sasaki, Y. (1992). Resolution of resistivity tomography inferred from numerical simulation. *Geophysical Prospecting, 40*, 453–464.

Schwartz, R. C., Casanova, J. J., Bell, J. M., & Evett, S. R. (2013). A reevaluation of time domain reflectometry propagation time determination in soils. *Vadose Zone Jounal, 13*, 1–13. https://doi.org/10.2136/vzj2013.07.0135.

Scollar, I., Tabbagh, A., Hesse, A., & Herzog, I. (1990). *Archaeological prospecting and remote sensing*. Cambridge: Cambridge University Press.

Tarantola, A. (1987a). *Inverse problem theory and methods for data fitting and model parameter estimation* (613 p.). Amsterdam: Elsevier.

Tarantola, A. (1987b). *Inverse problem theory* (613 pp.). Elsevier, New York.

Telford, W. M., Geldart, L. P., & Sheriff, R. E. (1990). *Applied geophysics*. New York: Cambridge University Press.

Tikhonov, A. N., & Arsenin, V. Y. (1977a). *Solutions of ill-posed problems*. Washington DC: Winston and Sons.

Tikhonov, A. N., & Arsenin, V. Y. (1977b). *Solutions of ill-posed problems*. New York: Wiley.

Tikhonov, A. N., Leonov, A. S., Yagola, A. G. (1998). *Nonlinear ill-posed problems* (Vols. 1 and 2). London: Chapman and Hall.

Topp, G. C., Davis, J. L., & Annan, A. P. (1980). Electromagnetic determination of soil water content: measurements in coaxial transmission lines. *Water Resources Research, 16*(3), 574–582.

Topp, G. C., Yanuka, M., Zebchuk, W. D., & Zegelin, S. (1988). Determination of electrical conductivity using time domain reflectometry: Soil and water experiments in coaxial lines. *Water Resources Research, 24*, 939–944.

Tripp, A. C., Hohmann, G. W., & Swift, C. M., Jr. (1984). Two-dimensional resistivity inversion. *Geophysics, 49*(10), 1708–1717.

Vasco, D. W., Datta-Gupta, A., & Long, J. C. S. (1997). Resolution and uncertainty in hydrologic characterization. *Water Resources Research, 33*(3), 379–397.

Yanuka, M., Topp, G. C., Zegelin, S., & Zebchuk, W. D. (1988). Multiple reflection and attenuation of time domain reflectometry pulses: Theoretical considerations for applications to soil and water. *Water Resources Research, 24*, 945–952.

Yungul, S. (1950). Interpretation of spontaneous-polarization anomalies caused by spherical ore bodies. *Geophysics, 15*, 237–246.

Zhdanov, M. S. (2002). *Geophysical inversion theory and regularization problems*. Amsterdam: Elsevier.

Chapter 5
Site Application: Forensic Civil Cases

Abstract In this chapter, some problems related to civil cases are addressed and solved. Many litigations in the courtrooms concern damage caused by neglect and lack of maintenance, for example of the underground water pipe and sewage systems, large infrastructures (the state of conservation of the irons, the quality of the concrete) etc. Therefore, some important cases will be illustrated to explain how to tackle and solve these important problems.

Keywords Civil cases · Water pipe loss · Concrete quality · Rebar conservation degree

5.1 Effects of Pipe Leaks

Detection and localisation of leaks are extremely important for the life of men and the stability of buildings. A pipe leak can cause several problems. These problems are the damage caused to the buildings that could range from structural to health issues around the home. Water can physically cause buildings structural damage that can be financially harmful. Health effects include mould and damp problems that could be very harmful to people with respiratory issues such as asthma.

The continuous introduction of water into the subsurface can lead to serious damage such as, for example, those of Gallipoli (in a village near Lecce in south Italy) and in Florence (north Italy) where the breaking of a water pipe has caused the sinking of the pavement road (Fig. 5.1).

Many peoples take civil actions against the institutions for the damage caused by water leaks. Based on these considerations, the experimental procedure was undertaken to identify water leaks. Three different techniques, Time Domain Reflectometry (TDR), Ground Penetrating Radar (GPR), and Electrical Resistivity Tomography (ERT) were simultaneously employed for the detection of water leaks in underground pipes, and their performances were compared.

At state of the art, TDR has been recently demonstrated to be a promising alternative for the individuation of leaks (Cataldo et al. 2012a, b), and also GPR and ERT has been successfully employed for this purpose (Demirci et al. 2012).

© Springer Nature Switzerland AG 2020

G. Leucci, *Advances in Geophysical Methods Applied to Forensic Investigations*,

https://doi.org/10.1007/978-3-030-46242-0_5

Fig. 5.1 The sinking of the pavement road in Florence (**a**), and Gallipoli (**b**)

To carry out a synoptic and comparative analysis, these techniques were tested on the experimental case in the laboratory on a plastic pipe buried under two different types of soil (one half of the pipe buried in silty soil, and the other half in clayey soil), and in which two leaks were intentionally provoked.

The TDR method can be used for leak detection and precisely for the inspection of *newly-installed* underground pipes made of any material (Cataldo et al. 2012a), and for the inspection of *pre-existing* underground metal pipes (Cataldo et al. 2012b).

Leak detection is related to the variation of the dielectric characteristics of the soil. The dielectric constant increase when the soil becomes moistened because of a water leak. The relative dielectric constant of water is approximately equal to 80, whereas the relative dielectric constant of dry soil is usually in the order of 3–4. As a result, the leaked water provokes a significant local increase of the permittivity (and, hence, a decrease of the electrical impedance) of the soil in the proximity of the leak point. This change of impedance is detected and localised through TDR measurement.

The experiment was carried out in a laboratory where the presence of a leaky underground pipe was reproduced as follows. Two adjacent wooden boxes were filled with two different types of soil (the first one clayey and the second one silty). The overall dimensions of the two boxes were 7.5 m × 1.2 m × 0.6 m. A 7.3 m-long plastic pipe was buried 45 cm below the top surface (Fig. 5.2).

For TDR measurements the HyperLabs HL1500 TDR unit was used.

Concerning Fig. 5.2, the holes were provoked at distances $L^{ref}_{EC1-1} = 2.5$ m and $L^{ref}_{EC1-2} = 5.5$ m, respectively, from the point of connection to the TDR instrument.

TDR results: the first step of the test was the measurement in dry condition. Successively water flows travel in the pipe, and TDR measurements were repeatedly performed while water escaped from the two leakage points, during the 20-minute test time. Results (Fig. 5.3) shows that in the acquired reflectogram, there are two dips in correspondence of approximately $d^{app} = 6.0$ m and $d^{app} = 11.5$ m. The dips became progressively more evident and wider after more time due to the increasing amount of leaked water. Figure 5.3b shows a zoom highlighting the two leak points.

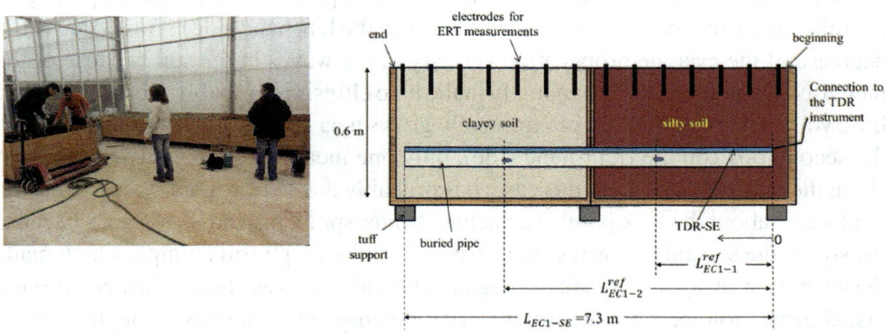

Fig. 5.2   Schematic representation of the test site

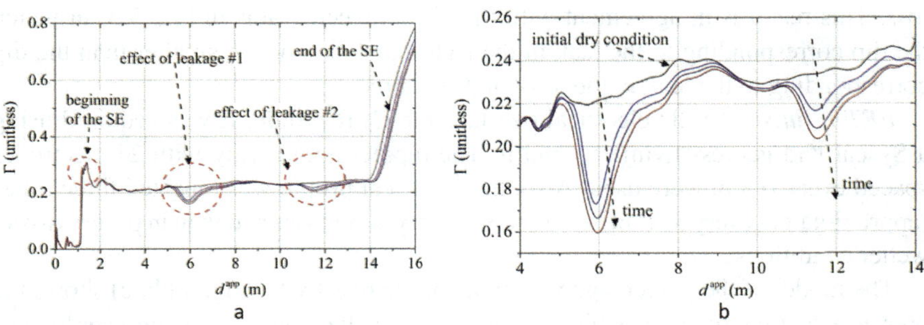

Fig. 5.3   **a** Reflectograms acquired for the test. **b** Zoom on the two leaks

Using a dedicated wavelet-based data-processing, as detailed in (Puust et al. 2010; Cataldo et al. 2012a, 2013), the resulting estimations for the two leak position were $L_{EC1-1}^{TDR} = 2.55$ m and $L_{EC1-2}^{TDR} = 5.51$ m, respectively. Therefore TDR technique identifies the leak positions with high accuracy. Moreover, it can also be seen that the different soils do not influence the performance of the TDR measurement system.

GPR results: The GPR survey was carried out with a Ris Hi-Mod system using a dual-band antenna, 200–600 MHz, manufactured by IDS.

GPR data were acquired (as TDR data) along a 7.3 m long line in a dry condition and for four different "wet soil conditions". Here the antenna at 600 MHz results is reported. In this case, migration based processing has been applied to the data. In particular, in all the images of Fig. 5.4, the vertical dashed yellow line represents the transition point from the first to the second wooden box, with a slight change of the propagation velocity (in particular, this involves a slightly shallower time depth of the return times from the bottom, as can be appreciated by the dashed black line). Compared to the GPR profile acquired in dry conditions (Fig. 5.4 upper panels), it is possible to notice the presence of anomalies caused by the two leaks. In particular, in the first box (on the left side of the figure), it is possible to note that, for increasing moisture levels, the anomalies (correspond to the leak zone) in red and blue ellipses expand and somehow move.

At the same time, it is possible to see also an apparent progressive deepening of the reflection from the bottom of the first box (on the left-hand side). This is due to the decrease of the average propagation velocity of the waves in the soil because of the increased water content. The anomaly in the blue ellipsis is probably more related to a little void that originates and progressively grows near the leakage point. Concerning the second box (on the right-hand side), only one meaningful anomaly can be seen about the leak point. Also, in this case, it is probably due to a progressively increasing void space about the leak point. The actual "water spot", instead, is not seen because the soil in the second box was characterised by quite tough soil clumps, which made the water flow away rather than impregnate the soil (this was also confirmed, through visual inspection, because the water started flowing out of the box through a hole at the bottom of the box).

The electromagnetic wave velocity analysis is shown in Fig. 5.5.

EM velocity analysis shows a strong decrease in correspondence of the water loss. This result is in agreement with the TDR reflectograms in Fig. 5.3, in which the dip corresponding to the leak in the right-hand side box is smaller than the dip corresponding to the leak in the left-hand side box.

ERT results: The 2D electrical overlaps the GPR profile. It was acquired using a Syscal Kid georesistivimetry, and it. The dipole-dipole array with 24 electrodes spaced every 0.3 m were used. Also, the electrical profile was repeated three times (approximately every 30 min), the first one in dry conditions and then in progressively wetter conditions.

The model of the percentage variations of the resistivity (Fig. 5.6b, c) shows the presence of two zones in which the resistivity values decrease significantly. The resistivity values decrease due to an increase in the volumetric water content. As a final note, it is important to point out that, also in this case, the evaluated leak

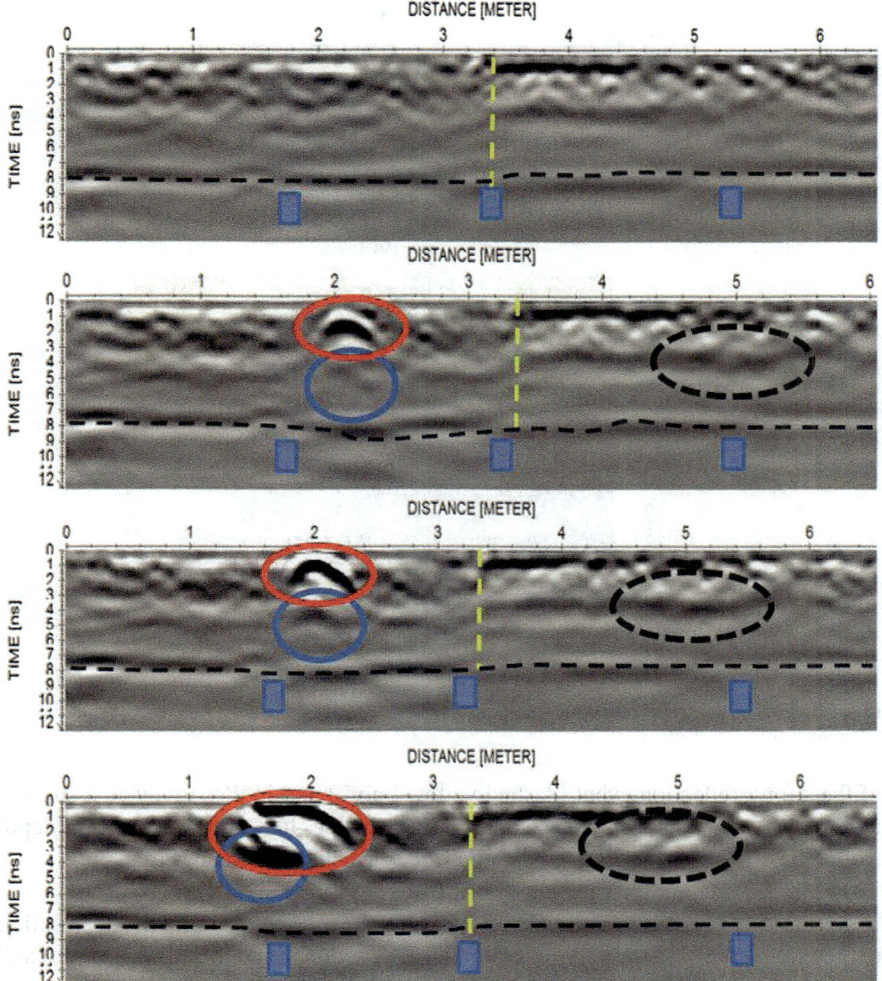

Fig. 5.4 Radar sections for the test site. From top to bottom: dry condition; slightly wet condition (water flown for 5 min); moderately wet condition (water flown for 10 min); and quite a wet condition (water flown for 15 min)

positions (2.6 and 5.3 m), are in good agreement concerning those evaluated through TDR, ERT and GPR results.

Successively GPR method was applied in the real water pipe where two pipe leaks were simulated. Data were acquired along a water pipe using the georadar Ris Hi-Mod with a dual-band antenna 200–600 MHz (Fig. 5.7).

A close examination of the raw data (Fig. 5.8) showed the presence of numerous reflection events that have the hyperbolae shape. Consider these reflection events is possible to estimate the EM wave velocity propagation. The estimated mean value of the EM wave velocity ranging from 0.07 to 0.12 m/ns. The shape and alignment

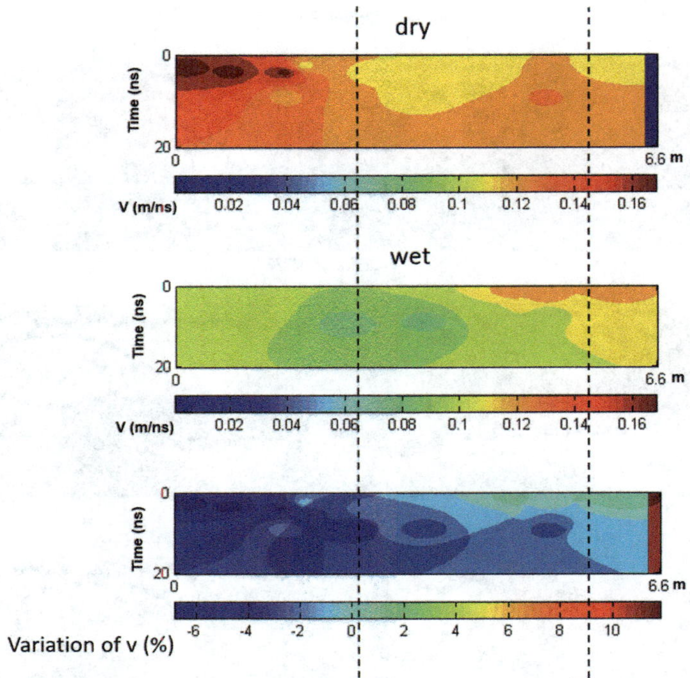

Fig. 5.5 EM velocity analysis

of the strong reflection event (dashed yellow line) found in the survey area suggest that they are related to the presence of the water pipe. It is posed at 1.5 m in depth (Fig. 5.8a). The losses points are labelled 1 and 2, respectively, in Fig. 5.8.

Figure 5.9a shows the range of the EM wave velocity distribution.

Figure 5.9 shows the comparison between the EM wave velocity distribution related to the profile acquired in dry conditions (without pipe losses) (Fig. 5.9a) and the EM wave velocity model related to the profile acquired after the first phase pipe loss (Fig. 5.9b) and after the second phase pipe losses (Fig. 5.9c). It is possible to note a decrease of the EM wave velocity in correspondence of the pipe losses.

The volumetric water content variation model in the subsoil after the pipe losses was estimated using the Topp formula (Topp et al. 1980) (Fig. 5.10).

Figure 5.10 show the increase of volumetric water content in the first and second phases of a water leak (Fig. 5.10b, c). In these zones, the volumetric water content is greater than 4 and 12% in comparison to the other zones (Fig. 5.11).

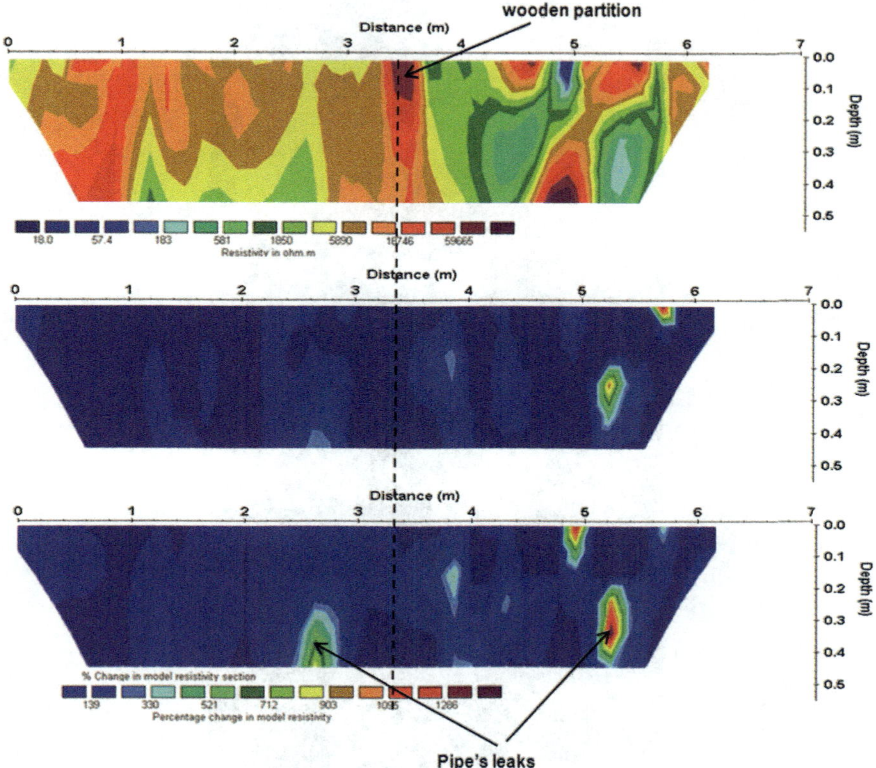

Fig. 5.6 2D resistivity models: **a** dry condition; **b** percentage variations of resistivity after 30 min; **c** percentage variations of resistivity after 60 min. Note that the percentage of variations of resistivity are negative

5.2   The Case of the City of Sant'Angelo (Sicily, South Italy)

The town of Sant'Angelo Muxare is a village located in the south part of Sicily (Fig. 5.12).

In the last few decades, the city has been affected by a series of subsidence events, which have, in some cases, resulted in the partial collapse of buildings and road surfaces. The causes of the ground subsidence events were studied using integrated 3D geophysical surveys. Results help the municipality to identify the response and to undertake the cause of compensation for the damages.

3D ERT data analysis: The surveyed area is shown in Fig. 5.13. To investigate the buildings, special ERT arrays were used. The ERT array was surrounding the buildings (Chavez et al. 2011; Argote-Espino et al. 2013; Tejero-Andrade et al. 2015). It was a dipole-dipole equatorial-parallel array. They are a series of L-profiles. Initially, a 2-D survey was conducted along each perpendicular line or transect. In the next step, the current electrodes remain at the end of one line, while the potential

Fig. 5.7 GPR data acquisition

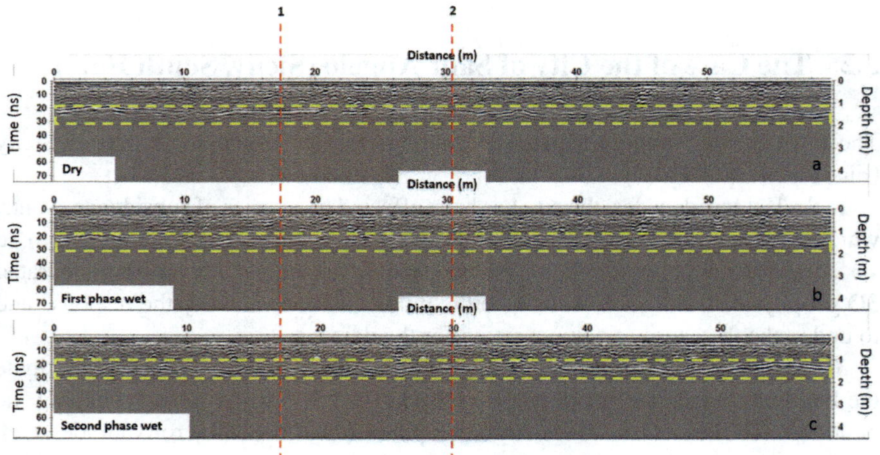

Fig. 5.8 Processed radar section acquired in the three phases: **a** dry; **b** first phase wet; **c** the second phase wet

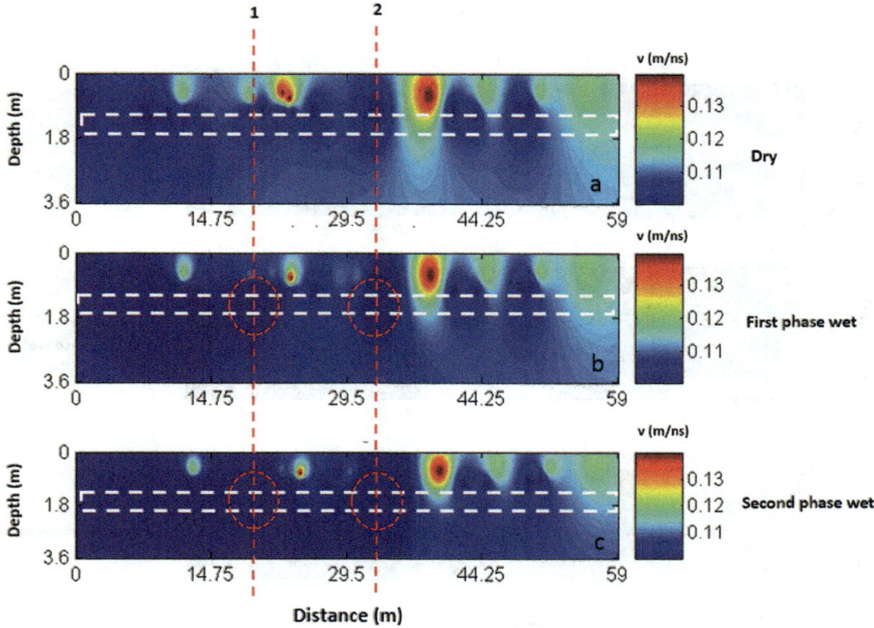

Fig. 5.9 Electromagnetic wave velocity distribution in the subsoil in the three phases: **a** dry; **b** first phase wet; **c** the second phase wet

is moved, along the line. Then the current electrodes move one electrode position, and the potential electrodes move as previously described.

The ABEM geo-resistivity meter supporting 64 electrodes was used for ERT data collection. The electrode separation for all arrays was 2 m (Fig. 5.13).

The working cube was computed using the software ErtLab (http://www.geostudiastier.it). The true resistivity model computed has an investigation depth of 6 m, which guarantees that the inverted true resistivity model is deeper than the expected building foundations. The foundations have an expected depth between 1 m and 2 m such revealed the GPR results. Figure 5.14 shows the resistivity depth slices from 0.5 to 6.0 m in depth. Here is possible to note the presence of a heterogeneous subsurface with resistivity values ranging from 100 to 2000 Ω per meter. Afterwards, it is possible to note the presence of:

(1) areas indicated with "A", with resistivity values between 1500 and 2000 Ω per meter; these values indicate the probable presence of areas where localised phenomena of instability are present. The relatively low resistivity values indicate that these anomalies are not attributable to the presence of empty volumes, but rather to incoherent materials, probably put there to fill up previous voids;

(2) areas indicated with "B", with resistivity values between 100 and 300 Ω per meter; these values indicate the probable presence of areas where a phenomenon of instability is present too. The low resistivity values indicate that

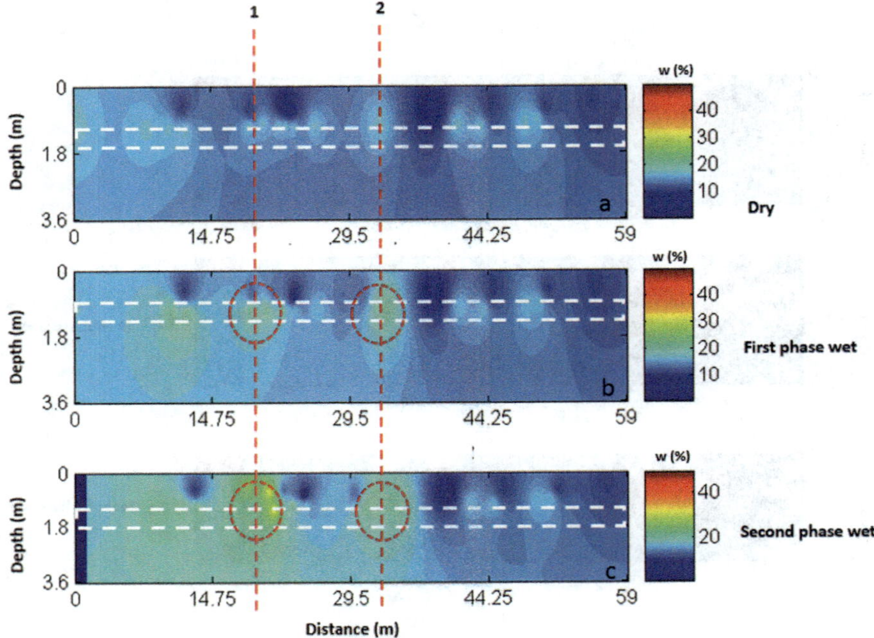

Fig. 5.10 Volumetric water content distribution in the subsoil in the three phases: **a** dry; **b** first phase wet; **c** the second phase wet

these anomalies are due to the potential presence of water-saturated incoherent materials;

(3) areas with resistivity values between 700 and 1200 Ω per meter; these values indicate the probable presence of subsoil zones free of in-homogeneities.

Figure 5.15 shows the low resistivity areas and the dashed dark arrows indicate a probable water path in the subsoil.

SP data analysis: self-potential anomalies are associated with water in subsurface structures and the flow of water through the ground (Telford et al. 1990; Lowrie 2007) that is usually indicated as a negative anomaly in the profile (Colangelo et al. 2006; Vichabian and Morgan 2002). The SP measurements covered the 3D ERT surveyed area. The self-potential signals were measured at the ground surface in a set of 432 measurement points located along the same ERT lines. Each electrode (stainless and nonpolarising) was placed inside a 10 cm bucket, filled with a moistened bentonite and gypsum mixture to ensure good contact between the electrode and the ground. Measurements of the self-potential signals were carried out with the same syscal kid, and nonpolarising Pb/PbCl2 (Petiau) electrodes (Perrier et al. 1997) were used. Processing of the SP data was a low-pass filter in the frequency domain to avoid edge effects of space domain filters so that high frequencies were eliminated and low frequencies were preserved (Aubanel and Oldham 1985). Figure 5.16 show the SP results build as maps at several depts. The result of this research shows that

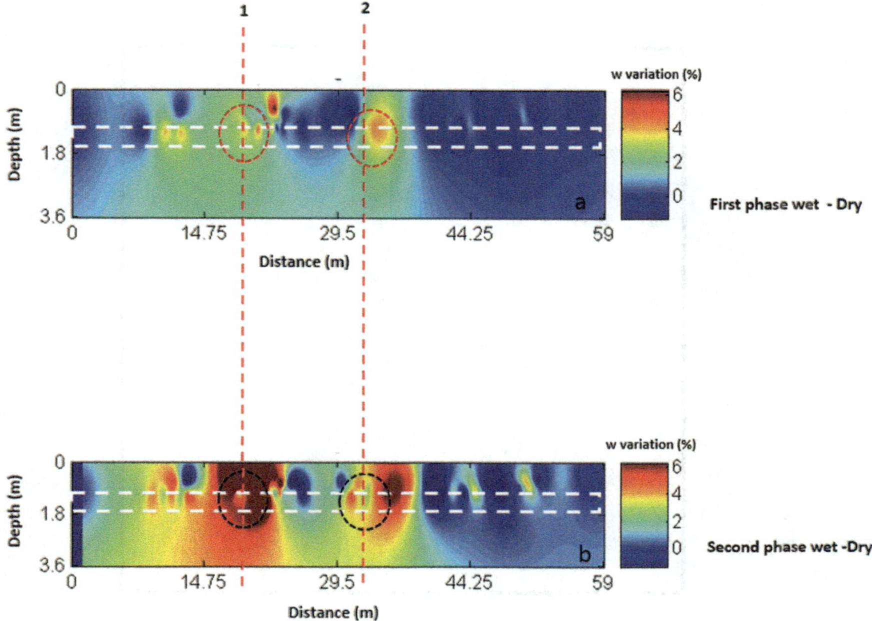

Fig. 5.11 Relative percentage changes in volumetric water content distribution in the subsoil:
a between the first phase wet and dry; **b** between second phase wet and dry

Fig. 5.12 The city of Sant'Angelo Muxaro (Sicily, Italy)

self-potential values vary between −20 and 50 mV. In Fig. 5.16 a quite uneven
distribution pattern of the self potentials is visible. In particular, it is possible to
note two points, indicated with "R", where there is a negative concentration of the
spontaneous potential (−20 mV). It is quite probable that, in these points, a flow
of materials waterborne, in the directions indicated by the arrows (that is, towards
positive SP values) was occurring at the moment of the measure.

Fig. 5.13 The geophysical surveyed area with the location of ERT profiles, red points represents the position of the electrodes

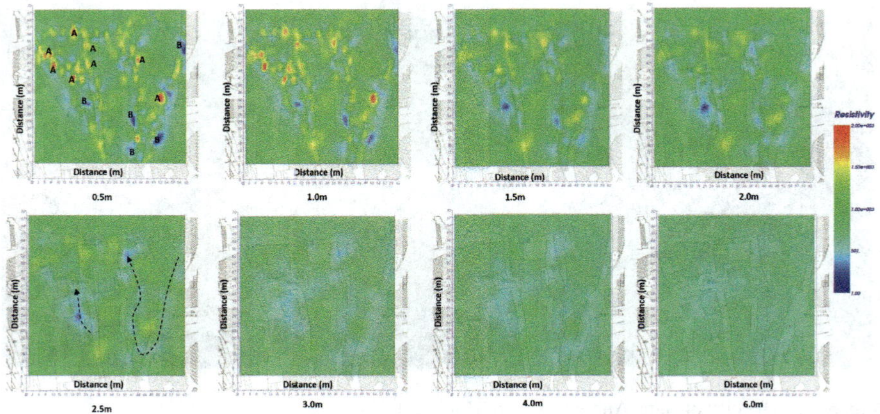

Fig. 5.14 The ERT depth slices

Subsidence phenomena occur in the investigated area in a zone indicate with R in Fig. 5.16 (Fig. 5.17).

GPR surveys were undertaken in some homes to understand the degree of damages (Fig. 5.17).

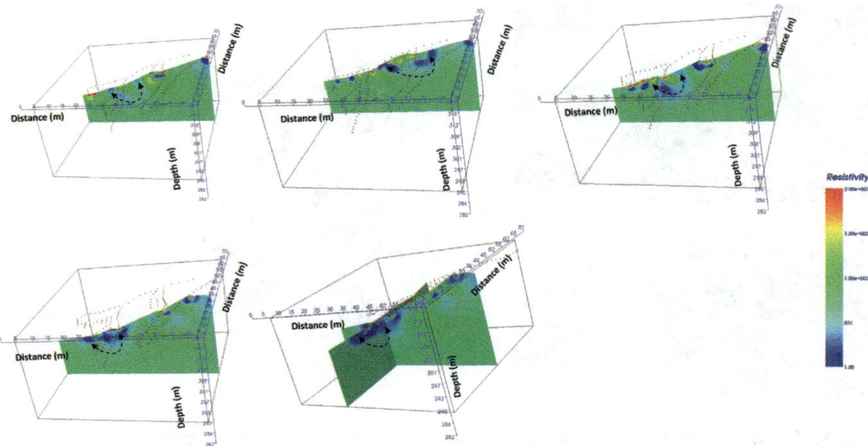

Fig. 5.15 The ERT pseudo-3D visualisation

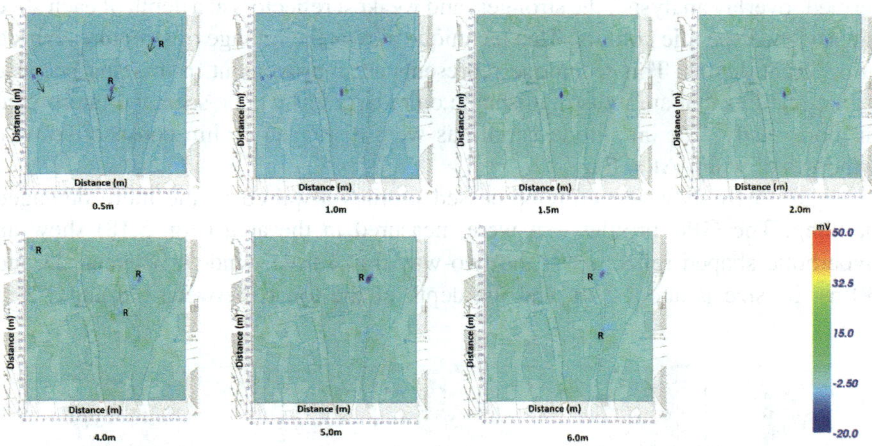

Fig. 5.16 The SP depth slices

GPR data analysis: A pulsed Opera-Duo system, manufactured by the Ingegneria dei Sistemi IDS-Corporation was used. The system is equipped with a dual-band antenna with central frequencies at 250 and 700 MHz, although only the results from the high-resolution 700 MHz survey are presented here. The data were subsequently processed using standard two-dimensional processing techniques using the GPR-Slice Version 7.0 software (Goodman 2013). The processing consists of the following steps: (i) frequency filtering; (ii) background removal to attenuate the horizontal banding in the deeper part of the sections (ringing); (iii) estimation of the average electromagnetic wave velocity by hyperbola fitting; (iv) Kirchhoff migration, using a constant average velocity value of 0.07 m/ns. Successively horizontal time slices

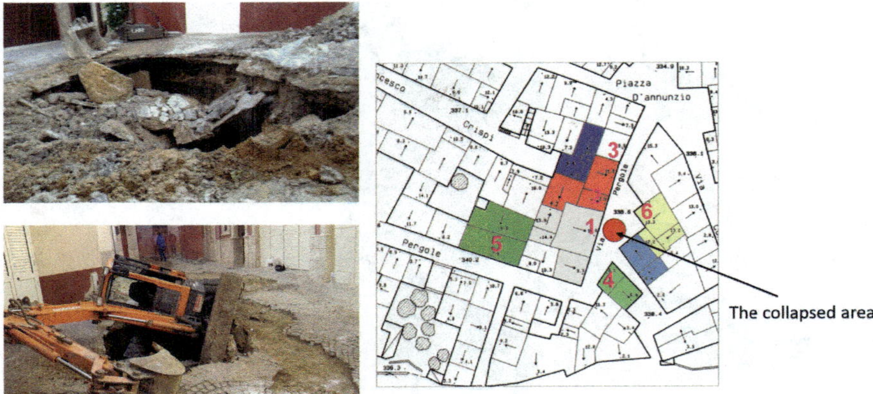

Fig. 5.17 The subsidence phenomena occur near some homes and the surveyed homes

were built (Conyers 2004). In a graphic method developed by Goodman et al. (2006), termed 'overlay analysis', the strongest and weakest reflectors at a depth of each slice are assigned specific colours. This technique allows the linkage of structures buried at different depths. This technique represents an improvement in imaging because subtle features that are indistinguishable on radargrams can be seen and interpreted in more easily. The amplitude variations were displayed within consecutive time windows of width $\Delta t = 5$ ns.

Figure 5.18 show the GPR processed profiles acquired in the more damaged building. The GPR profiles that were measured in the area (Fig. 5.18) show an hyperbolic shaped reflection at the two-way travel time window between 25 and 30 ns. Its size is about 3 m, and the depth of the top is between 0.8 and 1.2 m

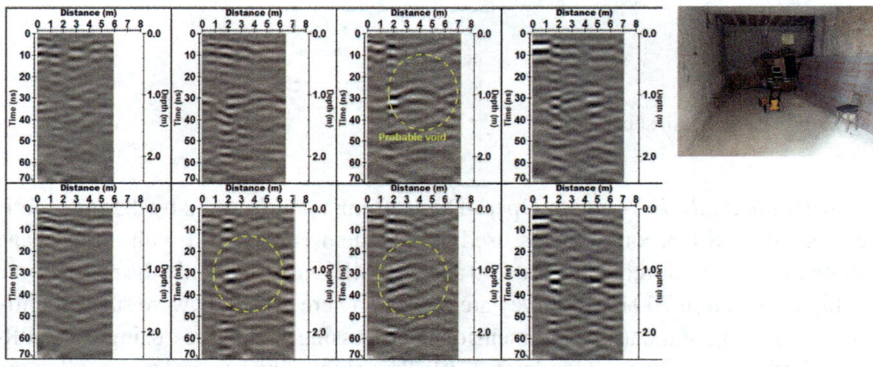

Fig. 5.18 The processed radar section related to one of the profiles acquired in a more damaged building

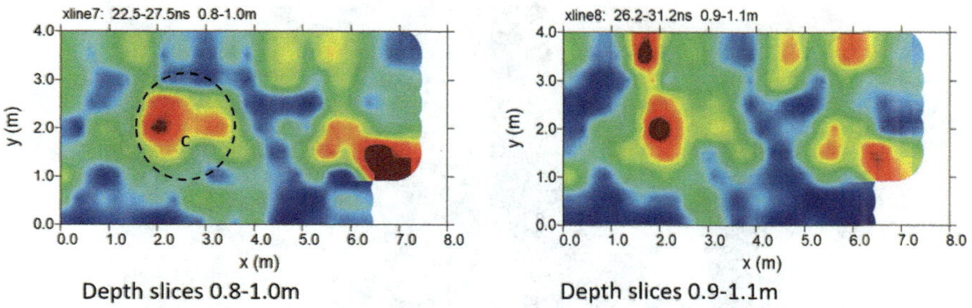

Fig. 5.19 GPR depth slices related to the more damaged building

(with an average electromagnetic wave velocity of 0.07 m/ns). This hyperbolic-shaped reflection was interpreted as a cavity. Time slices (Fig. 5.19) show the depth evolution of buried structures, including their size, shape and location.

In the slices ranging from 0.8 to 1.1 m depth, relatively high-amplitude anomalies (labelled C) are visible. Anomaly "C" was interpreted as a cavity. The iso-surfaces amplitude of EM wave (Fig. 5.20) shows the extension of the cavity C.

According to the achieved results, it is possible to deem that the instability events have been caused by losses within both the water supply networks and the drainage systems. Moreover, the excavation allows confirming the water pipe leaks (Fig. 5.21).

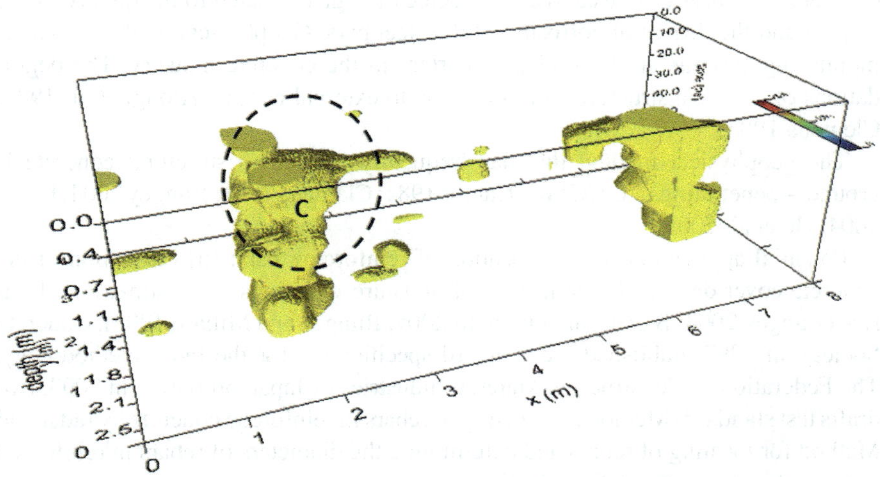

Fig. 5.20 Iso-surfaces amplitude of EM wave

Fig. 5.21 The old water pipe and the new pipe

5.3   The Corrosion of the Reinforced Bar in Concrete Structures

In many cases, it is necessary to establish the causes of the collapse of reinforced concrete structures or to establish the degree of neglect related to the quality of the cement and the degree of corrosion of the steel bars. Geophysical methods used as monitoring systems are becoming important in the concrete industry. The degradations of concrete structures are often due to external causes (Alongi et al. 1982; Clemena 1983).

The geophysical method that has frequently applied to structural concrete is ground—penetrating radar (GPR) (Cantor 1984; Clemena 1991; Bungey 2004; Forde 2004; He et al. 2009).

Potential applications are (i) location of reinforcing bars; (ii) estimation of the concrete cover depth; (iii) estimation of moisture variations: (iv) estimation of bar size (Bungey 2004; Barrile and Pucinotti 2005; Bungey and Millard 1993). Concrete Society, in 1997, published the standard specification for the radar methodology. The Federation of Construction Material Industries of Japan proposed, in 2003, two drafts test standard (Method for locating of rebars in reinforced concrete by radar and Method for locating of rebars and determining the diameters of rebars in reinforced concrete by electromagnetic induction).

One of the most widespread applications of GPR in Civil Engineering is the detection of reinforcing bars in concrete (Ulriksen 1982; Pucinotti and De Lorenzo 1994; Pucinotti and Barrile 2002; Barnes et al. 2008; Diamanti et al. 2008; Leucci 2012). Bar sizing is more difficult, and there is little evidence of industrial usage (Leucci 2012).

Leucci (2012) performed a study that allows founding a relationship between the relative dielectric constant and volumetric water content.

Fig. 5.22 Cubic concrete sample used for test measurement

$$w = 0.001k^3 + 0.1232k^2 - 0.043k + 3.3035$$

A combination of w (volumetric water content) and k (relative dielectric constant) could provide concise information for uniquely characterising concrete quality.

To test the validity of the proposed relationship a laboratory measurement, at two known volumetric water content (w = 4.90% and w = 5.98%) using the GPR, were performed on a concrete cubic sample (Fig. 5.22).

A Sir-3000 was used for two dimensional travel time tomography (TT). The results (Fig. 5.23) show low variations of k and therefore of the volumetric water content. Volumetric water content varies from about 5.8% to about 6.4% in the sample with known volumetric water content (w = 5.98%) and varies from about 4.3% to about 5.1% in the sample with known volumetric water content (w = 4.9%). The ranges of variations demonstrate the effectiveness of the relationship found by Leucci (2012).

For the rebar diameter, Leucci (2012) found a relationship

$$d = 4.482 \, A_r^{4.7241}$$

where d represent the rebar diameter, and A_r represents the amplitude ratio co-polarized/cross-polarized. The application of this equation in a controlled test (Fig. 5.24) show good results (Fig. 5.25).

The results are shown in Fig. 5.25.

Another important tool is the measure of the corrosion degree of the bar in the concrete. In this case, the SP methods can evidence the corrosion of reinforcement

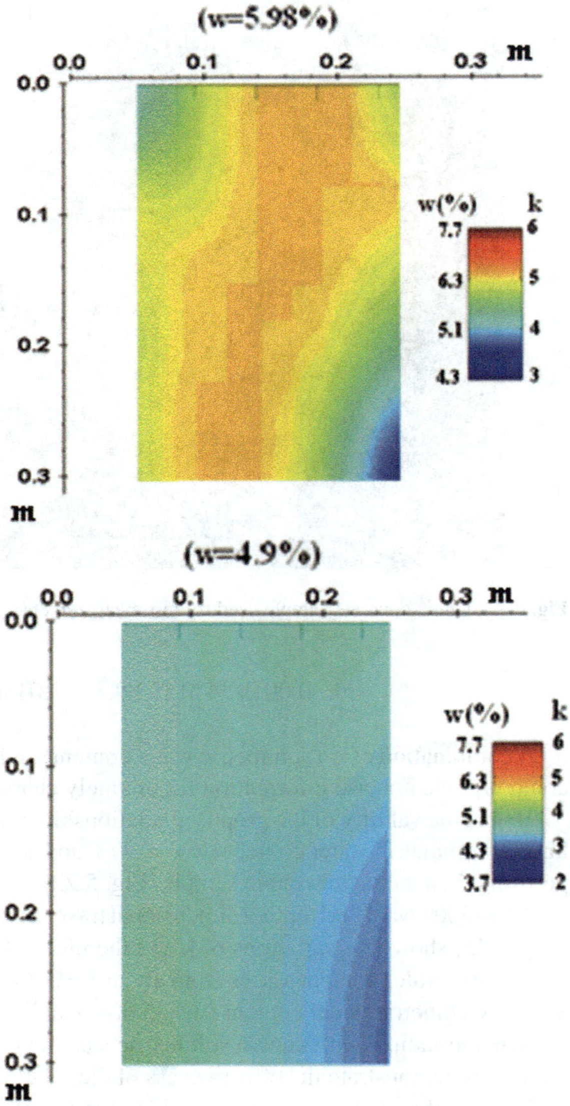

Fig. 5.23 GPR traveltime tomography relative dielectric constant (k) and volumetric water content (w) distribution models

steel. Under normal conditions, reinforcement steel is protected from corrosion by a thin, passive film of hydrated iron oxide. This passive film decomposes due to the reaction of the concrete with atmospheric carbon dioxide (CO_2), or by the penetration of substances aggressive to steel, in particular, chlorides from de-icing salt or saltwater. At the anode, ferrous ions (Fe^{++}) are dissolved, and electrons are set free. These electrons drift through the steel to the cathode, where they form hydroxide (OH^-) with the generally available water and oxygen. This principle creates a potential difference that can be measured by the half-cell method (Fig. 5.26a).

Fig. 5.24   The controlled test scheme

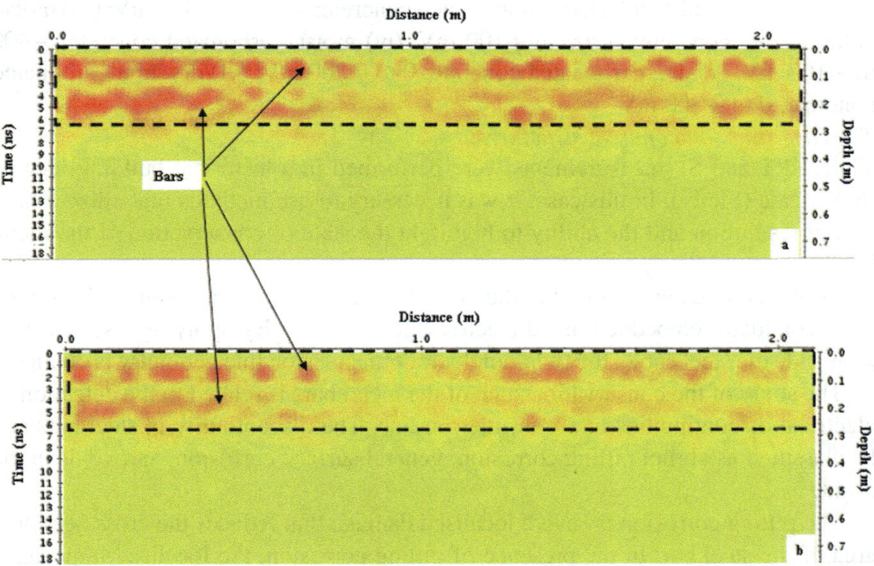

Fig. 5.25   The GPR processed data: **a** normal polarisation; **b** parallel polarisation

The basic idea of the SP measurement is to measure the potentials at the concrete surface to obtain a characteristic picture of the state of corrosion of the steel surface within the concrete. For this purpose, a reference electrode is connected via a high-impedance voltmeter to the steel reinforcement and is moved in a grid over the concrete surface (Fig. 5.26b).

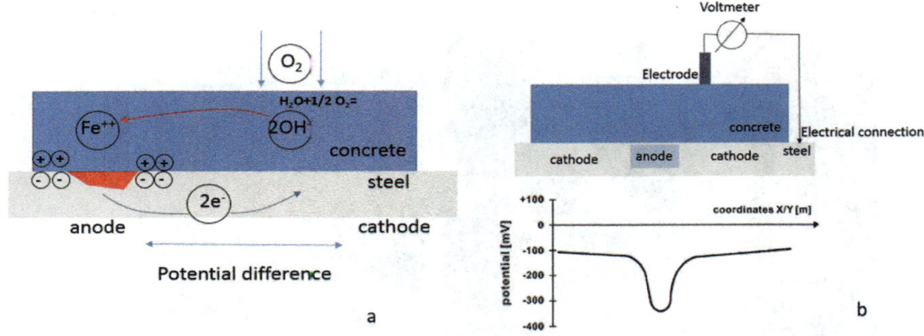

Fig. 5.26 **a** Principle of steel corrosion in concrete; **b** typical SP curve for moist carbonated concrete

The reference electrode is a Cu/CuSO$_4$ half-cell. It consists of a copper rod immersed in a saturated copper-sulphate solution, which maintains a constant, known potential. Typical orders of magnitude for the half-cell potential of steel in concrete measured against a Cu/CuSO$_4$—reference electrode are in the following ranges (Leucci et al. 2017; Leucci, 2019): (i) water-saturated concrete without O$_2$: -1000 to -900 mV; (ii) moist, chloride contaminated concrete: -600 to -400 mV; (iii) moist, chloride-free concrete: -200 to $+100$ mV; (iv) moist, carbonated concrete: -400 to $+100$ mV; (v) dry, carbonated concrete: 0 to $+200$ mV; (vi) dry, non-carbonated concrete: 0 to $+200$ mV.

The case study on concrete structures:

TDR, ERT and SP measurements were performed in a historical building located in Acireale (Sicily). In this case, it was necessary to use methods that allow a high spatial resolution and the ability to highlight the state of conservation of the metal bars.

TDR technique has applied to the bars in the concrete. Hera the metal bars are a good conductor embedded in a dielectric (the concrete). By applying a sensor wire alongside the steel cable, the twin-conductor transmission line geometry is obtained.

The study of the conservation state of the metal bars is related to the detection of electrical discontinuities in a transmission line. The discontinuity in steel bars can be classified as abrupt pitting corrosion, general surface corrosion, and voids in the grout.

The pitting corrosion is severe localised damage that reduces the cross-sectional area of the steel bar. In the presence of pitting corrosion, the localised impedance should increase abruptly and therefore a positive reflection from the site of pitting corrosion is expected. As seen in Chap. 3, the evaluation of the EM transit time allows the location of the corrosion point and the reflection amplitude indicates the magnitude of the damage.

The surface corrosion tends to reduce the radius of the steel bar on the order of a few per cent over a part of the length of the line. In this case, the application of TDR allows determining the extent and length of the corrosion. The surface corrosion is detectable from the magnitude and duration of the EM signal reflection, respectively.

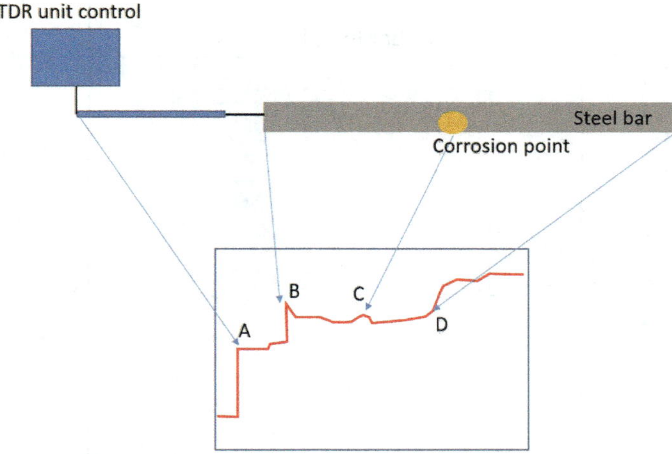

Fig. 5.27   TDR return of a 3-m rebar specimen with pitting corrosion in the middle

The voids in grout leave a section of the steel bar vulnerable to corrosion. The presence of void in the grout will change the dielectric constant in the sense that the void tends to reduce the dielectric constant and therefore increase the characteristic impedance. Furthermore, the presence of voids will change the velocity of propagation in the transmission line.

An experimental test was performed on a 3-m long steel bar sample (Fig. 5.27). The sample has pitting corrosion in the middle part of the bar. In point A will generate a step waveform.

A coaxial cable is used to connect the sample to the measuring system. The wave is launched into a coaxial cable and cause a positive reflection at the beginning of the sample (point B). The wave velocity propagation in the transmission line is v_p. In point C, a wave reflection related to the physical damage is observed. The time, $T = T_C - T_B$, is the transit time from point B to the mismatch and back again. At the end of the sample, the wave goes up because the line is terminated by an open circuit (point D).

Based on these results, TDR was performed on the steel bar part of the foundation base. The TDR signal is noisy due to the presence of other electrically connected conductors. Figure 5.28 shows TDR readings on 10 m length of the steel bar. The end of the bar can be easily identified, indicating that energy loss for the embedded transmission line is not a major problem. The signal between point A and B allow identifying a significant corrosion state of the bar.

Furthermore, ERT and SP measurements were performed. Therefore, 64 electrodes with variable interelectrode distance were used in a non-conventional array (Fig. 5.29). The distributions of the parameters' electrical resistivity and spontaneous potentials were estimated.

Resistivity maps were constructed using ERTLab software and a special algorithm implemented in the Matlab environment (Fig. 5.30).

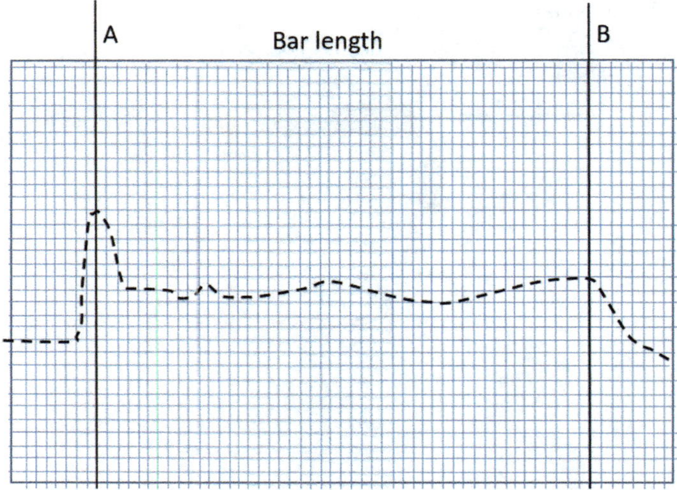

Fig. 5.28   TDR results obtained from a 10 m steel bar

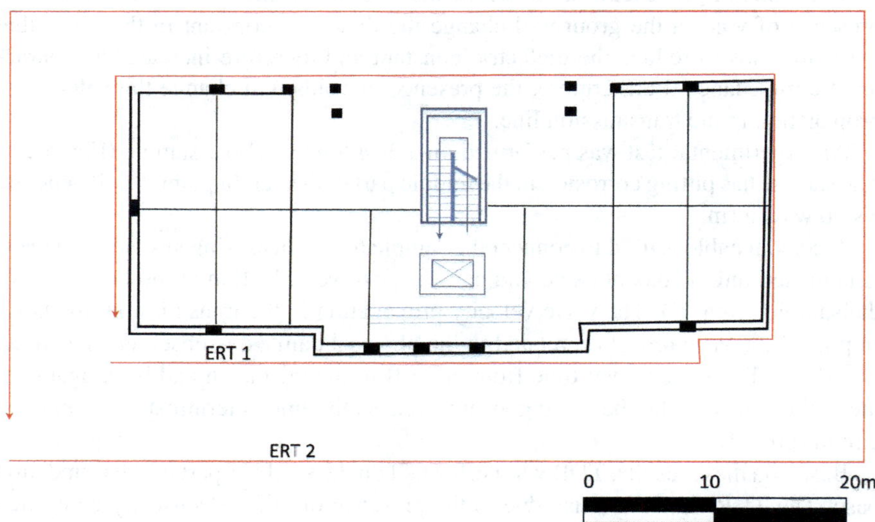

Fig. 5.29   The location of ERT and SP profiles

From the resistivity distribution model, the presence of a heterogeneous structure with resistivity values between 10 and 6000 Ω m is evident. Note the presence of an area (blue colour "B") with low resistivity values between 10 and 50 Ω m, where there may be high water content. High resistivity values 6000 Ω m (indicates with the arrows in Fig. 5.30) can be related to the foundation of the structure that has a good degree of conservation. The 2D distribution of the resistivity around the

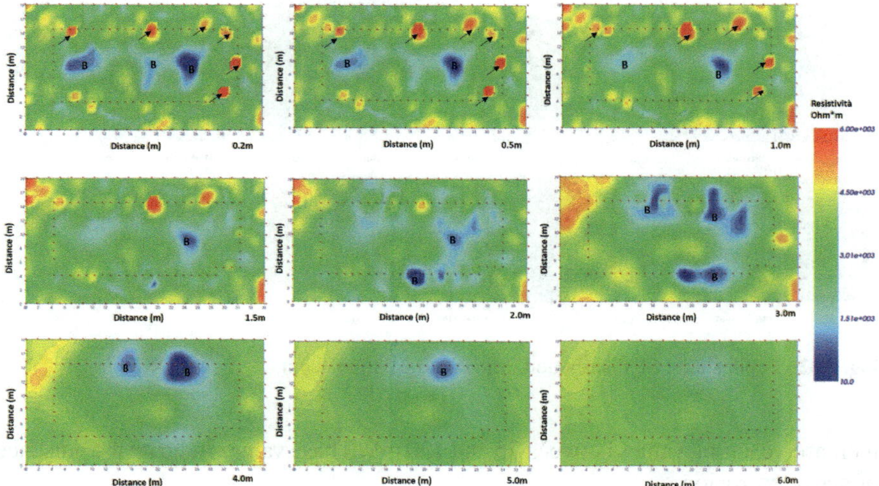

Fig. 5.30 Depth slices: resistivity distribution from 0.2 to 6.0 m in depth

building show (Fig. 5.31) that foundation leans on incoherent materials with a strong presence of water.

In the SP distribution maps (Fig. 5.32), there are three points (blue) in which there is a concentration of negative potentials (-50 mV) which indicates the presence of water-saturated concrete. Around these areas (green colour), the values of SP

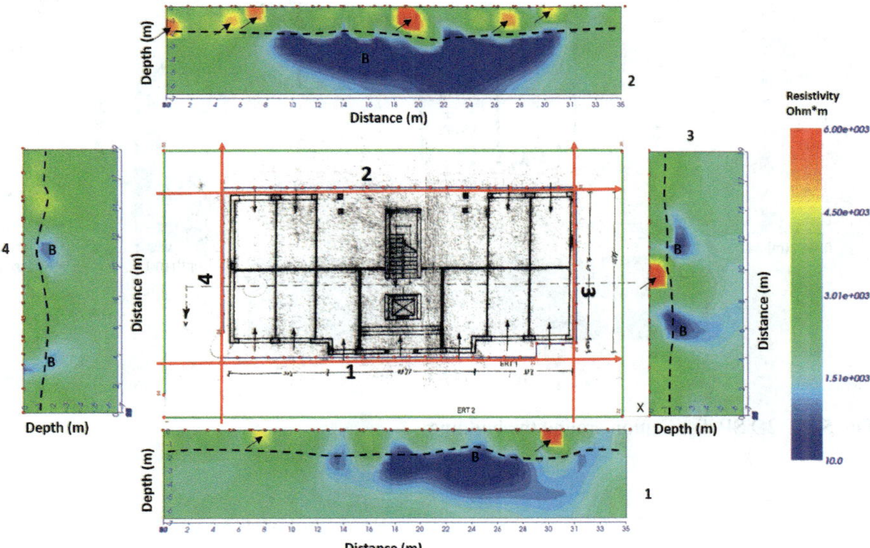

Fig. 5.31 2D resistivity distribution around the building

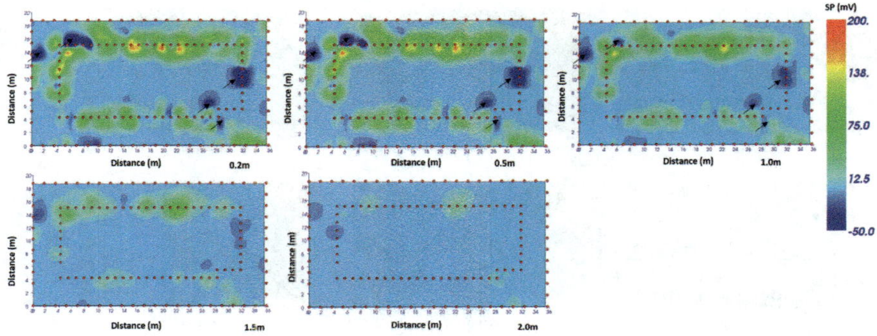

Fig. 5.32 Depth slices: SP distribution from 0.2 to 2.0 m in depth

increase, reaching values between 75 and 130 mV. These values indicate the presence of active corrosion.

The 2D distribution of the SP around the building show (Fig. 5.33) the part of the foundation where it is possible to found high corrosion phenomena. This phenomenon has been confirmed by direct observation (Fig. 5.34).

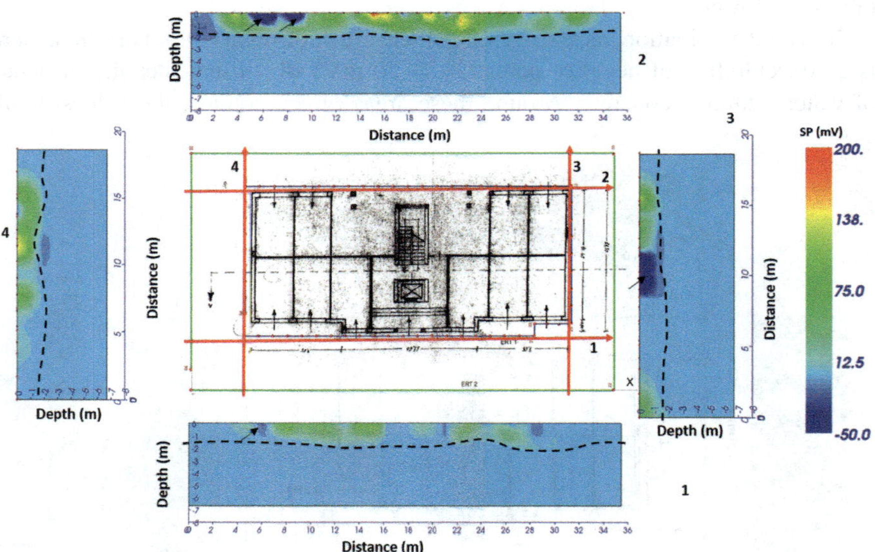

Fig. 5.33 2D SP distribution around the building

Fig. 5.34 Photo of the degraded area

5.4 The Damage Related to the Fall of Trunks

Trees can cause direct damage to human-made structures (Fig. 5.35) and in some cases, even the death of people. Causes of the tree damage may be a result of the decay of the biomechanical characteristics related to the presence of voids or disease presents in the trunk (Yau 1991; Moore 1995).

Fig. 5.35 Photo of some damages linked to trees collapse

Damage can be caused by the distribution of root zones in the subsoil that causes subsidence phenomena (Bradshawa and Hunt 1995; Mattheck and Breloer 1998; Lawson 1998).

A series of experiments were undertaken to understand the best geophysical method can be used for the tree trunk and root conduction studies.

The first experiment was conducted in the laboratory on a trunk using ground-penetrating radar, electrical resistivity tomography and sonic tomography geophysical methods (Fig. 5.36).

The trunk sample presents a degraded central area and a compact peripheral area (Fig. 5.37).

Fig. 5.36   The test performed on a trunk using **a** ERT; **b** GPR; **c** ST

Fig. 5.37   The sample trunk

The ERT Test The electrical resistivity technique using a ring array was applied to image the internal electrical structures of the trunk. The dipole-dipole array was chosen with a 1 cm electrodes space. Measured electrical resistivity data were inverted using a 2D iterative finite-element algorithm which incorporates the cylindrical geometry of the trunk. The technique is successfully tested and show that the resolution obtained when anomalous mapping zones inside the trunk are higher for the dipole-dipole configuration. In this preliminary study using ERT to obtain images of sample trunk, it was apparent that the internal water content was measured at 2%. The ERT surveys were carried out with the Syscal Kid georesistivitymeter with 24 electrodes arranged in a non-conventional array (Fig. 5.38).

Laboratory tests allowed to define the order of magnitude of the electrical characteristics of wood with different degrees of decay. The estimation of the behaviour of the resistivity of the trunk sample drove a better understanding of the trunk conservation state. The resistivity tomography results (Fig. 5.39) show a distribution of the resistivity inside the trunk.

Figure 5.39 shows a relationship between trunk decay and high electrical resistivity corresponds to the presence of probably voids spaces.

Fig. 5.38 The ERT measurements on the trunk sample

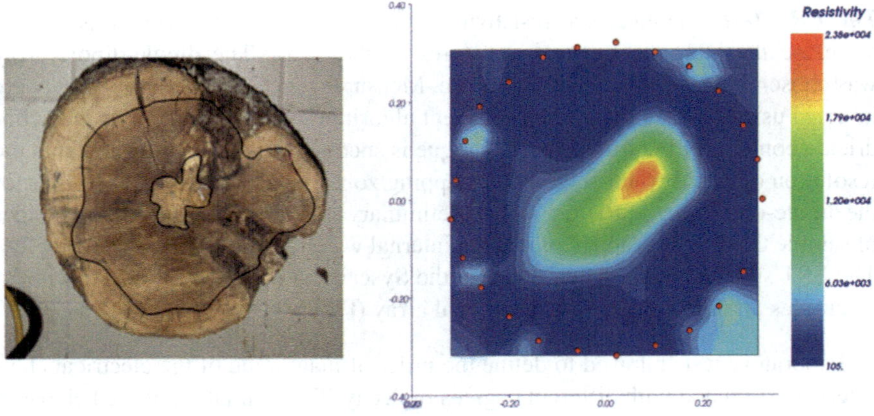

Fig. 5.39   The resistivity distribution inside the trunk

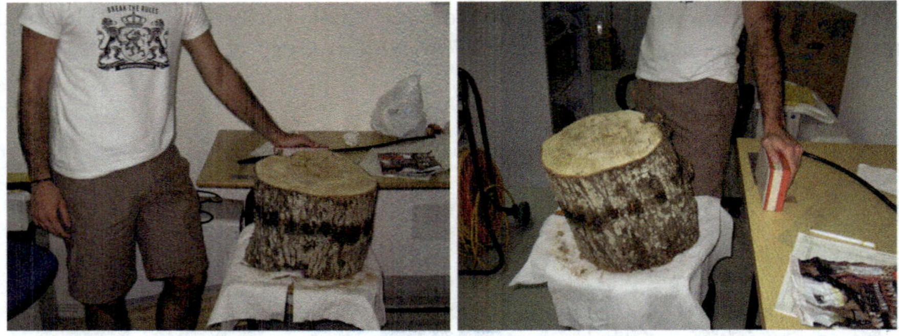

Fig. 5.40   The GPR measurements on the trunk sample

The GPR Test   The GPR test was performed using a Sir 3000 georadar system with the 1.5 GHz antenna (Fig. 5.40).

The processed radar section (Fig. 5.41) show very good correspondence with the central part of the trunk. It is possible to see also the trunk bark.

The ST Test   The ST test was undertaken using a sledgehammer as an energy source, with 24 piezoelectric geophones (4 kHz) with 1 cm spacing, and a shot point for each geophone position. The measurements were collected using a Geometric Strataview Seismograph (model Nimbus 1220; Geometrics, Silicon Valley, CA, USA) with 24 active channels (Fig. 5.42). Using travel time measurements of the seismic waves passing within the trunk is possible to obtain the seismic P-wave velocity (V_p) distribution in the trunk. The software Reflexw 8.0 version, developed by Sandmeier (2018), was used to perform the data processing and to interpret the seismic direct wave data.

The results of the inversion of seismic data are shown in Fig. 5.43.

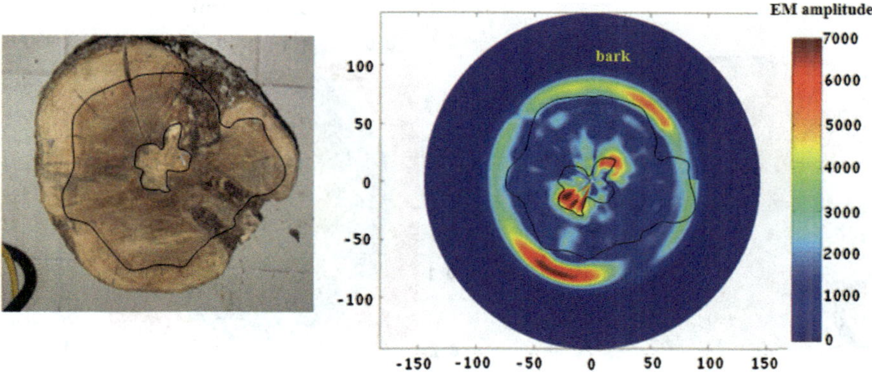

Fig. 5.41 The processed radar section

Fig. 5.42 The ST measurements on the trunk sample

The seismic tomography survey points out that the trunk may be divided into three main zones. A main low Vp zone (300 < Vp < 400 m/s); a medium Vp zone (1000 < Vp < 1100 m/s) and an higher Vp zone (1600 < Vp < 1900 m/s). The first zone is interpreted to be related to the trunk bark. The increase of Vp is probably related to a more compact material.

Discussion on results The integration of the results allow us to understand that the central part of the trunk seems to be more compact. High resistivity values associated with higher Vp values and high electromagnetic amplitude reflections are therefore associated with a compact zone and not to decay zone.

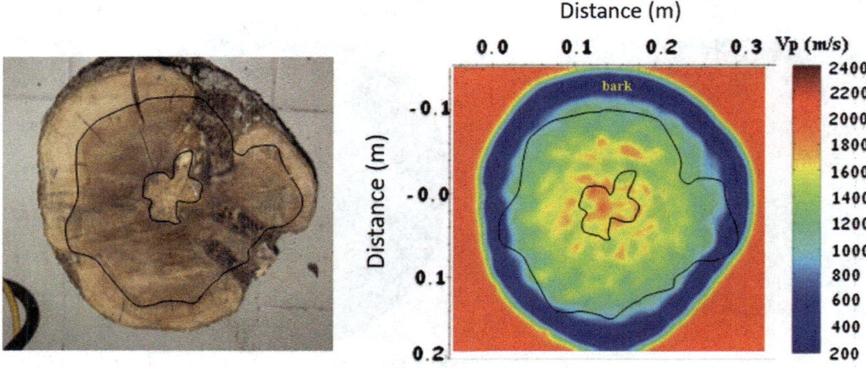

Fig. 5.43   The Vp distribution inside the trunk sample

Test in situ   The reliability of the GPR, ERT and seismic geophysical methods, in the tree trunk studies, was tested on eucalyptus trees in an urban environment. The studied trees are located in a former quarry inside a rock-cut (Fig. 5.44).

The tests on the trunk   A more degraded trunk (apparently on the surface) was chosen to investigate the inside degree of damages (Fig. 5.45). For this purpose, ST, ERT and GPR methods were used.

Fig. 5.44   The studied trees

Fig. 5.45 Photo of the acquired lines on the eucalyptus

ST data acquisition and analysis on the trunk The ST measurements in situ were performed using a sledgehammer as an energy source, with 24 piezoelectric geophones (4 kHz) with 5 cm spacing, and a shot point for each geophone position. The measurements were collected using a Geometric Strataview Seismograph with 24 active channels (Fig. 5.46).

The software Reflexw 8.0 version, developed by Sandmeier (2018), was used to perform the data processing and to interpret the seismic direct wave data. Results are shown in Fig. 5.47.

The seismic tomography survey points out that the trunk may be divided into three main zones. A main low Vp zone ($300 < Vp < 400$ m/s); a medium Vp zone ($800 < Vp < 1000$ m/s) and an higher Vp zone ($1200 < Vp < 1300$ m/s). The first zone is interpreted to be related to the trunk bark. The increase of Vp is probably related to a more compact material.

The GPR data acquisition and analysis on the trunk The GPR test was performed using a Sir 3000 georadar system with the 1.5 GHz antenna. The processed radar sections are shown in Fig. 5.48.

Results show some high EM amplitude anomalies related to the trunk bark (the outer one); the more internal anomalies could be related to some problems inside the trunk.

Fig. 5.46 Photo of the geophones located on the tree trunk

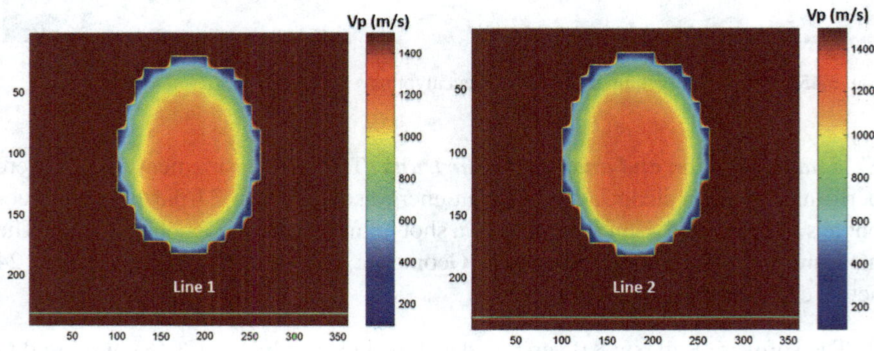

Fig. 5.47 Vp distribution inside the trunk for the acquired line 1 and line 2

The ERT data acquisition and analysis on the trunk The ERT test was performed using a Syscal Kid system 24 electrodes 3 cm spaced. The processed data are shown in Fig. 5.49.

Results show some high resistivity anomalies related to the trunk bark (the outer one); the more internal anomalies could be related to some problems inside the trunk.

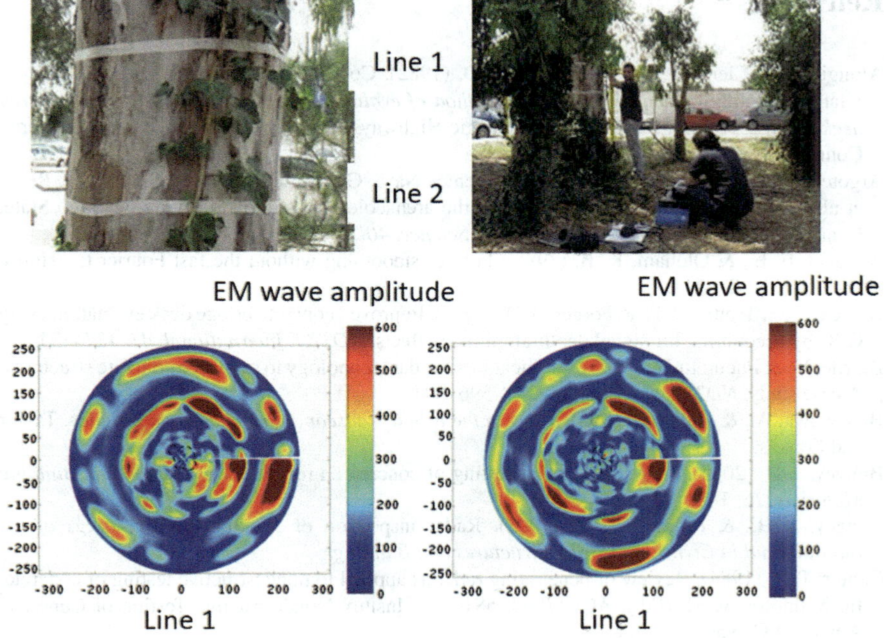

Fig. 5.48   The processed radar sections

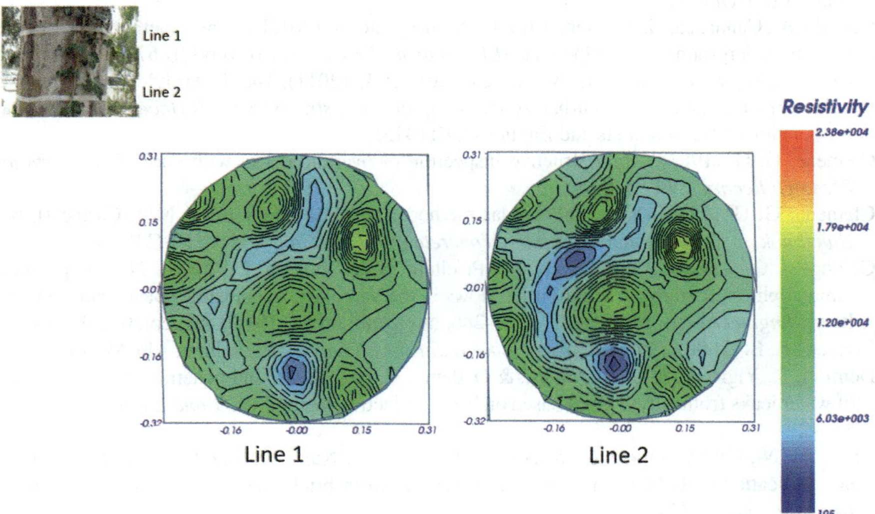

Fig. 5.49   The processed ERT 2D data

References

Alongi, A. J., Clemefia, G. G., & Cady, P. D. (1982). Condition evaluation of concrete bridges relative to reinforcement corrosion. In *Method of evaluating the condition of asphalt-covered decks* (vol. 3). Report SHRP-S-325. Strategic Highways Research Program, National Research Council, Washington DC.

Argote-Espino, D., Tejero-Andrade, A., Cifuentes-Nava, G., Iriarte, L., Farías, S., Chávez, R. E., et al. (2013). 3D electrical prospection in the archaeological site of El Pahñú, Hidalgo State, Central Mexico. *Journal of Archaeological Science, 40*(2), 1213–1223.

Aubanel, E. E., & Oldham, K. B. (1985). Fourier smoothing without the fast Fourier transform. *Byte, 10*(2), 207–218.

Barnes, C. L., Trottier, J. F., & Forgeron, D. (2008). Improved concrete bridge deck evaluation using GPR by accounting for signal depth-amplitude effects. *NDT&E International, 41*, 427–433.

Barrile, V., & Pucinotti, R. (2005). Application of radar technology to reinforce concrete structures: A case study. *NDT&E International, 38*, 596–604.

Bradshawa, A., & Hunt, B. (1995). *Trees in the urban landscape, Principles and practice*. Taylor and Francis.

Bungey, J. H. (2004). Sub-surface radar testing of concrete: a review. *Construction and Building Materials, 18*, 1–8.

Bungey, J. H., & Millard, S. G. (1993). Radar inspection of structures. *Proceedings of the International in Civil Engineering Structures and Buildings, 173*.

Cantor, T. R. (1984). Review of penetrating radar as applied to nondestructive testing of concrete. In: Malhotra, V. M. (Ed.), ACI *SP-82*, 581–602, Insituy Nondestructive Testing of Concrete, American Concrete Institute.

Cataldo, A., Cannazza, G., De Benedetto, E., & Giaquinto, N. (2012a). A TDR-based system for the localization of leaks in newly-installed, underground pipes made of any material. *Measurement Science & Technology, 23*(10), 1–9.

Cataldo, A., Cannazza, G., De Benedetto, E., & Giaquinto, N. (2012b). A new method for detecting leaks in underground water pipelines. *IEEE Sensors Journal, 12*(6), 1660–1667.

Chavez, G., Tejero, A., Alcantara, M. A., & Chavez, R. E. (2011). The 'L-Array', a tool to characterize a fracture pattern in an urban zone. In *Expanded Abstracts: Near Surface 2011*, European Association of Geoscientists and Engineers, P114155.

Clemena, G. G. (1983). Nondestructive inspection of overlaid decks with GPR. *Transportation Research Record, 899*, 21–32.

Clemena, G. G. (1991). Short pulse radar methods. In V. M. Malhotra & N. J. Carino (Eds.), *Handbook on non-destructive testing of concrete* (pp. 253–274). Boston: CRC Press.

Colangelo, G., Lapenna, V., Perrone, A., Piscitelli, S., & Telesca, L. (2006). 2D self potential tomographies for studying groundwater flows in the Varco d'Izzo landslide (Basilicata, southern Italy). *Engineering Geology, 88*(3), 274–286. https://doi.org/10.1016/j.enggeo.2006.09.014.

Conyers, L. B. (2004). *Ground—penetrating radar for archaeology*. Lanham: Alta Mira Press.

Demirci, S., Yigit, E., Eskidemir, I. H., & Ozdemir, C. (2012). Ground penetrating radar imaging of water leaks from buried pipes based on back-projection method. *NDT and E International, 47*, 35–42.

Diamanti, N., Giannopoulos, A., & Forde, M. C. (2008). Numerical modelling and experimental verification of GPR to investigate ring separation in brick masonry arch bridges. *NDT&E International, 41*, 354–363.

Forde, M. C. (2004). Ground penetrating radar, Introduction to nondestructive evaluation technologies for bridges. In *Transportation Research Board Pre-conference Workshop*.

Goodman, D. (2013). *GPR Slice Version 7.0 Manual*. http://www.gpr-survey.com. Accessed June 2013.

Goodman, D., Steinberg, J., Damiata, B., Nishimure, Y., Schneider, K., Hiromichi, H., & Hisashi, N. (2006). GPR overlay analysis for archaeological prospection. In *Proceedings of the 11th International Conference on Ground Penetrating Radar*, Columbus, Ohio, CD-rom.

He, X.-Q., Zhu, Z.-Q., Liu, Q.-Y., & Lu, G.-Y. (2009). Review of GPR rebar detection. In *PIERS Proceedings* (pp. 804–813), Beijing, China, March 23–27, 2009.

Lawson, M. (1998). Peer review of tree root damage to buildings (P G Biddle). *Arboricultural Journal, 22*, 4–13.

Leucci, G. (2012). The use of gpr to estimate volumetric water content and reinforced bar diameter in concrete Structures. *Journal of Advanced Concrete Technology, 10*, 411–422.

Leucci, G. (2019). *Nondestructive testing for archaeology and cultural heritage: A practical guide and new perspective* (pp. 217). Springer, ISBN 978-3-030-01898-6.

Leucci, G., De Giorgi, L., Gizzi, F. T., Persico, R. (2017). Integrated geo-scientific surveys in the historical centre of Mesagne (Brindisi, Southern Italy). *Natural Hazards, 86*, 363–383. https://doi.org/10.1007/s11069-016-2645-x.

Lowrie, W. (2007). *Fundamentals of geophysics*. Cambridge: Cambridge University Press.

Mattheck, C., & Breloer, H. (1998). *La stabilità degli alberi.* Il Verde editoriale, Roma.

Moore, G. M. (1995). *Realities of street tree planting in relation to built structures, from trees in the urban environment* (p. 11). RAIPR: Adelaide.

Perrier, F. E., Petiau, G., Clerc, G., Bogorodsky, V., Erkul, E., Jouniaux, L., et al. (1997). A one-year systematic study of electrodes for long period measurements of the electric field in geophysical environments. *Journal of Geomagnetism and Geoelectricity, 49*(11–12), 1677–1696. https://doi.org/10.5636/jgg.49.1677.

Pucinotti, R., & Barrile, V. (2002). L'utilizzo di tecniche radar per le indagini non distruttive sulle opere in c.a. In *Atti del 148 Congresso C.T.E.* (Vol. 1, pp. 147–156). Mantova.

Pucinotti, R., & De Lorenzo, R. A. (1994). Nondestructive in situ testing for the seismic damageability assessment of ancient r/c structures. In *Book of Proceedings, Third International Conference on NDT* (p. 189), Chania, Crete, Greece.

Puust, R., Kapelan, Z., Savic, D., Koppel, T. (2010). A review of methods for leakage management in pipe networks. *Urban Water Journal, 7*(1), 25–45.

Sandmeier, K. J., (2018). Reflexw 7.0 manual, sandmeier software, zipser strabe 1. *D-76227 Karlsruhe*, Germany.

Tejero-Andrade, A., Cifuentes, G., Chavez, R. E., Lopez Gonzalez, A., & Delgado-Solorzano, C. (2015). "L" and "Corner" arrays for 3D electrical resistivity tomography: An alternative for urban zones. *Near Surface Geophysics, 13*, 1–13. https://doi.org/10.3997/1873-0604.2015015.

Telford, W. M., Geldart, L. P., & Sheriff, R. E. (1990). *Applied geophysics*. Cambridge: CambridgeUniversity Press.

Topp, G. C., Davis, J. L., & Annan, A. P. (1980). Electromagnetic determination of soil water content: Measurements in coaxial transmission lines. *Water Resources Research, 16*(3), 574–582.

Ulriksen, P. (1982). *Application of impulse radar to civil engineering*. Ph.D. Lund University of Technology, Lund, Coden, Lutvdg (TVTG-1001), Sweden.

Vichabian, Y., & Morgan, F. D. (2002). Self potentials in cave detection. *The Leading Edge, 21*(9), 866–871. https://doi.org/10.1190/1.1508953.

Yau, P. (1991). Urban tree impact on building structures. In *Proceedings Royal Australia Institute of Parks and Recreation, State Conference*, Melbourne.

Chapter 6
Site Application: Forensic Crime Cases

Abstract Objects of the forensic investigations vary from illegally buried weapons and explosives, landmines and improvised explosive devices, drugs and weapons caches to clandestine graves of murder victims and mass genocide graves, environmental crime (illegal dumps, a spill of pollutants, food sophistication, etc.). Various techniques are available for forensic crime search teams. These techniques allow detecting a hidden object successfully. Near-surface geophysical search methods have been dominated by ground-penetrating radar but recently other techniques, such as electrical resistivity, magnetometric, time domain reflection have become more common. This chapter discusses some case history related to various research studies applied to crime forensic science. Will be examined the application of TDR to evidence food sophistication, the application of ERT to evidence the pollution related to dumps and hydrocarbons, the application of GPR, ERT and magnetic for the hidden bodies, clandestine graves and the search of hidden arms.

Keywords Geophysics · Food sophistication · Subsoil pollutants · Clandestine graves · Hidden bodies · Arms

6.1 TDR to Evidence Food Sophistication

To certify and guarantee the geographical origin of typical food with excellent characteristics, the European Union (EU) identifies specific food quality denominations, based on their geographical origin: the Protected Geographical Indication (PGI), the Protected Designation of Origin (PDO) and the Traditional Speciality Guaranteed (TSG) (EC Reg. 510/2006). Since the PDO and PGI certifications for EVOOs usually define the area of origin of each product, the varieties of used olives, as well as some specific guidelines for the production, the extra virgin olive oil (EVOO), is included in the same PDO share some common characteristics, including both sensory properties and specific chemical composition. Sensory analysis is an essential tool for evaluating the quality of oils obtained by olives.

The investigation of the dielectric properties of EVOO can done good results since really different levels of dielectric behaviour characterise oil and water, with

© Springer Nature Switzerland AG 2020

G. Leucci, *Advances in Geophysical Methods Applied to Forensic Investigations*,
https://doi.org/10.1007/978-3-030-46242-0_6

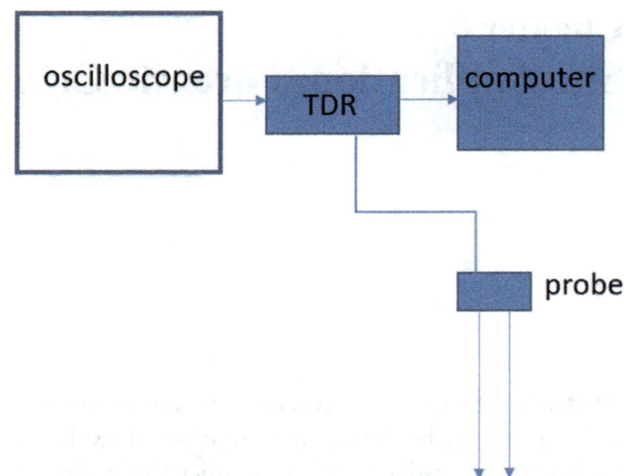

Fig. 6.1 Layout of the TDR instrumental

dielectric constant ε values ranging from about 3–3.2 for edible oils and about 80 for water (measured at 1 MHz and 25 °C) (Lizhi et al. 2008).

Literature proves the effectiveness of the non-destructive measurement for moisture content applied for the analysis of food dielectric properties (Nelson 1991; Sumnu et al. 2005; Tang 2005; Sosa-Morales et al. 2010). The TDR was successfully used in the last 20 years for the assessment of soil water content and salinity (Topp et al. 1982; Dalton and Van Genuchten 1986; Noborio 2001; Fellner-Feldegg 1968; Van Loon et al. 1995; Nozaki and Bose 1990; Pettinelli and Cereti 2002; Cataldo et al. 2009; Miura et al. 2003; Kent et al. 2004).

TDR measurements: The layout of the TDR instrumental is shown in Fig. 6.1.

TDR measurements were performed on a sample of EVOO. A step-like voltage signal at a frequency of 15 GHz was generated in a coaxial probe designed and realised to ensure a stable 50 Ω impedance (Cataldo et al. 2009).

The probe (length 66 mm, external diameter $a = 2$ mm and inner diameter $b = 4.4$ mm) is filled with the liquid under test.

TDR signals were analysed by the classical theory of the time domain response to assess the water content (WC) of the EVOO samples. The dielectric constants of the system characterised by the EVOO samples in contact with the glass pipes covering the 2-terminal metallic wires were calculated by analysing the voltage reflection coefficient (q) waveforms. These waveforms were obtained from the TDR signals according to the following equation:

$$\rho = V_r / V_i$$

where V_i (V) is the incident voltage (measured at the air-oil interface) and V_r (V) is the reflected voltage.

By analysing the reflection coefficient waveforms the dielectric constant ε of the dielectric system characterised by oil and materials surrounding the conductive probe terminal was calculated:

$$\varepsilon = (ct/2L)^2$$

where c is the velocity of the electromagnetic wave in free space (0.3 m/ns), t is the travel time of the wave (from the interface air–oil to the end of the probe), and L is the probe length immersed in the oil. The travel time of the reflection (t) was obtained according to the method described in Chap. 4 of this book, as showed in Fig. 6.2.

The length (L) of the probe portion inserted in the oil, was calculated by averaging eight measures conducted in the intersection points between the concave oil meniscus and the probe.

TDR signals, both for measurements without and with oil, are plotted, for all the used oils, in Fig. 6.3.

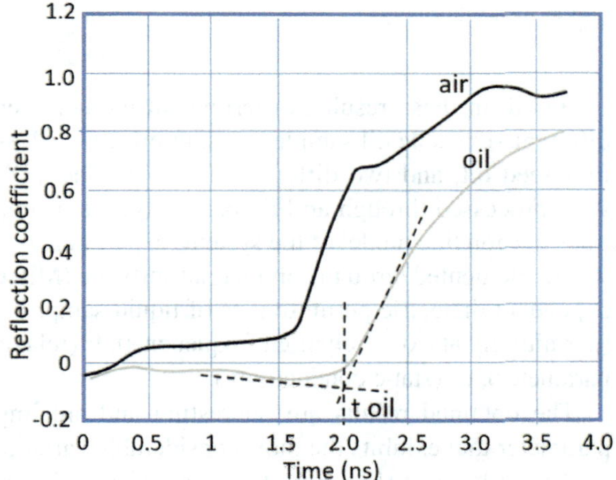

Fig. 6.2 Estimation of the travel time of the reflection (t) from the rise of the reflection and from the use of tangent lines to the waveform (in the graph only the first 4 ns of the signal are shown)

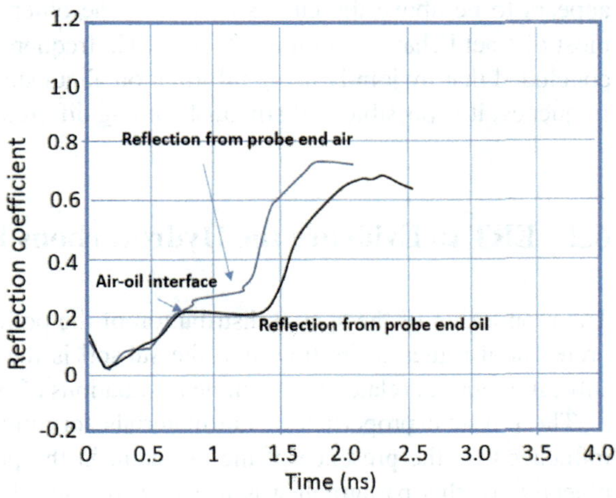

Fig. 6.3 TDR signals for measurements with air (grey) and with EVOO samples (black)

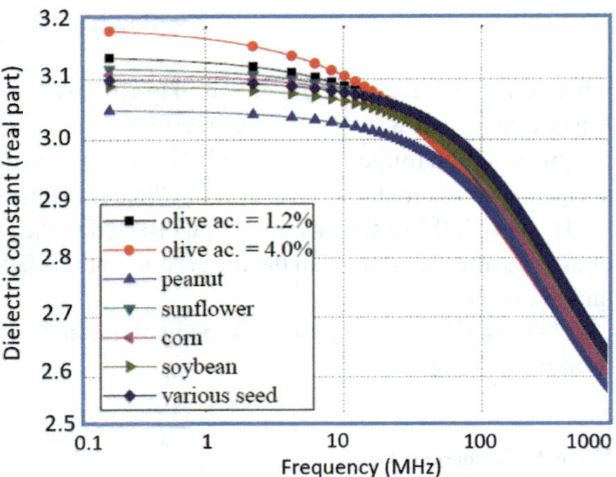

Fig. 6.4 Real part of the relative permittivity for the seven types of oil (Cataldo et al. 2009 modified)

Based on these results, experimental measurements were performed on seven different vegetable oil samples: peanut oil, corn oil, sunflower oil, soybean oil, various seed oil, and two different olive oils. The TDR collected time-domain data were processed through an FFT-based algorithm (implemented in MATLAB). The transmission line model of the system (experimental setup, probe, and filling liquid) was implemented through commercial software (Microwave Office). The frequency-dependent dielectric permittivities of liquid samples are ε_s (static permittivity), ε_∞ (permittivity at extremely high frequencies), fr (relaxation frequency), β (dispersion parameter), σ_s (static conductivity).

The obtained results are interesting and are important to note that fr is the parameter that exhibits the most considerable variations among the considered oils.

As can be seen (Fig. 6.4), there are specific frequency ranges in which the permittivities of the different oils can be discriminated. The permittivity of olive oils appears to be above the curves related to the other oils. Vegetable oils show the most distinct behaviour in the 0.2–100 MHz frequency range. Altogether, it can be concluded that by jointly using information about static permittivity and relaxation frequency, it is possible to distinguish among different oils.

6.2 ERT to Evidence the Hydrocarbons Pollution

The monitoring of the spatial distribution of the pollutant and the identification of preferential routes of its flow into the subsoil is the objective of the study of the subsoil pollution, related for example to situations of sites polluted by hydrocarbons.

The physical properties of the materials forming the subsoil are significantly influenced by the presence of the pollutant in the pores and the cracks. The first objective of this paragraph was to show to determine the presence of pollutants, spatial distribution, quantity, and what could be its migration as a function of time.

In this work, the possibility offered by the geophysical methods ERT in identifying and mapping the presence of hydrocarbons in the subsoil is taken into consideration. For this purpose, an application of this methodology is presented near a (disused) fuel distributor.

The identification of a pollutant in the subsoil is generally achieved through direct methods (such as coring). These methods can give remarkable results with the drawback, however, of being limited to the point of application only. The extension of these methods to a larger or smaller area, in addition to the high cost, could cause more damage than the pollutant itself. The methods of geophysical investigation allow a rapid qualitative analysis of the investigated structures and, in many cases, are a valid alternative to the methods of direct investigation. The physical properties of a medium, expressible through measurable geophysical parameters such as the electrical resistivity, the velocity of propagation of seismic and electromagnetic waves in the medium itself, are influenced by the presence of the pollutant. Therefore, the information obtainable with the use of geophysics should be of particular utility for the characterisation and evaluation of the integrity of large areas.

Experimentally, the possible response, in terms of variation of the physical parameters involved, of a site polluted by hydrocarbons to a possible geophysical survey was taken into consideration. Therefore geophysical surveys have been carried out in controlled situations. The data were acquired in two different phases: the first in the absence of a pollutant; the second immediately after the hydrocarbon introduction into the subsoil. The data were collected using the dipole-dipole electrode arrangement on a linear array of 24 equispaced electrodes of 0.4 m. After five iterations, the subsoil resistivity model was obtained, shown in Fig. 6.5.

The distribution pattern of the underground resistivity shows an almost horizontal stratigraphy, with resistivity increasing with depth from values lower than 80 Ωm, characterising the superficial layer from about 0 to 0.5 m of depth, at values close to 700 Ωm, characterising the next layer (Fig. 6.5a). In the two profiles acquired before and after the introduction of the pollutant (Fig. 6.5a and b) no essential differences are noted. However, considering the variation model, in percentage, of the resistivity (Fig. 6.5c), obtained from the difference of the models in Fig. 6.5a and b respectively are possible to note an area, precisely at the point of injection of the pollutant, in which the increase of resistivity varies from 3.0 to 5.0%.

For the real field measurements, the area of a disused fuel distributor was considered (Fig. 6.6a). To understand whether the area, at a distance of 15 years from the closure, is or is not affected by hydrocarbon pollution, 3D ERT survey was carried out.

The 3D resistivity measurements were acquired on a square area in a grid of 23 m × 23 m. A dipole-Dipole array was used with 24 parallel profile 1 m spaced. The electrodes distance was 1 m. The results of some 2D profiles are shown in Fig. 6.6b. Their interpretation will be useful to understand the 3D results better.

The 2D models showed in Fig. 6.6b have highlighted the presence of (i) two anomalies (indicated with Cisterna1 and Cisterna2) whose roof is placed at a depth of about 0.5 m, and resistivities above 800 Ωm. These anomalies have a regular shape of about 4 m of length, 4 m of width and extend to the depth of about 2.5 m. Their

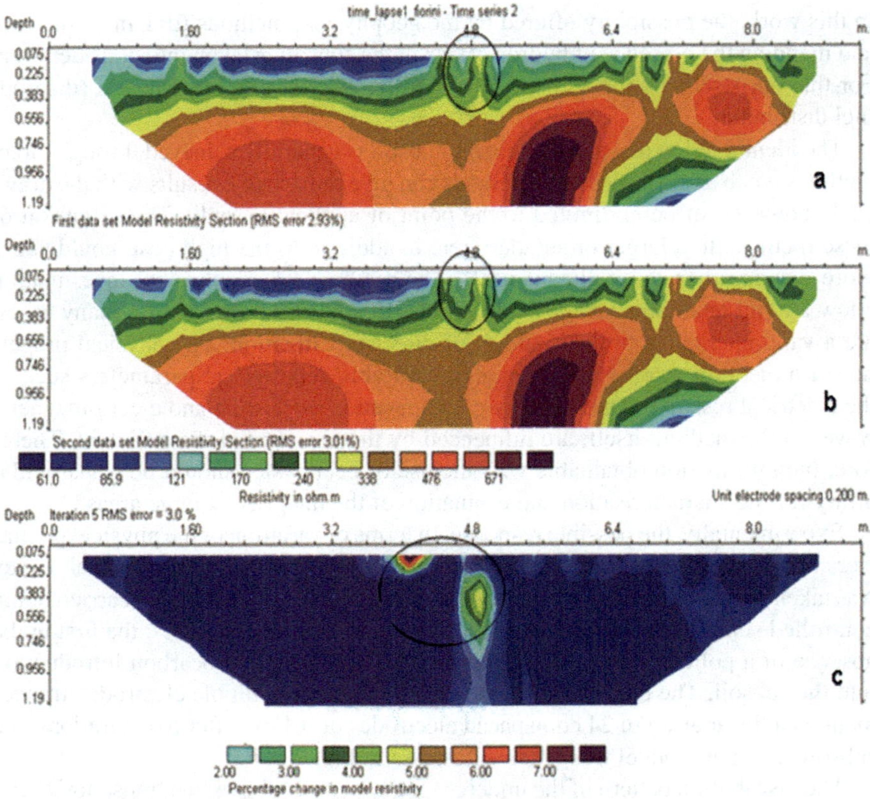

Fig. 6.5 Model of distribution of the resistivity in the subsoil; dipole-dipole electrode device: **a** without pollutant; **b** with pollutant; **c** percentage variation of resistivity

shape and size suggest the presence of the tanks that contained the fuel. The fact that the resistivity values are low compared to those expected from the presence of voids (greater than 10,000 Ωm) suggests that the cavity is not empty but partially filled with wet debris; (ii) a layer from the surface to about 7 m in depth with resistivity values ranging between 30 and 50 Ωm. These resistivity values suggest the presence of geological deposits typical of the area; (iii) an anomaly (indicated by C), at a depth of about 2.5 m, immediately below the anomaly called Cisterna1 with a resistivity of about 80 Ωm. The fact that the resistivity values are slightly higher than the surrounding medium reinforces the hypothesis of a probable presence of fuel in the subsoil. Figure 6.6c shows the resistivity depth slices that describe the variations in resistivity in the subsoil at various depth levels. In the resistivity slices are evident the main alignment of the resistivity anomaly (about 80 Ωm) indicated with C, in the slices from 1 to 5 m in depth, indicate the presence of pollutant. Furthermore, the presence of three cisterns is well highlighted. If the threshold is decreased, using

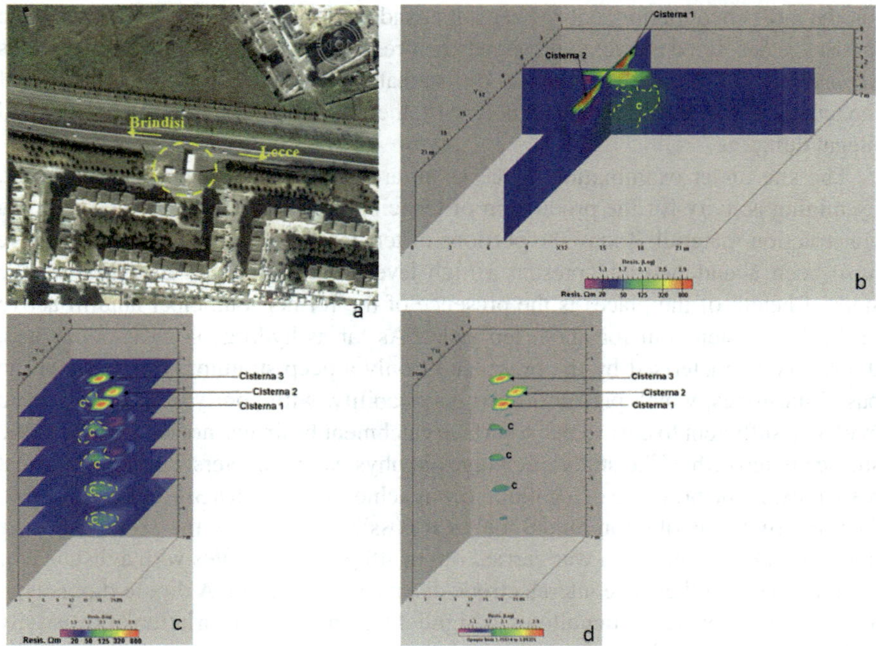

Fig. 6.6 **a** Site investigated with geophysics. **b** 2D models of the distribution of the resistivity in the subsoil relative to two profiles acquired within the grid. **c** 3D model of resistance variation in the subsoil at various depths. **d** 3D visualisation of the resistivity anomalies greater than 800 Ωm

resistivity values greater than 70 Ωm it is possible to highlight the distribution of the pollutant in the subsoil (Fig. 6.6d).

6.3   ERT to Evidence the Pollution Related to Dumps

The results of a geophysical survey carried out in an area affected by the presence of a former municipal landfill are reported. The presence in the area of the former municipal landfill creates a scenario with a potentially significant impact on the environment. The possible lack of waterproofed side and bottom barriers allows landfill leachate to enter the subsoil, with a high probability of pollution of the main groundwater zone.

The effects of this phenomenon can be observed for decades, so it is clear what the risk of a hygienic-sanitary nature to which the resident populations near the landfill have been (but still are) has been. In this regard, it is sufficient to think that for conformation the area is suitable for the development of the agricultural activity, with consequent use of water potentially contaminated by potentially toxic substances. The scenario that can be observed along the roads surrounding the area is not the most encouraging: entire heaps of abandoned waste form the backdrop to the landscape described above; they range from tires to car batteries, from durable goods of various types to production waste. Unfortunately, we talk about peripheral areas of the territory, easily accessible and poorly lit, therefore the ideal site for uncontrolled illegal dumping.

The site under examination concerns an area interested, above all in the past, by mining activity for the production of tenacious calcarenite rock to be used as a construction material. Today the portions of territory that are no longer productive have been abandoned and present a high level of degradation. To aggravate the state of health of the places is the presence of the former municipal landfill active in the last millennium for about ten years. As far as hydrogeology is concerned, the site is characterised by the presence of only a deep stratum, circulating in the base limestones, which present mixed permeability, with poorly developed karstic evidence sufficient to ensure that a surface catchment basin has not developed on the surface noteworthy. The study of changes in physical parameters, such as electrical resistivity, in the presence of a pollutant, was achieved through a 3D data acquisition. The resistivity distribution model makes it possible to identify the presence of the pollutant. Data acquisition was carried out on 48 parallel profiles with a distance of 10 m from each other. The interelectrode distance is also 10 m. A dipole-dipole type array was used, which better allows to highlight the horizontal variation of resistivity and therefore highlight any spills present in the subsoil.

The first slice, located at a depth of about 1 m from the ground level, is characterised by the presence of material with resistivity values typical of debris covers (Fig. 6.7a). Already at this depth begin to see higher values represented by the yellow colour: here is present a structure with high resistivity such as the geomembrane. In fact, at a depth of about 8 m, the landfill body wrapped by the waterproofing membrane is completely highlighted by a colour ranging from yellow to dark orange, this shade representing high resistivity values typical of insulating material. A much lower resistivity value appears in the middle of the structure with a green tone. Probably in this point, the geomembrane has been torn apart, leaving this "hole" of resistivity.

The consequence of this aspect can be noted in the lower layers of the sections of Fig. 6.7a, where a body with a lower resistivity value appears, represented by a blue tone: it is the leachate that comes out of the body of the landfill and goes to spread in the adjacent subsoil. The deeper it goes down, the more the percolate spot widens, highlighting a diffusion phenomenon that over time will favour a precise direction based on what are the geological and hydrogeological characteristics of the area. The software used made it possible to obtain a three-dimensional model of the landfill in question, and through a resistivity filter operation, it was possible to highlight the hypothetical form of the main structures of the former municipal landfill (Fig. 6.7b). It began to highlight what could be the waste body housed in the landfill; it was sufficient to isolate an iso-resistivity surface between 107.7 and 109.9 Ωm, which represents a typical range of a landfill body insulation material, as can be seen in Fig. 6.8b. Subsequently, even higher resistivity values were taken into consideration, to identify what could be the shape of the residual geomembrane of the landfill (Fig. 6.7c). From the image, it is evident the fragmentation of this

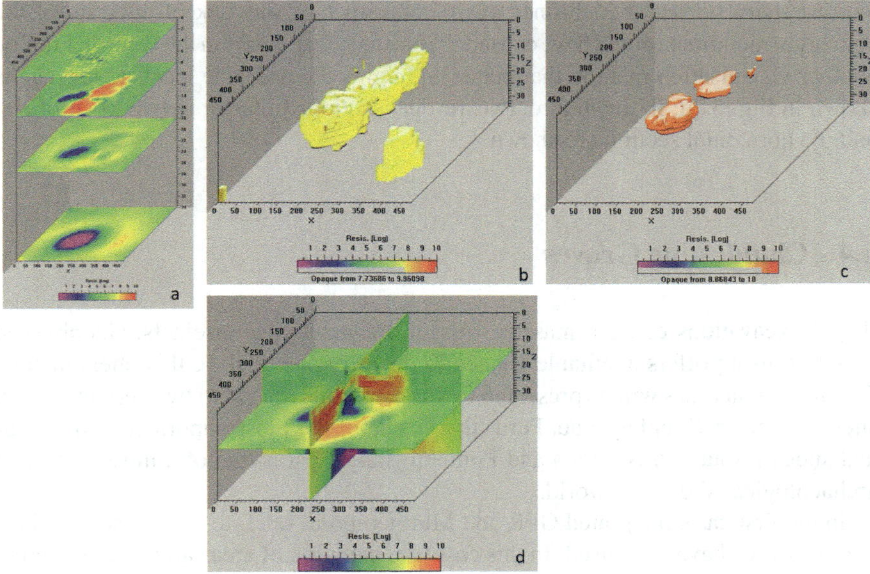

Fig. 6.7 a Distribution of resistivity at various depths, obtained from data processing; **b** iso-resistivity surface that highlights the body of the landfill; **c** iso-resistivity surface that shows the damaged geomembrane; **d** distribution of resistivity on two vertical and one horizontal section

Fig. 6.8 1955 aerial image of the Ceremonial Centre of Cahuachi. The photo shows a wide area affected by circular holes made by grave robbers (Lasaponara et al. 2014)

iso-resistivity surface, which once again confirms how the lack of integrity of the insulation membrane could have caused the leachate leaking from the landfill. Finally, another visualisation method that better highlights the probable leachate leakage is shown in Fig. 6.7d. In this figure, two vertical sections (relative to two 2D profiles) and the horizontal section are shown.

6.4 Clandestine Graves

Illegal excavations cause a massive loss of archaeological artefacts. Geophysical remote sensing offers a suitable chance to quantify and analyse this phenomenon. Two important cases will be presented in this paragraph related to two very important sites: Ventarron (Lambayeque, Peru) that is one of the most important archaeological sites in Southern America and Pompeii (Italy) that is one of a most important archaeological site in the world.

In the first case, integrated GPR and Multitemporal satellite images acquired for the study area have been used. In this case, the mapping of areas affected by looting offered the opportunity to investigate such areas not previously systematically documented. To this purpose, Ground Penetrating Radar prospections were conducted in some looted sites. In the second case, integrated GPR and ERT measurements were performed.

South America case: in Peru, illegal excavations date back to the Spanish colonial period (Silverman 1993) and have had a strongly increased in the twentieth century (Alva 2001). Figure 6.8 show the dimension of this problem.

Lambayeque region, in northern Peru, represents an emblematic case for the illegal excavations and rescue archaeology in Peru (Fig. 6.9).

The well-known discovery of the Royal Tombs of Sipan in 1989 was made after a "disagreement amongst looters" involved in an illegal excavation. This quarrel caused the intervention of the police and the involvement of local archaeologists, Walter Alva and Susana Menezes, to examine the looted objects (see Watson 1999). As a consequence, archaeological rescue started, and the following systematic researchers and excavations enabled the discovery of one of the richest tomb area in South America. The study area is located in Ventarron, at about 20 km from Sipán (Fig. 6.10).

Here satellite imagery and geophysical techniques were used. The mainly research focused on areas close to Arenal and Cafetal (Fig. 6.11).

The two areas (Cafetal and Arenal, named A and B in Fig. 6.11) chosen for the survey, had been affected by intense excavation in the last decade. Figure 6.12a shows two satellite images both related to the Area A in Cafetal acquired in 2003 and 2010, respectively. In the two images are visible the circular holes.

In Arenal (see Fig. 6.13), the selected test site (Area B) puts into evidence significant changes in the pattern linked to plundering.

The excavation area was enlarged of about 36% (0.74 ha in 2003, 1.01 ha in 2010).

In Cafetal, the total extension of the surfaces affected by plundering is 2.39 ha. In Arenal, the extension of looted areas is 3.15 ha. The archaeometry analysis of

Fig. 6.9 Typical looting patterns in Lambayeque from an aerial view (courtesy by Ignacio Alva)

Fig. 6.10 The investigated area

Fig. 6.11 Arenal, Ventarron and Cafetal. A and B indicate the areas located in Cafetal and Arenal, respectively, investigated by processing VHR satellite imagery. In B some circular holes have been investigated by georadar

the circular holes put into evidence different sizes for the two investigated sites: about 2.5–4 m in Cafetal and larger than 4 m in Arenal. This is due to different soil granulometry (mailingly sand in Arenal and silty sand in Cafetal) which determines different stability conditions of the cavities.

GPR data, acquired with the Ris Hi mod georadar equipped with the dual-band antenna 200–600 MHz, show meaningful returns from meaningful depths (a depth of some returned echoes up to 5 m was estimated) but also from quite shallow targets in other points. Good penetration is because the area is semi-desertic. From the hyperbolic tails of the GPR data, the EM wave propagation velocity was estimated. It is ranging from about 7 cm/ns to about 12 cm/ns, corresponding to a relative dielectric permittivity ranging from 19.62 to 6.25.

GPR data were acquired in two selected areas. In the first area located in Cafetal (Fig. 6.14), GPR data were acquired directly on the excavation to verifying the presence of archaeological remains.

Only two GPR profiles were acquired (Fig. 6.14b) because the objective difficult related to the irregular excavations (Fig. 6.14a). After the processing data has been corrected according to the topography of the surface that, of course, is not flat because the prospection follows the residual excavation (and soil deposit) profile left by the huajeros. This poses a theoretical problem about the correctness of the processing because some operations (in particular the migration) presuppose a flat interface. However, the average curvature ray of the hole is quite large with respect the central wavelength, and this makes us reasonably confident about the fact that a "usual" GPR

Fig. 6.12 a Area A in Cafetal from 2003 Google Earth scene (left-hand panel) and 2010 GeoEye imagery (right-hand panel). The comparison of the two scenes puts into evidence illegal excavations carried out between 2003 and 2010; b zoom on 2010 image

processing is licit also in the cases at hand. However, the final topographic correction drove us to change the interpretation of some of the focused anomalies. In particular, in Figs. 6.15 and 6.16, the achieved results are shows.

In both Figs. 6.15a and 6.16a refer to the result achieved without topographic correction and Figs. 6.15b and 6.16b refer to the result after topographic correction. A hyperbolic shaped reflection labelled "A" at two-way travel time window between 10 and 30 ns has a size of about 0.5 m, and depth of the top is between 0.4 and 1.2 m (with an average electromagnetic wave velocity of 0.1 m/ns). Even so, the topographic correction revealed to be essential to reduce the "false alarm ratio". In particular, in Figs. 6.15a and 6.16a the anomaly labelled "A" might be interpreted as buried targets from the non-corrected profile. However, the topographic correction reveals that they are quite probably just related to the curved stratification of the excavated area (Figs. 6.15b and 6.16b). Unfortunately, the acquisition of just two GPR profiles allows, only, to create a pseudo-3D visualisation of the acquired data. Data were displayed in a cube in which they appear in the position where they were

Fig. 6.13 Area B in Arenal, from 2003 Google Earth scene and 2010 GeoEye RGB, respectively. The multitemporal observation puts into evidence the increasing of looting activity in the southwestern part of this area (Lasaponara et al. 2014)

acquired (Fig. 6.17). This allows us to understand the development in a 3D way of the anomaly highlighted in the 2D profiles.

Another aspect appreciable at this stage is the level of the apparent maximum excavation depth. For example, in Fig. 6.16b seems that the maximum excavated depth has reached the time depth of about 1.2 m below the soil level. However, these aspects deserve further investigations.

In the second area located in Arenal (Fig. 6.18), the GPR prospecting was carried out near an abusive excavation to verify the existence of archaeological feature that could be preserved. Data were acquired in continuous mode along 0.5 m spaced

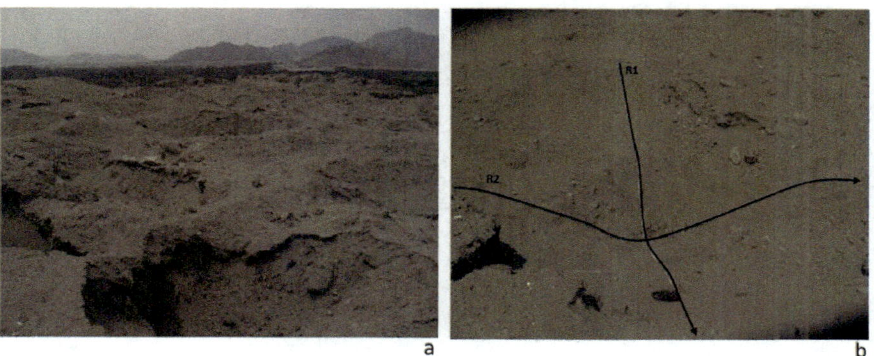

Fig. 6.14   Looted area in Cafetal investigated by GPR; **a** detail of the holes; **b** GPR profiles location

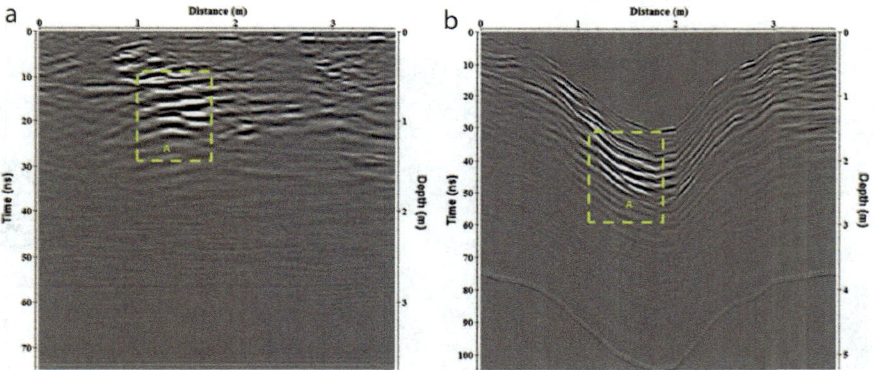

Fig. 6.15   Processed R1 GPR radar section in an alleged looted tomb, in Arenal (area B). **a** Without topographic correction; **b** with topographic correction

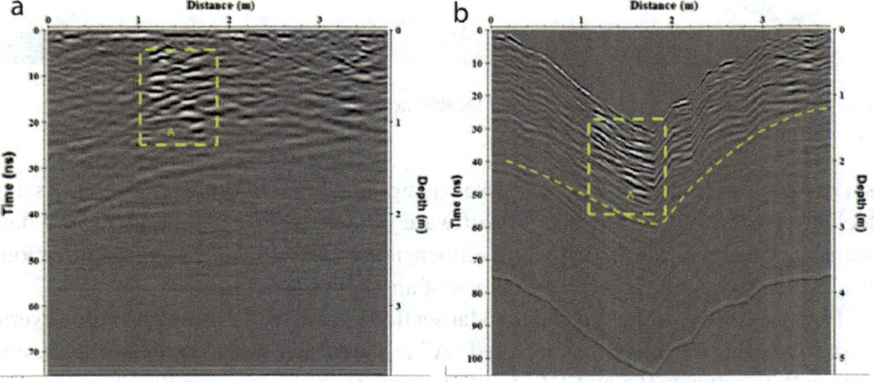

Fig. 6.16   Processed R2 GPR radar section in an alleged looted tomb, in Arenal (area B). **a** Without topographic correction; **b** with topographic correction

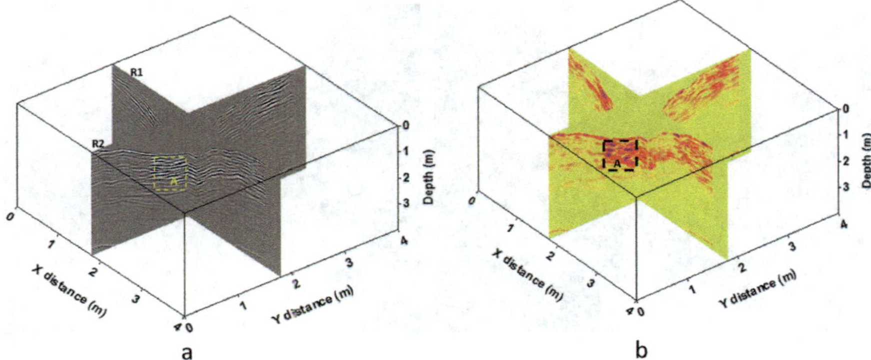

Fig. 6.17 The pseudo-3D visualisation of the 2D processed radar sections labelled R1 and R2:
a radar sections; **b** enveloped radar sections

Fig. 6.18 The 3D surveyed area in Arenal located near an abusive excavation

survey lines and subsequently processed using standard two-dimensional processing
techniques with the GPR-Slice 7.0 software (Goodman 2013). The processed data
were subsequently merged into three-dimensional volumes and visualised in various
ways to enhance the spatial correlations of anomalies of interest.

Figure 6.19 shows the processed radar section related to the second profile. Several
hyperbolic shaped reflections labelled "A" are present. Its size is about 0.5 m, and
the depth is between 0.5 and 1.17 m (with an average electromagnetic wave velocity
of 0.078 m/ns). Furthermore a continuous and slightly undulating reflector appears
strong and irregular and reaching a maximum depth below the ground surface ranging
from 1.1 to 1.3 m (dashed yellow labelled "S").

Fig. 6.19 Processed R2
GPR radar section

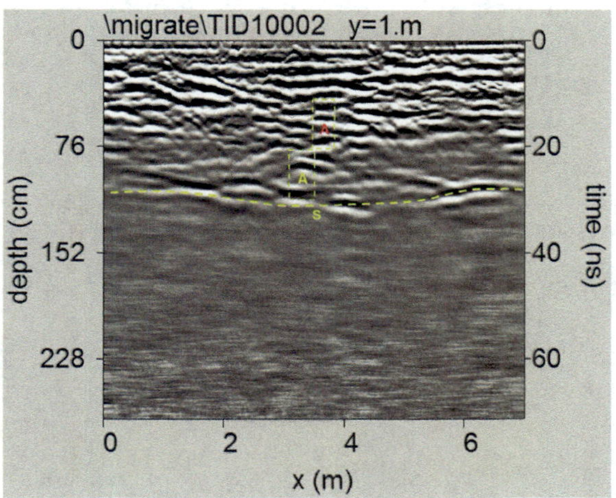

Time slices using the overlay analysis (Goodman et al. 2006; Goodman and Piro 2013) were built (Fig. 6.20).

Depth slices (0.54–1.28 m depth) show the anomaly A, while depth slice ranging from 1.22 show the anomaly S. The anomalies labelled A can be related to the adobe walls, while the anomaly labelled S, is related to the ancient living surface. The visualisation of the GPR data as isoamplitude surfaces (Fig. 6.21) allows the archaeological interpretation (Conyers 2004). Relatively strong continuous reflections are visible on the threshold volumes. In this case, is visible the 3D development of the anomalies labelled "A" and "S" respectively.

Pompeii case: The suburb of the ancient city of Pompeii was populated by numerous settlement complexes scattered over the territory that responded to both productive (farms for the production of wine and oil) and residential or seasonal needs. The protection activity carried out by the Archaeological Superintendence of Pompeii and now by the Archaeological Park of Pompeii has allowed outlining a rather complex and articulated picture, with the identification of various "villas", placed in the territory of competence. The current excavation campaign, in the area of Civita Giuliana, an area about 700 m^2 northwest of the walls of ancient Pompeii, as well as confirming these data, highlighted the productive sector—servile of a large villa already, in part, investigated at the beginning of the twentieth century and the area (south and south-west of the structure) intended for agricultural use.

This area in recent decades has been affected by clandestine excavations. A geophysical campaign was undertaken in an area of about 7000 m^2 (Fig. 6.22). Successively the need to definitively interrupt these criminal actions of the impoverishment of the national archaeological heritage has determined the need to carry out a new excavation campaign through a synergic operation between the Archaeological Park of Pompeii and the Procura della Repubblica of Torre Annunziata.

Two geophysical methods were used to evidence clandestine excavation: the GPR and ERT (Fig. 6.23).

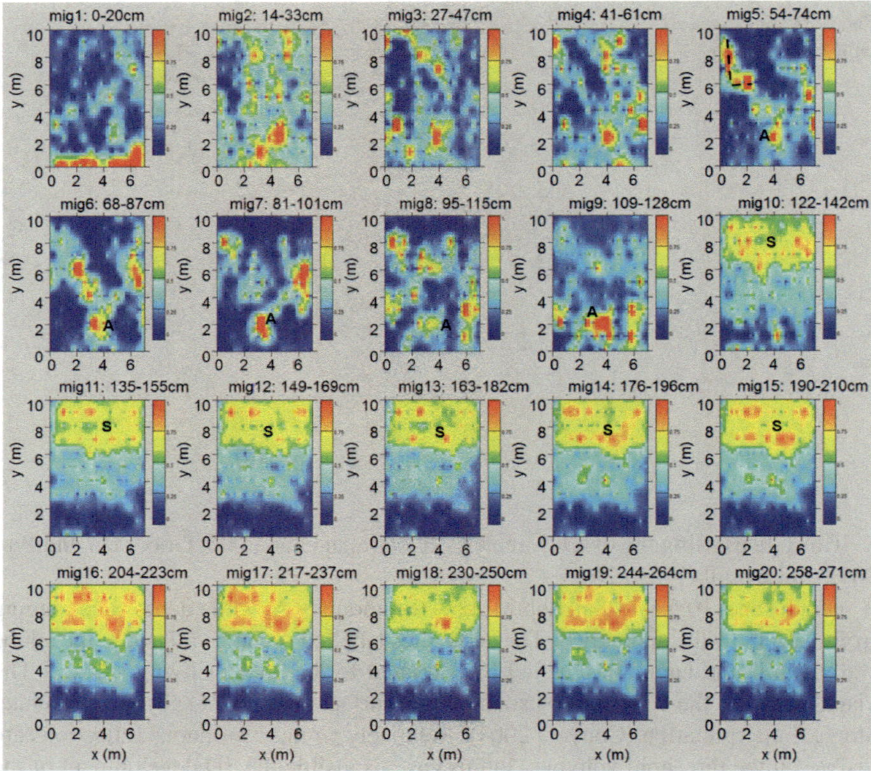

Fig. 6.20 Depth slices

Fig. 6.21 Examples of 3D visualisations using iso-amplitude surfaces of the complex trace amplitude

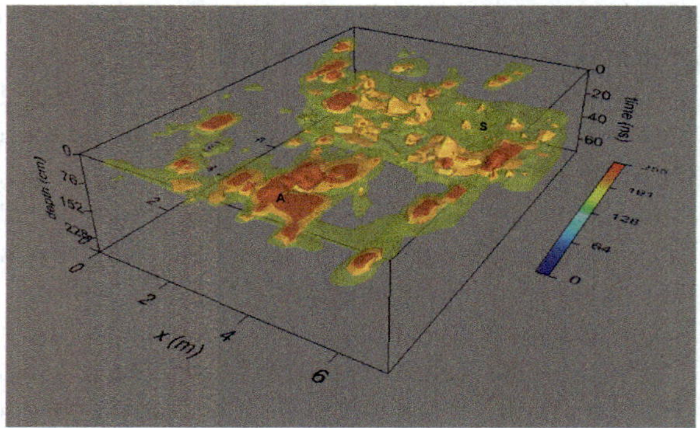

GPR data analysis: For GPR measurements a georadar Ris Hi mod with the dual-band 200–600 MHz antenna were used. Due to the presence of several obstacles, the surveyed area was divided into several areas (Fig. 6.24). In these areas, GPR data were acquired in a grid with parallel and orthogonal profile spaced 0.5 m.

Fig. 6.22 The surveyed area in Civita Giuliana Street (Pompeii, Italy)

Fig. 6.23 Phases of geophysical data acquisition

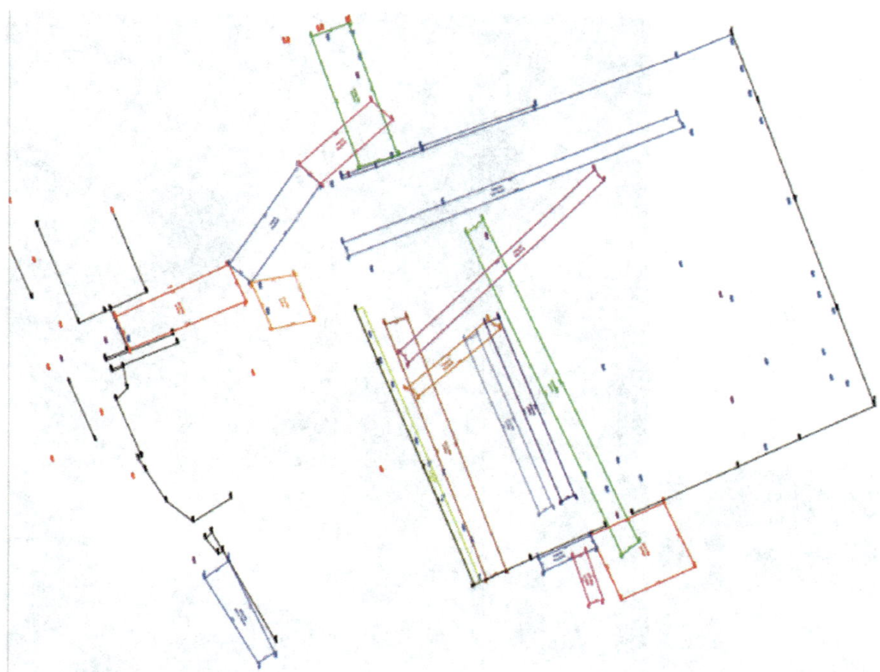

Fig. 6.24 Areas surveyed with GPR the Ground-penetrating radar surveyed areas (colored rectangulars)

The quality of the campaign data was discreet thanks to a series of measures adopted in the acquisition phase. However, to attempt to eliminate a noise component, however present in the data, and allow the simple interpretation of the data itself, processing was carried out. The processing steps were: (i) zero time filter; (ii) background removal filter; (iii) migration. The analysis of the processed profiles acquired with the 600 MHz antenna in the area A shown (Fig. 6.25):

- The first layer of variable thickness between 0.6 and 0.8 m;
- The presence of some reflected events at a depth of about 3.0–3.5 m (dashed yellow line). These events are visible between the abscissas 0 and 14 m and could be related to the presence of a tunnel;
- The presence of other reflected events related to probable structures of archaeological interest that can be found starting from the depth of 1.8 m.

The profiles acquired with the 200 MHz antenna (Fig. 6.26) allow to investigate up to a depth of about 8 m and do not show different anomalies concerning the profiles acquired with the 600 MHz antenna.

The planimetry of the profiles, acquired in a step grid equal to 0.5 m, allowed the anomalies present on each section to be spatially correlated, using the analysis of the amplitude of the events reflected within assigned time intervals (time slices).

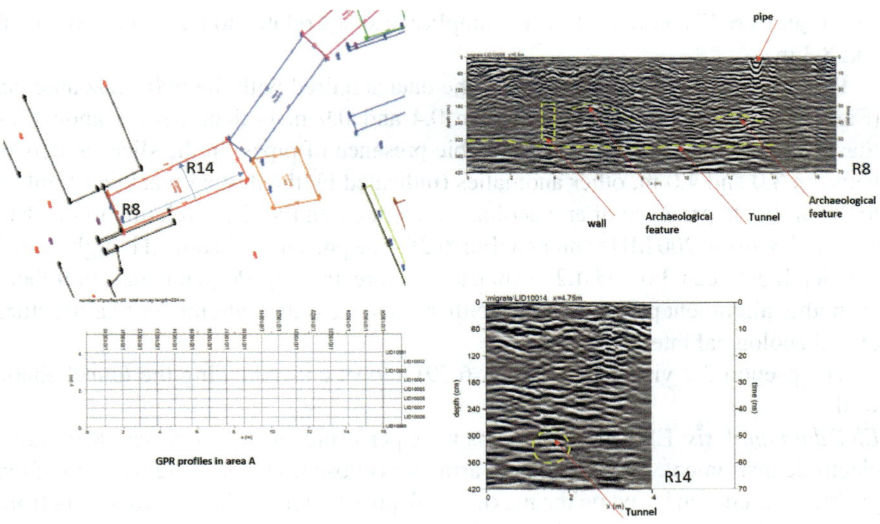

Fig. 6.25 Area: processed GPR profiles related to the 600 MHz antenna

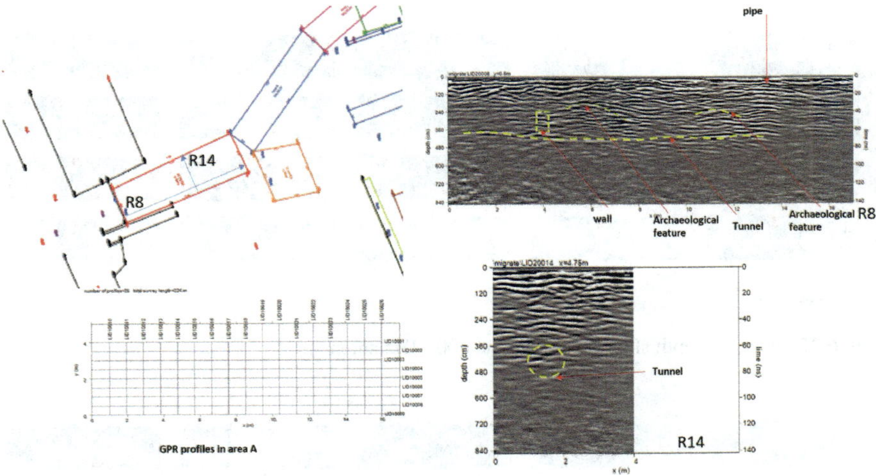

Fig. 6.26 Area: processed GPR profiles related to the 200 MHz antenna

Amplitude slices were constructed at time intervals of 5 ns; each slice corresponds to a soil thickness of about 0.3 m. The blue colour indicates the weak amplitude of the reflected signal (subsoil consisting of substantially homogeneous material); the colours from light blue to the most intense red indicate variations in amplitudes of the reflected signal and therefore the presence of significant electromagnetic discontinuities. The variations in amplitude (and therefore in colour) in the same slice are an indication of horizontal variations in the electromagnetic characteristics of the soil. Figures 6.27 and 6.28 show the amplitude slices related to a depth between 0.0 and 8.3 m.

Particularly in the slices related to the data acquired with the 600 MHz antenna (Fig. 6.27) is possible to see, between 0.4 and 0.9 m in-depth, some anomalies (dashed black line) linked to the probable presence of pipes. In the slices at depths between 3.0 and 4.0 m, other anomalies (indicated by the dashed black line) linked to the probable presence of archaeological features. In the slices relating to the data acquired with the 200 MHz antenna (Fig. 6.28), the presence of a tunnel is highlighted at a depth between 3.0 and 4.2 m (black dotted rectangle). Perpendicular to it there is another alignment probably linked either to an excavation attempt or to a structure of archaeological interest.

The pseudo-3D visualisation (Fig. 6.29) allows understanding the tunnel shape well.

ERT data analysis: ERT measurements were performed using the Syscal Kid multi-electrode instrument. A dipole-dipole array was chosen. The maximum length of the profiles was chosen based on the maximum depth of interest (the first 10 meters from the ground level) and the probable resolution requested. Therefore, 24 electrodes with

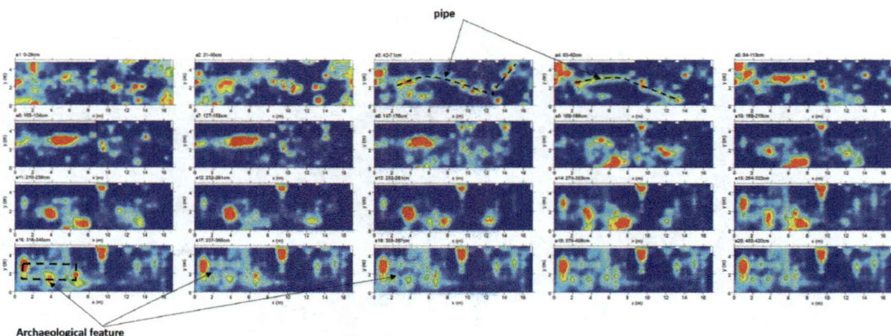

Fig. 6.27   Area A: depth slices related to the 600 MHz antenna

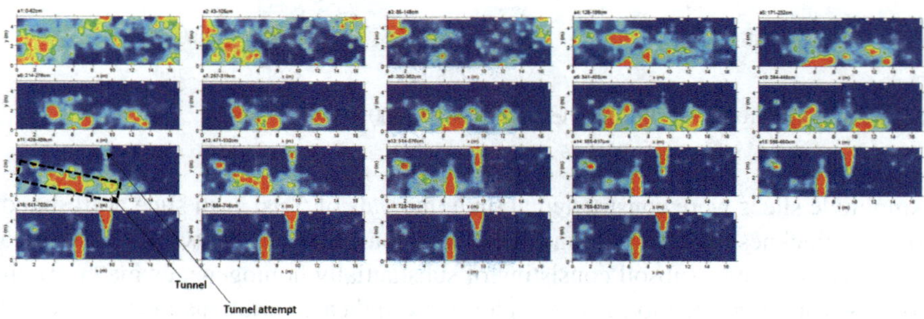

Fig. 6.28   Area A: depth slices related to the 200 MHz antenna

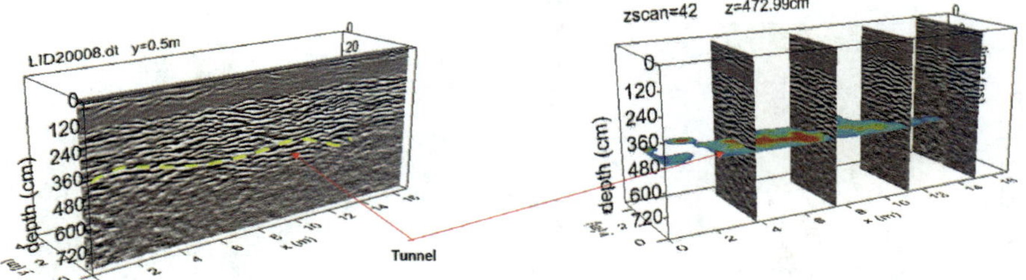

Fig. 6.29 Area A: pseudo-3D visualisation

an interelectrode distance variable from 1 to 2 m were used. Non-standard acquisition geometry was used which provides for the provision on the ground of an electric line that follows the so-called snake path (Fig. 6.30). Several profiles, in a roll along with mode, were acquired to cover the entire area. Resistivity maps were constructed using the ERTLab software.

The electrical resistivity distribution models at various depths are shown in Fig. 6.31.

From the resistivity distribution model (Fig. 6.31) is evident the presence of a heterogeneous subsoil with resistivity values between 3000 and 10,000 Ωm. Particularly it is possible to note the presence of:

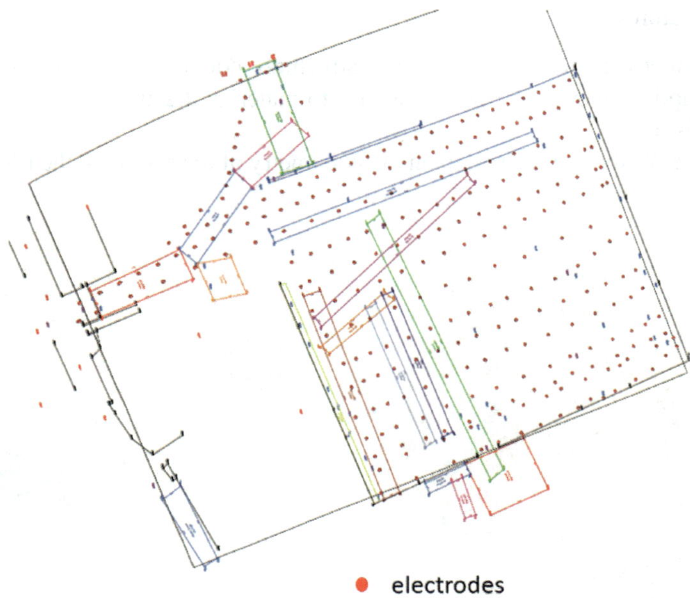

Fig. 6.30 The electrodes location in ERT survey

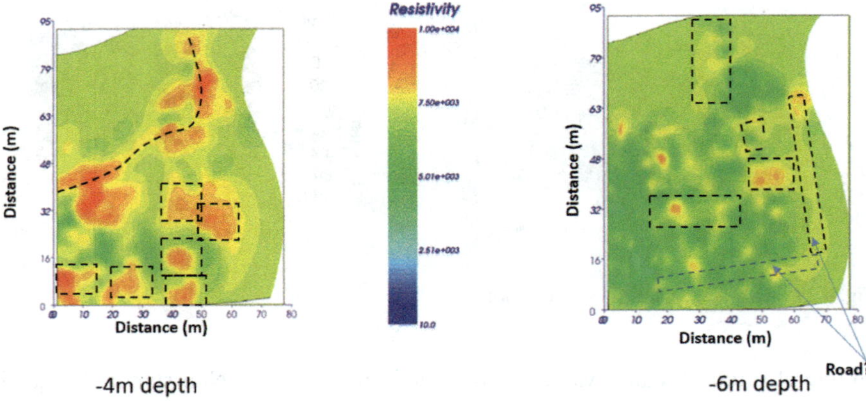

Fig. 6.31   Resistivity distribution at 4 and 6 m depth

1. areas (in red), in the slice at 4 m depth, with resistivity values between 8000 and 1000 Ωm; these values indicate the probable presence of areas in which a void is located. Here the black dotted line indicates the probable path of a tunnel. Still, at the same depth, the anomalies within the dashed black boxes could be related to the probable presence of structures of archaeological;
2. areas inside the black boxes outlined in the slice at 6 m depth, with resistivity values between 6000 and 7500 Ωm; these values indicate the probable presence of structures.

The imperfect identification of the structures could be because, at the time of the eruption, they were incorporated into materials having the same electrical characteristics.

Figure 6.32 shows the overlap on the planimetry of the results of the ERT survey.

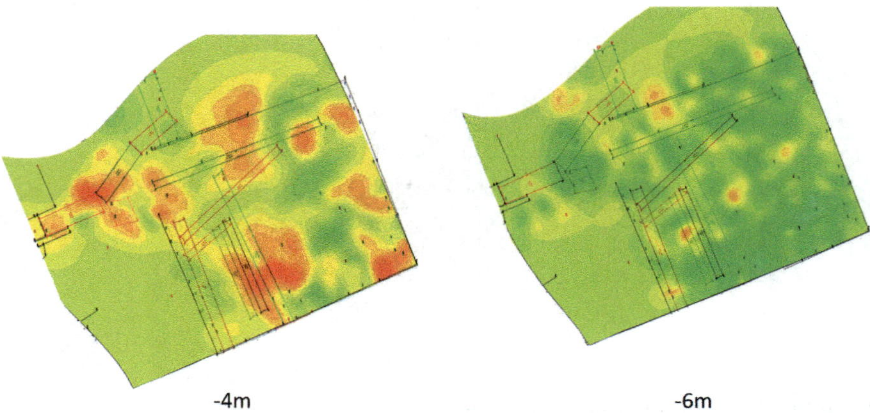

Fig. 6.32   Resistivity distribution at 4 and 6 m depth overlapped to the planimetry of the investigated area

The successive excavation has revealed the presence of several structures (Fig. 6.33) and the tunnel (Fig. 6.34).

Fig. 6.33 Photo of the excavation

Fig. 6.34 Photo of the excavation

6.5   Search of Hidden Human Body

Ground-penetrating radar mapping allows for three-dimensional analysis of archaeological features within the context of landscape studies. The method's ability to measure the intensity of radar reflections from deep in the ground can produce images and maps of buried features not visible on the surface. A study was conducted in the *Basilica della Trinità* in Venosa (south Italy) (Fig. 6.35) to study the buried human remains inside a sarcophagus. Here it is assumed that pouring of cement buried the human remains during the restoration work of the foundations of the same Basilica.

The ground-penetrating radar analysis showed them to be small anomalies likely associated with human remains allowing to recover them without damaging them.

Several authors (Conyers 2004; Freeland et al. 2002; Miller et al. 2002; Davenport et al. 1988, 1990; France et al. 1992, 1997; Leucci 2019) have demonstrated the efficacy and applicability of GPR in the detection of buried bodies. Conclusions from these studies indicate that GPR was the most important tool used to delineate graves and practical application by law.

The GPR survey was carried out with the IDS Hi-Mod georadar with 2000 MHz high-resolution antennae. Data were acquired in continuous mode along 0.05 m spaced survey lines (longitudinal and transversal profiles), using 512 samples per trace, 16 ns time range and a manual time-varying gain function.

The investigated sarcophagus was divided into two areas labelled respectively, area A and area B (Fig. 6.36).

Data were subsequently processed using standard two-dimensional processing techniques using the GPR-Slice Version 7.0 software (Goodman 2013). The processing flow-chart consists of the following steps (Leucci 2019): (i) header editing for inserting the geometrical information; (ii) frequency filtering; (iii) manual gain,

Fig. 6.35   The Basilica della Trinità in Venosa (South Italy)

Fig. 6.36   Surveyed areas inside the sarcophagus

to adjust the acquisition gain function and enhance the visibility of deeper anomalies; (iv) customised background removal to attenuate the horizontal banding in the deeper part of the sections (ringing), performed by subtracting in different time ranges a 'local' average noise trace estimated from suitably selected time–distance windows with low signal content (this local subtraction procedure was necessary to avoid artefacts created by the classic subtraction of a 'global' average trace estimated from the entire section, due to the presence of zones with a very strong signal); (v) estimation of the average electromagnetic wave velocity by hyperbola fitting (Conyers 2006; Leucci 2019); (vi) Kirchhoff migration, using a constant average velocity value of 0.1 m/ns.

The migrated data were subsequently merged into three-dimensional volumes and visualised in various ways to enhance the spatial correlations of anomalies of interest. In the present work, the time-slice technique has been used to display the amplitude variations within consecutive time windows of width $\Delta t = 1$ ns. To define the depth of human buried remains the electromagnetic (EM) wave velocity, using the characteristic hyperbolic shape of a reflection from a point source (diffraction hyperbola), was used.

Area A: 27 GPR profiles were acquired in the area A (Fig. 6.37).

The GPR profiles that were measured in the area A (Figs. 6.38 and 6.39) show different reflectors. A hyperbolic shaped reflection labelled "A" at the two-way travel time window between 3 and 6 ns is visible in radar sections. Its size is about 16 cm, and the depth of the top is between 0.20 and 0.21 m (with an average electromagnetic wave velocity of 0.1 m/ns). This reflection event was interpreted as probably due to human rests.

The hyperbolic-shaped reflection labelled "O" at two-way travel time window between 4 and 6 ns is visible in the radar section (with a size of about 5–6 cm and

Fig. 6.37 Area A: GPR profiles location

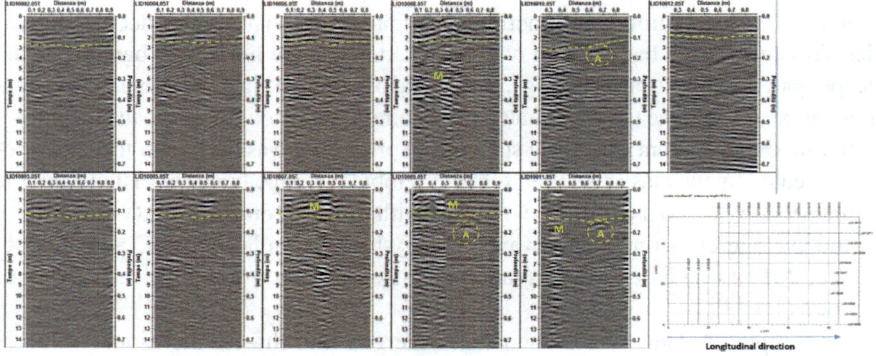

Fig. 6.38 Area A: longitudinal processed radar sections

a depth between 0.24 and 0.31 m). This reflection event was interpreted as probably due to the bones.

On each of the GPR records the lowest (dashed yellow) continuous and slightly undulating reflector appears strong and irregular and reaches a maximum depth below the surface ranging from 0.12 to 0.16 m. It probably represents the concrete boundary.

To identify the depth evolution of buried human remains, including their size, shape and location, time slices using the overlay analysis (Goodman et al. 2006; Goodman and Piro 2013) were built (Fig. 6.40). The time slices show the normalised amplitude using a range defined by blue as zero and red as 1. In the slice ranging from 6.1 to 8.6 cm depth, the relatively high-amplitude anomaly is visible (Fig. 6.41).

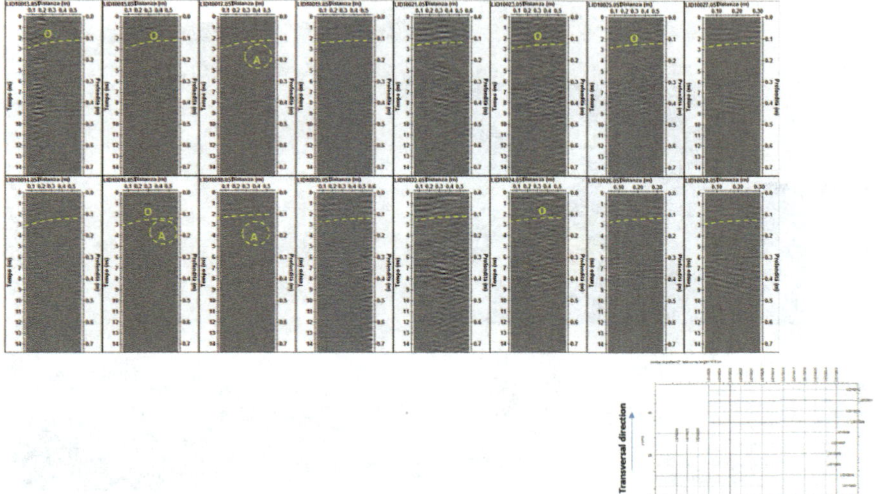

Fig. 6.39   Area A: transversal processed radar sections

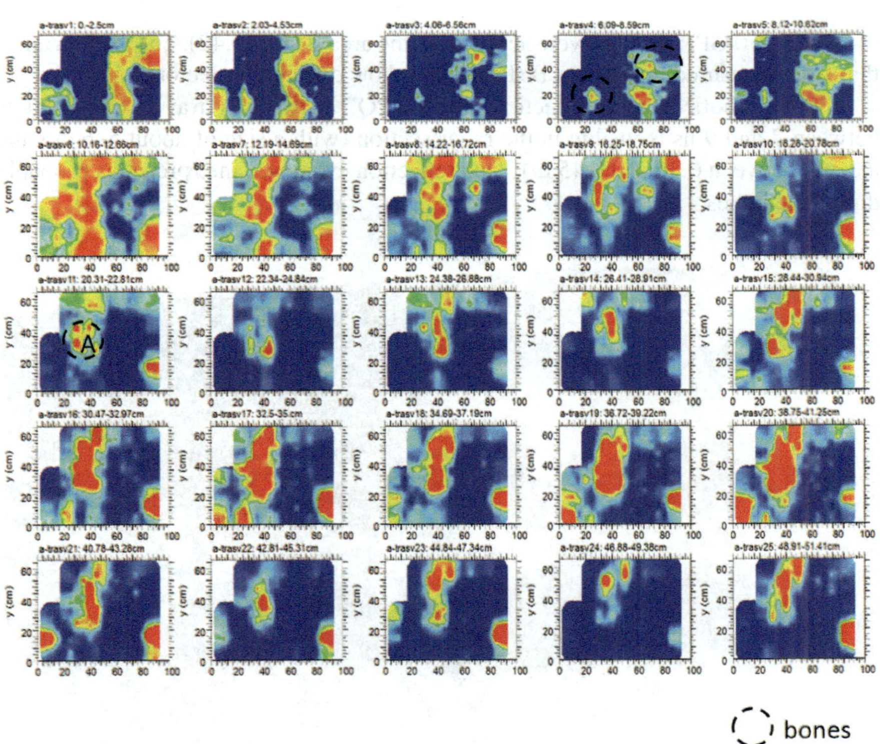

Fig. 6.40   Area A: depth slices

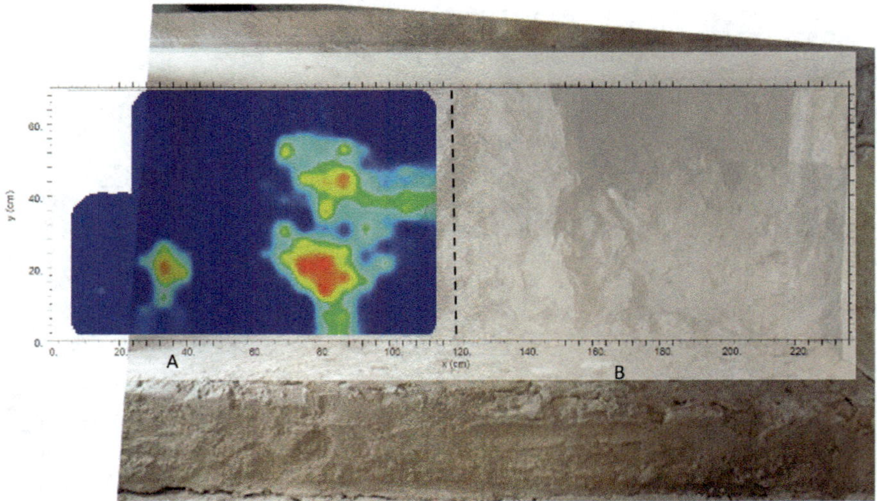

Fig. 6.41 Area A: depth slices 6.1–8.6 cm

Area B: 50 GPR profiles were acquired in the area B (Fig. 6.42). The GPR profiles that were measured in the area B (Fig. 6.43) show different reflectors.

The hyperbolic-shaped reflection labelled "O" at two-way travel time window between 7 and 9 ns is visible in the radar section (with a size of about 5–6 cm and a depth between 0.36 and 0.45 m). This reflection event was interpreted as probably due to the bones.

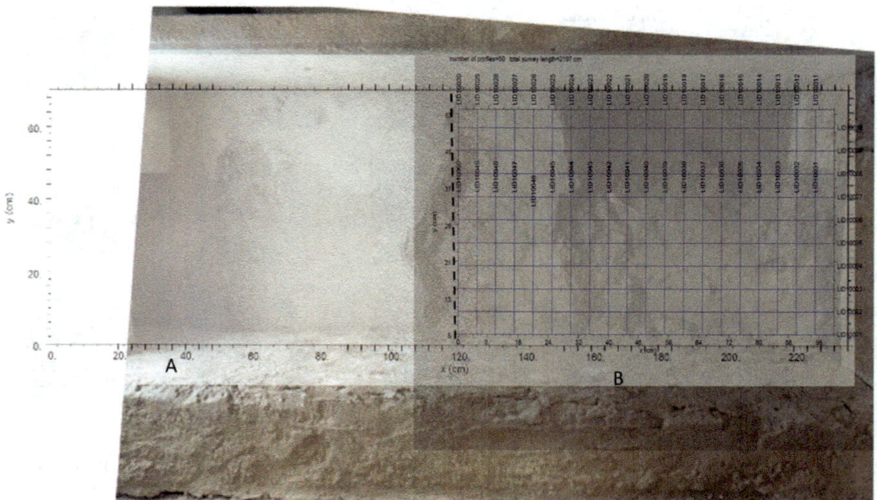

Fig. 6.42 Area B: GPR profiles location

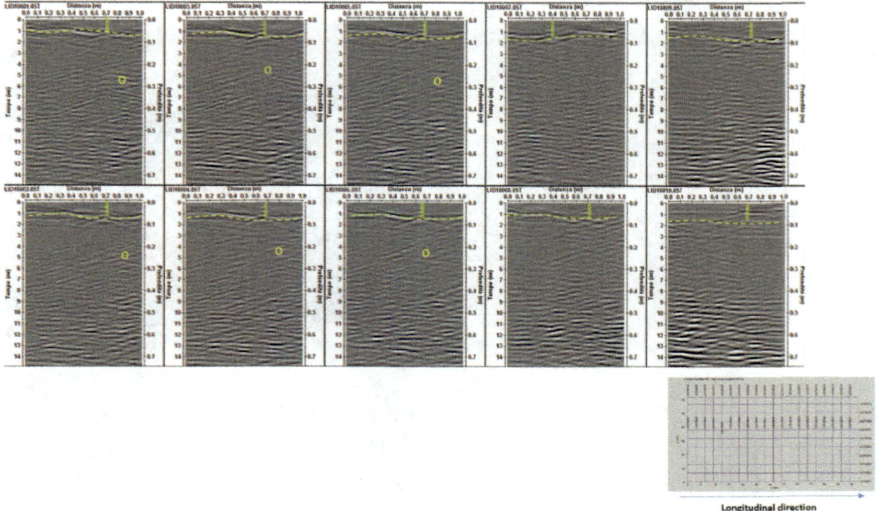

Fig. 6.43   Area B: longitudinal processed radar sections

On each of the GPR records the lowest (dashed yellow) continuous and slightly undulating reflector appears strong and irregular and reaches a maximum depth below the surface ranging from 0.08 to 0.1 m. It probably represents the empty space boundary.

Also, in this case, the time slices were built (Fig. 6.44). In the slice ranging from 3.1 to 3.5 cm depth (below the empty space), relatively high-amplitude anomaly (dark dashed circle) is visible (Fig. 6.45).

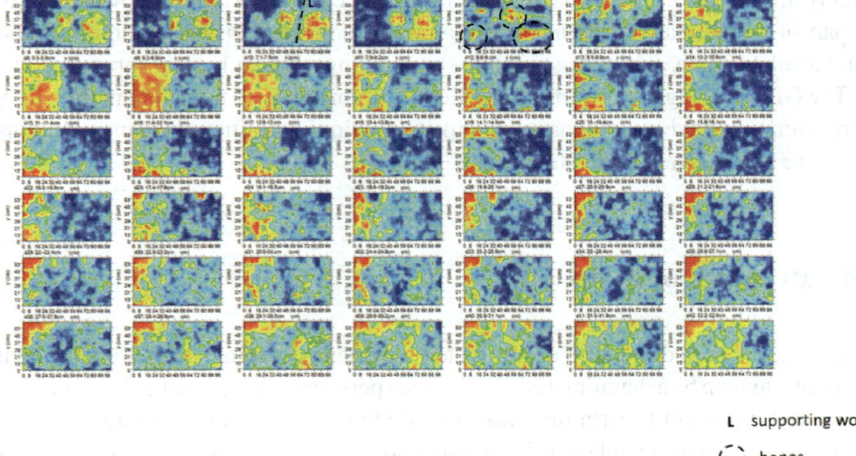

Fig. 6.44   Area B: depth slices

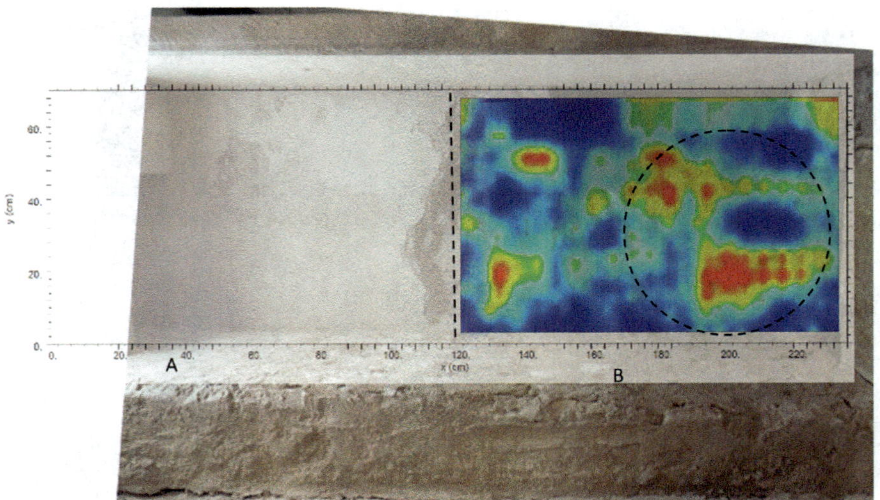

Fig. 6.45 Area B: depth slices 4 cm

The request for GPR measurements was due to provide accurate information about the exact location of buried human remains. Human remains may need to be removed due to associate a name with those remains. The benefits of GPR survey, in this case, were the possibility to use compact, portable, rugged GPR instruments; the non-invasively and rapid acquisition of the data; the onsite and real-time location and identification information on the zone of interest. Results are interesting and can help to develop an understanding of GPR variability detecting human remains in similar forensic cases, on the other hand, the results can give the possibility both to focus precisely the conservation/restoration plans and to enrich the cultural activities. Finally, the results have emphasised the importance for this kind of GPR method to be part of an integrated program of research that considers historical, archival, and other available data, mitigating possible shortcomings of geophysical survey results.

The GPR investigations allowed the identification of buried human remains which were overlapped with sarcophagus photos to obtain the position of human remains that were successively excavated (Fig. 6.46).

6.6   Research of Arms

Searching for buried metallic or not metallic evidence at crime scenes or potential disposal sites can be a daunting task for forensic personnel. In particular, it is common to search for a small firearm that was discarded or buried by the perpetrator. When performing forensic searches, it is recommended first to use non-invasive methods such as geophysical instruments to minimise damage to evidence and the crime scene.

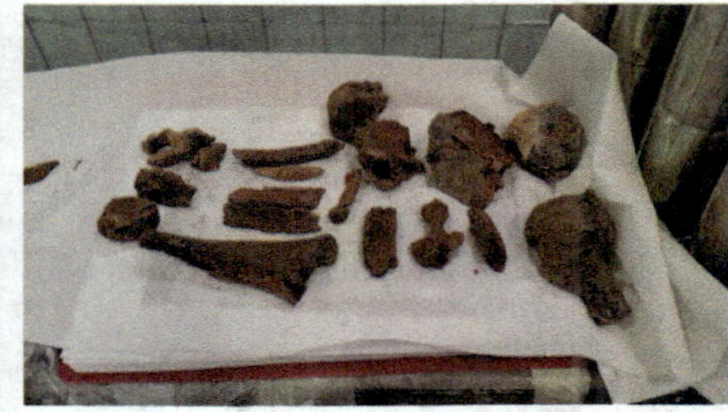

Fig. 6.46   Photo relative to the excavation results

Geophysical tools are used to pinpoint small areas of interest across a scene that will be invasively tested later.

To investigate the ability of GPR in hidden arms detection, a laboratory test was performed using two toy guns, one in metal and one in plastic (Fig. 6.47).

The two guns were hidden first individually and then together. Subsequently, a GPR survey was carried out using the Ris Hi mod georadar with the 2 GHz antenna. Parallel and orthogonal profiles have been made in a grid of 2 cm (Fig. 6.48). *Metal toy gun*: GPR data were processed as follow:

Zero time correction (Fig. 6.49);
Background removal filter (Fig. 6.50).

Analysis of processed data shows some reflection events from the air (Fig. 6.50). These events were related to the corner of the tank that contains the gun. For this reason, another step is needed in the elaboration of the data—the FK filter (for more see Chap. 4, p. 250) (Fig. 6.51).

Fig. 6.47   Photo of the two toy gun: **a** metal; **b** plastic

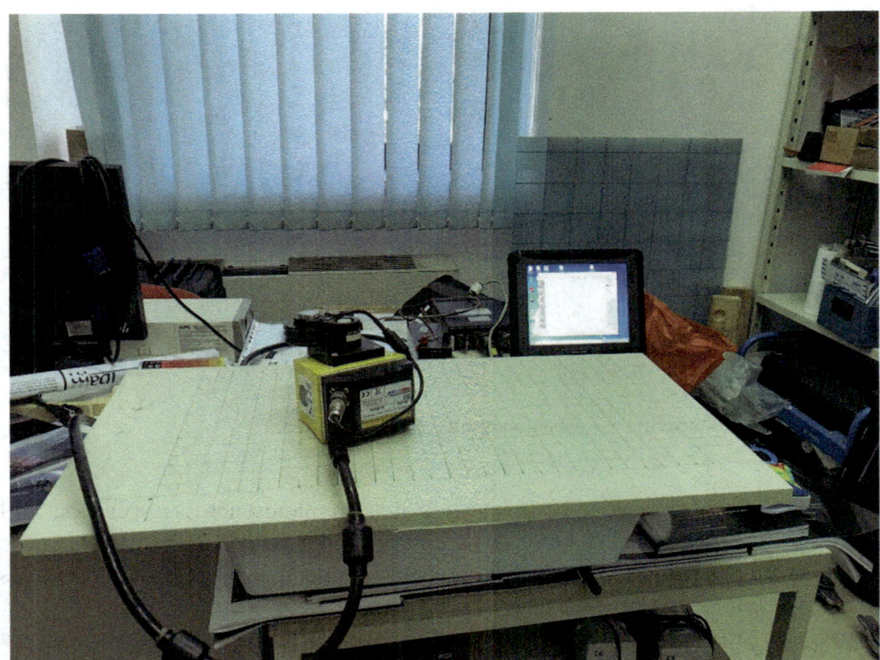

Fig. 6.48 Photo of the acquisition phases

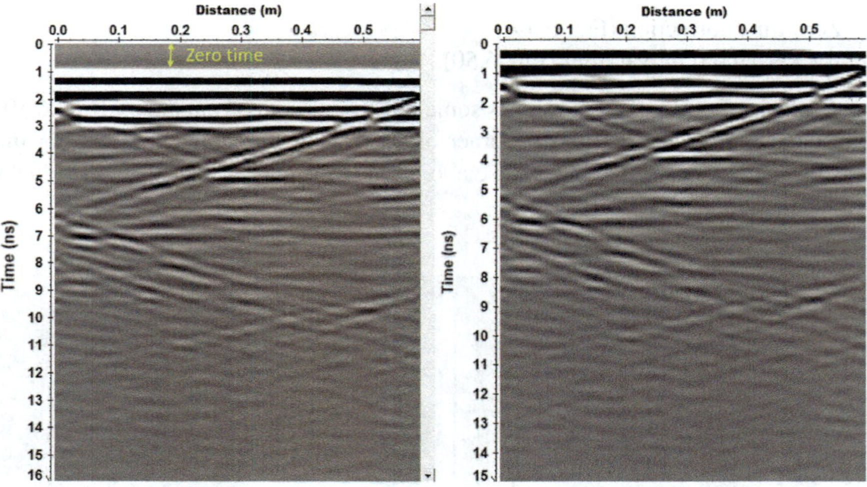

Fig. 6.49 Zero time correction

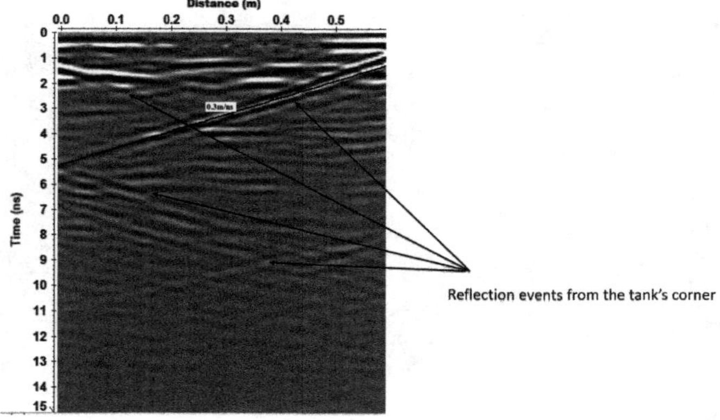

Fig. 6.50 Radar section processed with the background removal filter. Reflection events from the tank's corners are evidenced

Fig. 6.51 Radar section FK filtered

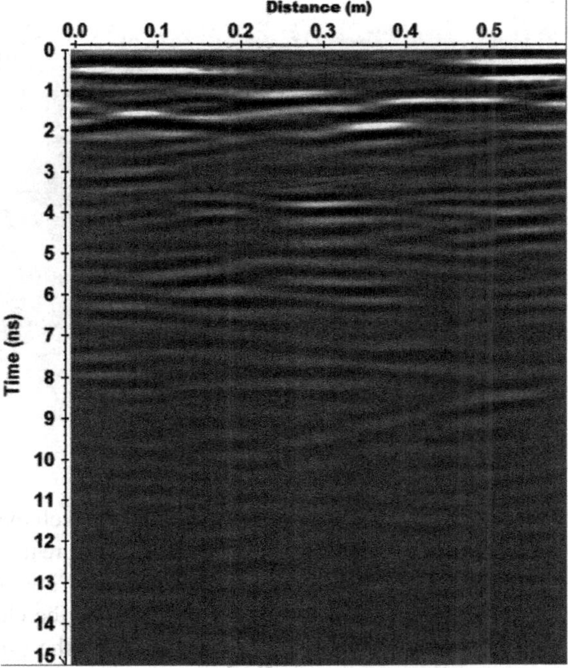

The analysis of the data was successively performed in a time slice way. Results show good evidence of the metal toy gun (Figs. 6.52 and 6.53).

Plastic toy gun: GPR data were processed as the data related to the metal toy gun (Fig. 6.54).

Also, in this case, the analysis of the data was successively performed in a time slice way. Results are shown in Figs. 6.55 and 6.56.

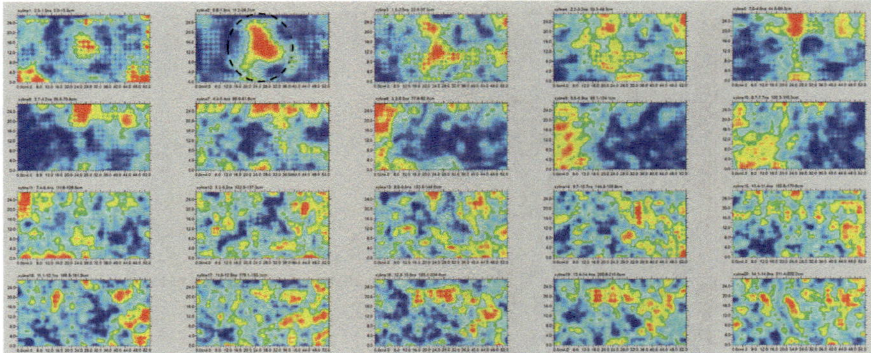

Fig. 6.52 Depth slices

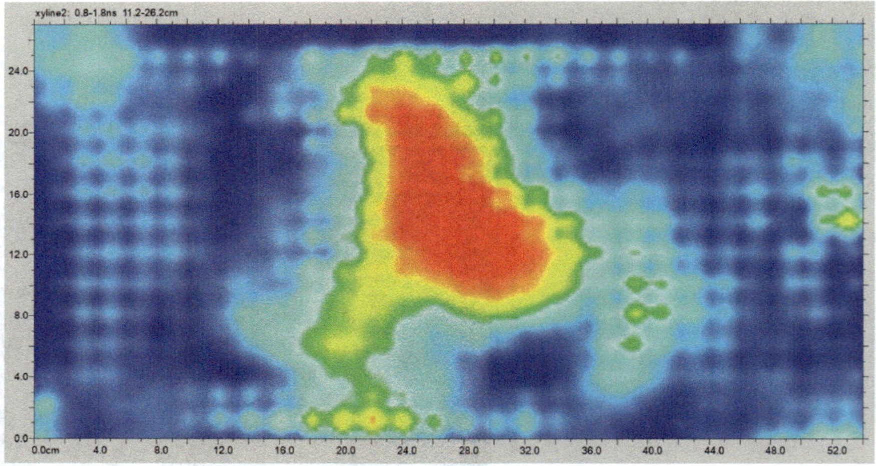

Fig. 6.53 Slice at 11.2–26.8 cm depth

Also, in this case, seem that the toy gun is well evidenced.

Several cases related to the crime forensic were taken into consideration. An alternative method for monitoring characteristics of oils was investigated. The proposed approach investigate on a quality indicator of the characteristics of vegetable oils. This approach preserves low costs (thanks to the use of instrumentation operating in the time domain) and, additionally, anticipates the possibility of in-situ real-time monitoring (thanks to the possibility of using portable TDR instruments). Results showed that different oils could be discriminated in terms of dielectric parameters.

Other aspects are related ERT method applied to problems related to the presence of pollutants in the subsoil. The first case evidenced hydrocarbon materials in the subsoil. The second case highlighted a problem common to most landfills built in the second half of the last millennium: waterproofing. Leaking leachate makes the aquifer

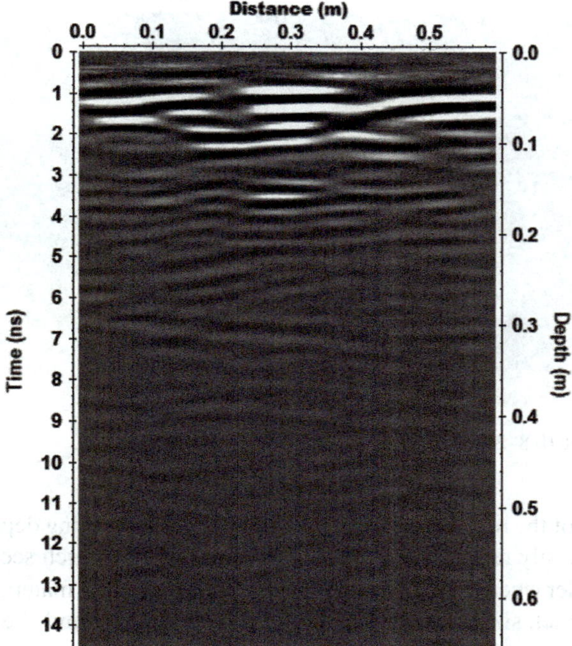

Fig. 6.54 Processed radar section

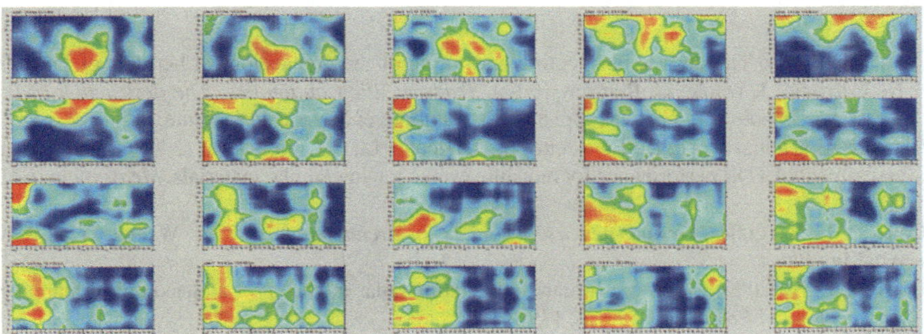

Fig. 6.55 Depth slices

water hardly usable for irrigation purposes. In the case of the study, a dangerous leakage of leachate was detected.

The plundering of archaeological sites, fed by the illicit trade of antiquities, affects cultural heritage and, mainly, causes irreversible losses of knowledge about the human past. In this case, the methodological approach proposed to allow evidence of the clandestine excavation.

Geophysical methods are also a useful tool for hidden bodies and arms discovery.

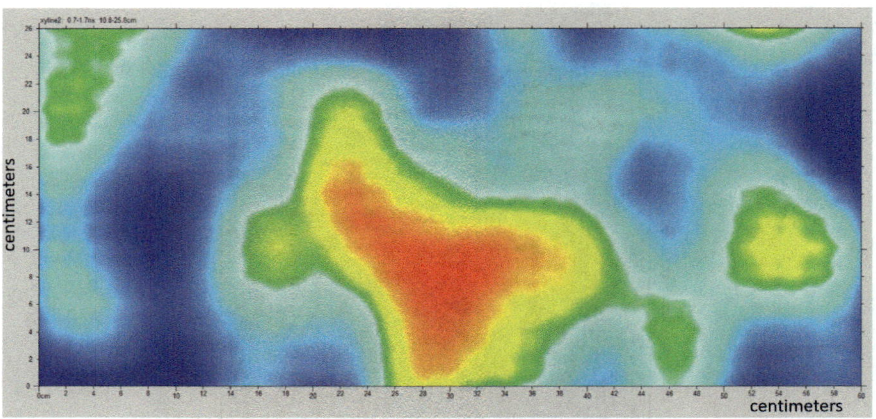

Fig. 6.56   Slice at 10.8–25.8 cm depth

The results of the research illustrated in this chapter providing dependable results which can be easily replicated by investigators during the search scenarios. Several issues to consider when choosing which geophysical tool to use materials of the target, depth of the target, size of the target, size of the search area, and site conditions.

References

Alva, W. (2001). The destruction, looting and traffic of the archaeological heritage of Peru. In N. J. Brodie, J. Doole, C. Renfrew (Eds.), *Trade in illicit antiquities: The destruction of the world's archaeological heritage* (pp. 89–96). Cambridge: McDonald Institute.

Cataldo, A., Cannazza, G., De Benedetto, E., Tarricone, L., & Cipressa, M. (2009). Metrological assessment of TDR performance for moisture evaluation in granular materials. *Measurement, 42,* 254–263.

Conyers, L. B. (2004). *Ground-penetrating radar for archaeology* (p. 255). Walnut Creek, CA: AltaMira Press.

Conyers, L. B. (2006). Innovative ground-penetrating radar methods for archaeological mapping. *Archaeological Prospection, 13*(2), 139–141.

Dalton, F. N., & Van Genuchten, M. Th. (1986). The time-domain reflectometry method for measuring soil water content and salinity. *Geoderma, 38,* 237–250.

Davenport, C. G., Lindemann, J. W., Griffin, T. J., & Borowski, J. E. (1988). Geotechnical applications 3: Crime scene investigation techniques. *Geophysics: The Leading Edge of Exploration, 7*(8), 64–66.

Davenport, C. G., Griffin, T. J., Lindemann, J. W., & Heimmer, D. (1990). Geoscientists and law officers work together in Colorado. *Geotimes, 35*(7), 13–15.

Fellner-Feldegg, H. (1968). The measurement of dielectrics in the time domain. *Journal of Chemical Physics, 73,* 616–623.

France, D. L., Griffin, T. J., Swanburg, J. G., Llindemann, J. W., Davenport, G. C., Trammell, V., et al. (1992). A multidisciplinary approach to the detection of clandestine graves. *Journal of Forensic Science, 37,* 1445–1458.

France, D. L., Griffin, T. J., Swanburg, J. G., Llindemann, J. W., Davenport, G. C., Trammell, V., et al. (1997). NecroSearch revisited: Further multidisciplinary approaches to the detection of clandestine graves. In W. D. Haglund & M. H. Sorg (Eds.), *Forensic taphonomy: The postmortem fate of human remains* (pp. 497–509). Boca Raton, FL: CRC Press.

Freeland, R. S., Yoder, R. E., Miller, M. L., & Koppenjan, S. K. (2002). Forensic application of sweep-frequency and impulse GPR. In *Ninth International Conference on Ground Penetrating Radar*, Santa Barbara, CA, USA, April 30–May 2 (pp. 533–538).

Goodman, D. (2013). GPR Slice V.7.0 manual. Available from http://www.gpr-survey.com. June 2013.

Goodman, D., & Piro, S. (2013). GPR remote sensing in archaeology. In *Geotechnologies and the environment* (Vol. 9, p. 233). New York: Springer.

Goodman, D., Steinberg, J., Damiata, B., Nishimure, Y., Schneider, K., Hiromichi, H., et al. (2006). GPR overlay analysis for archaeological prospection. In *Proceedings of the 11th International Conference on Ground Penetrating Radar*, Columbus, OH, USA. CD-rom.

Kent, M., Oehlenschlager, J., Mierke-Klemeyer, S., Manthey-Karl, M., Knöchel, R., Daschner, F., et al. (2004). A new multivariate approach to the problem of fish quality estimation. *Food Chemistry, 87,* 531–535.

Lasaponara, R., Leucci, G., Masini, N., & Persico, R. (2014). Investigating archaeological looting using satellite images and georadar: The experience in Lambayeque in North Peru. *Journal of Archaeological Science, 42,* 216–230.

Leucci, G. (2019). *Nondestructive testing for archaeology and cultural heritage: A practical guide and new perspectives.* Cham, Switzerland: Springer.

Lizhi, H., Toyoda, K., & Ihara, I. (2008). Dielectric properties of edible oils and fatty acids as a function of frequency, temperature, moisture and composition. *Journal of Food Engineering, 88,* 151–158.

Miller, M. L., Freeland, R. S., & Koppenjan, S. K. (2002). Searching for concealed human remains using GPR imaging of decomposition. In *Ninth International Conference on Ground Penetrating Radar*, Santa Barbara, CA, USA, April 30–May 2 (pp. 539–544).

Miura, N., Yagihara, S., & Mashimo, S. (2003). Microwave dielectric properties of solid and liquid foods investigated by time-domain reflectometry. *JFS: Food Engineering and Physical Properties, 68,* 1396–1403.

Nelson, S. O. (1991). Dielectric properties of agricultural product measurements and applications. *IEEE Transactions on Electrical Insulation, 26,* 845–869.

Noborio, K. (2001). Measurement of soil water content an electrical conductivity by time domain reflectometry: A review. *Computers and Electronics in Agriculture, 31,* 213–237.

Nozaki, R., & Bose, T. K. (1990). Broadband complex permittivity measurements by time-domain spectroscopy. *IEEE Transaction on Instrumentation and Measurement, 39,* 945–951.

Pettinelli, E., & Cereti, A. (2002). Time domain reflectometry: Calibration techniques for accurate measurement of the dielectric properties of various materials. *Review of Scientific Instruments, 73,* 3553–3562.

Silverman, H. (1993). *Cahuachi in the Ancient Nasca World.* Iowa City: University of Iowa Press.

Sosa-Morales, M. E., Valerio-Junco, L., López-Malo, A., & García, H. S. (2010). Dielectric properties of foods: Reported data in the 21st century and their potential applications. *LWT—Food Science and Technology, 43,* 1169–1179.

Sumnu, G., Raghavan, G. S. V., & Datta, A. K. (2005). Dielectric properties of foods. In M. A. Rao, A. K. Datta, & S. S. H. Rizvi (Eds.), *Engineering properties of foods* (3rd ed.). Boca Raton, FL: CRC Press by Taylor and Francis Group, LLC.

Tang, J. (2005). Dielectric properties of food. In H. Schubert & M. Regier (Eds.), *Microwave processing of food* (pp. 22–40). Cambridge: CRC Press, Woodhead Publishing Limited.

Topp, G. C., Davis, J. L., & Annan, A. P. (1982). Electromagnetic determination of soil water content using TDR: I. Applications to wetting fronts and steep gradients. *Soil Science Society of America Journal, 46,* 672–678.

Van Loon, W. K. P., Smulders, P. E. A., van den Berg, C., & van Haneghem, I. A. (1995). Timedomain reflectometry in carbohydrate solutions. *Journal of Food Engineering, 26,* 319–328.

Watson, P. (1999). The lessons of Sipan: Archaeologists and huajeros. Culture without context. *The Newsletter of the Illicit Antiquities Research Centre, 4,* 15–20.

Chapter 7
Conclusions

Abstract A new approach of the geophysical methods to the forensic sciences has been developed in this book. The basic knowledge of geophysical methods that could help the investigators has expatiated. Not all investigators are familiar with geophysical methods and this can create misunderstandings in the interaction between investigators and experts in geophysical methods. Therefore the author of this book hopes to have contributed to a dialogue about how law enforcement and forensic experts can improve investigative methods and techniques in the search, discovery, understand the several problems that imposed to the civil and criminal cases.

Keywords GPR · ERT · SP · Seismic · Magnetic · Gravimetry · Forensic geophysics

Several new methods not previously employed in civil and crime were proposed and underline that interdisciplinary research and field investigations hold the key to the of the investigator's ability to respond more effectively. Hope that the readers understand and appreciate that each case is different, and jurisdictions across the country have different policies, personnel, needs, and budgets. In response, law enforcement professionals and forensic experts are motivated to improve their investigative methods by striving for excellence/success, knowing that each new case will scientifically contribute to our knowledge base.

In the last decade the noninvasive in situ techniques such as seismic (sonic and ultrasonic techniques), ground-penetrating radar (GPR), electrical methods (electrical-resistivity tomography-ERT-and self-potential-SP), time-domain reflectometry (TDR), microgravimetry and magnetometry has had a development in terms of technologies (improvement of performance and information resolution of sensors and devices) and user-friendly software for data analysis, processing, and interpretation. This development in hardware and software have increased the interest of the investigator towards the use of these technologies.

Seismic investigations use elastic waves and allow to determine the 2D and/or 3D velocity distribution in the studied materials, including mechanical characteristics and the state of cracking, fractures, and other discontinuous elements. It can be used in the forensic field because it provides information about the degradation of the element and allows to locate areas of decay and structural weakness.

© Springer Nature Switzerland AG 2020

G. Leucci, *Advances in Geophysical Methods Applied to Forensic Investigations*,

https://doi.org/10.1007/978-3-030-46242-0_7

Ground-penetrating radar (GPR) is one of the most significant geophysical techniques applied in forensic science as a consequence of its high resolution and applicability. Its portability in terms of weight and size makes access to various zones easier.

Active and passive geo-electrical methods are newly applied in forensic and help to identify anomalies related to the presence of high volumetric water content, water flow and pollutant flow eventually related to the soil pollution.

TDR method helping in the evaluation of food sophistication and helping in the analysis of collapsed structures.

Microgravimetry allows the study of dangerous areas where carsick problems are presents.

Magnetic methods help in identify illegal landfills and hidden bodies.

A combination of the geophysical methods treated in this book could offer more informative interpretations and help in the accurate analysis of the possible uncertainties. Combining methodologies is a strategy used to corroborate results and to obtain accurate interpretations. As seen in Chaps. 5 and 6, the joint application of these methods have yielded the best results in the field of civil and criminal forensic investigations.

Finally, for almost ten years the attention of the media has been of the country focused on the so-called "forensic sciences" also thanks to the support, the introduction and dissemination in Italy of a whole series of fictions coming from the United States, which were then re-proposed in an "Italianized" version. But this has led to the creation of a series of "false myths" and to the belief that science and scientific analyses are the only way to conclude an investigation, or that they are the only ones able to give answers. This is not the case, or rather it should not be; analyses, tools and the contribution of knowledge scientific should help the professionals working in the field to facilitate them in conducting investigations, but they are not and should not be perceived as the solution to the question.

9783030462413